JOHN W. THOMSON is a member of the Department of Botany at the University of Wisconsin and author of *The Lichen Genus* Cladonia *in North America* and *The Lichen Genus* Physcia *in North America*.

Lichens are the predominant vegetation in the arctic environment, but the literature on them has remained scattered until now. This broad compendium, based on extensive field investigations, covers more than 500 species of lichens occurring on the North Slope of Alaska. It gives keys to the genera and species as well as succinct descriptions of the species, including information on the ecology and range of each. The book covers a specific area but, as the majority of species included are circumpolar, it will be useful in investigations extending far beyond the borders of Alaska.

This is the only detailed lichen guide and key for any region of arctic North America and will be an indispensable reference work for lichenologists, botanists, mycologists, and ecologists alike.

Lichens of the Alaskan Arctic Slope

JOHN W. THOMSON

UNIVERSITY OF TORONTO PRESS

Toronto Buffalo London

© University of Toronto Press 1979
Toronto Buffalo London
Printed in Canada
Reprinted in 2018
ISBN 978-1-4875-8253-1 (paper)

Library of Congress Cataloging in Publication Data

Thomson, John Walter, 1913–
  Lichens of the Alaskan arctic slope.

  Bibliography: p.
  Includes index.
  1. Lichens–Alaska–North Slope—Identification.
  I. Title.
  QK587.5.A4T47   589'.1'097987   78-31899
  ISBN 0-8020-5428-5

Contents

Preface

This volume was originally intended as a report of the lichens which were collected along the North Slope of Alaska during the summer of 1958. I soon realized, while working on the material, that there existed no assembled keys and descriptions of arctic lichens which would assist me in making the identifications. In order to make the report usable to others studying the lichen vegetation of this area I decided to add keys and descriptions. This process added two years to the writing. Although the fieldwork took a comparatively short time, the study of the material collected has been time consuming, especially that of the crustose species, which gave many difficulties. This volume is part of a larger study encompassing the lichens of the American Arctic. I hope eventually to publish in their entirety the results of this study.

The fieldwork of 1958 was made possible by joint grants from the Arctic Institute of North America and the National Science Foundation (G4315) to work at the Arctic Research Laboratory supported at Point Barrow by the Office of Naval Research. The research since has been supported by the National Science Foundation in a renewed series of grants (G8781, G19443, GB2702, GB7461) and by the Research Committee of the University of Wisconsin Graduate School.

Study of material in European herbaria was absolutely essential for the classifying of the material I had found, and in the intervening years I had two opportunities to do so. I was fortunate in being an exchange professor to the University of Helsinki during 1966, an opportunity which permitted study in the incomparable herbaria of Nylander, Räsänen, Acharius, and Vainio. A study trip to Europe during the spring and summer of 1961 supported by the National Science Foundation permitted me to study material in the vast collection of arctic lichens accumulated by Bert Lynge at the University of Oslo.

This book was possible only with the help of many individuals. I am indebted above all to my friend and esteemed colleague, Dr Sam Shushan of the University of Colorado, who helped to make the field collections. I am also indebted to Barbara Murray for permission to incorporate reports based on her collections made at Prudhoe Bay and Beaufort Lagoon. Dr A.J. Sharp of the University of Tennessee provided invaluable specimens which he collected along Mancha Creek and the Firth River. The collections of Dr Thomas H. Nash III of Arizona State University made at Anaktuvuk Pass added a great many records and species to the listing. We were able to use our field time very efficiently because of the assistance of Dr Max Brewer, director of the Arctic Research Laboratory at Point Barrow, Alaska. Supplies and transport and the scheduling of trips to good sampling areas were all arranged by Dr Brewer and his assistant Paul Tietjen, releasing the investigators to pursue their interests to the utmost. I would also like to thank the pilots, Robert J. Fisher of the Laboratory staff and Porter Lockhart of Wien Airlines, who so ably placed us in remote areas for research.

The material collected on the expeditions of 1958 was packaged and processed at the University of Wisconsin, and a full set of material is preserved in the herbarium there. A nearly complete set of duplicates has been deposited at the US National Herbarium, Smithsonian Institution. The specimens collected by Dr Nash are filed at Arizona State University, and many duplicates are at the University of Wisconsin. Many other institutions have had less complete duplicate sets of this arctic material made available to them. Exsiccat reference sets, Lichenes Arctici, currently containing ninety-six numbers, have been distributed to forty institutions or students of lichens throughout the world, disseminating as widely as possible the results of this Alaskan work.

For comments upon and assistance with the classifying of certain groups I am indebted to a number of individuals: to Dr Hannes Hertel for certain *Lecideae* and for help in better understanding the genus *Lecidella*; to Dr I.M. Lamb for part of the genus *Stereocaulon*; to Dr W.L. Culberson for describing the species *Cetrelia alaskana*; to Dr Teuvo Ahti for assistance with the loan of *Candelariella* as well as his great help while I was at the University of Helsinki; to Dr E.D. Rudolph for help with the Caloplacaceae; to Dr Irwin Brodo for help with the genus *Alectoria* and a careful reading of the manuscript; to Dr W.P. Jordan for help with the genus *Lobaria*; and to Dr Martyn Dibben for help with the Pertusariaceae. Naturally, however, any errors remain my responsibility. I thank Miss Renee Vanderlois for typing the manuscript. My wife, Olive, helped immensely in the preparation of the manuscript.

This report contains 504 species, many not hitherto reported from Alaska and some not previously reported from North America. Only five species described from the Alaskan material are new to science: *Cetrelia alaskana* (W. Culb. & C. Culb.) W. Culb. & C. Culb., *Lecidea shushanii* Thoms., *Lecidea carbonoidea* Thoms., *Lecanora concinnum* Thoms., and *Rhizo-*

carpon cumulatum Thoms. This scarcity of new species is not surprising as so many of the arctic lichens are circumpolar. Many species of the Beringian element in the Alaskan lichen flora have already been described from material collected in Kamchatka during the past century. A word of warning is in order. Although this work on the North Slope of Alaska may prove useful over much of the Arctic to identify many species, it covers only about one-quarter of the approximately 2000 species which have been reported from the arctic regions. While the more comprehensive work on lichens of the American Arctic is in progress, it is hoped that the present volume will assist those who are working on the vegetation of the north.

In constructing the keys I have occasionally introduced an error recovery system so that if an error occurs in keying out a specimen, the species recurs in a position which my lichenological colleagues will recognize as erroneous. I hope they will forgive this practice and understand the attempt to aid the student. The descriptions are not full but are intended to give sufficient diagnostic information to permit reasonably certain identification.

Publication of this book has been assisted by grants from the National Research Council and from the Publications Fund of the University of Toronto Press.

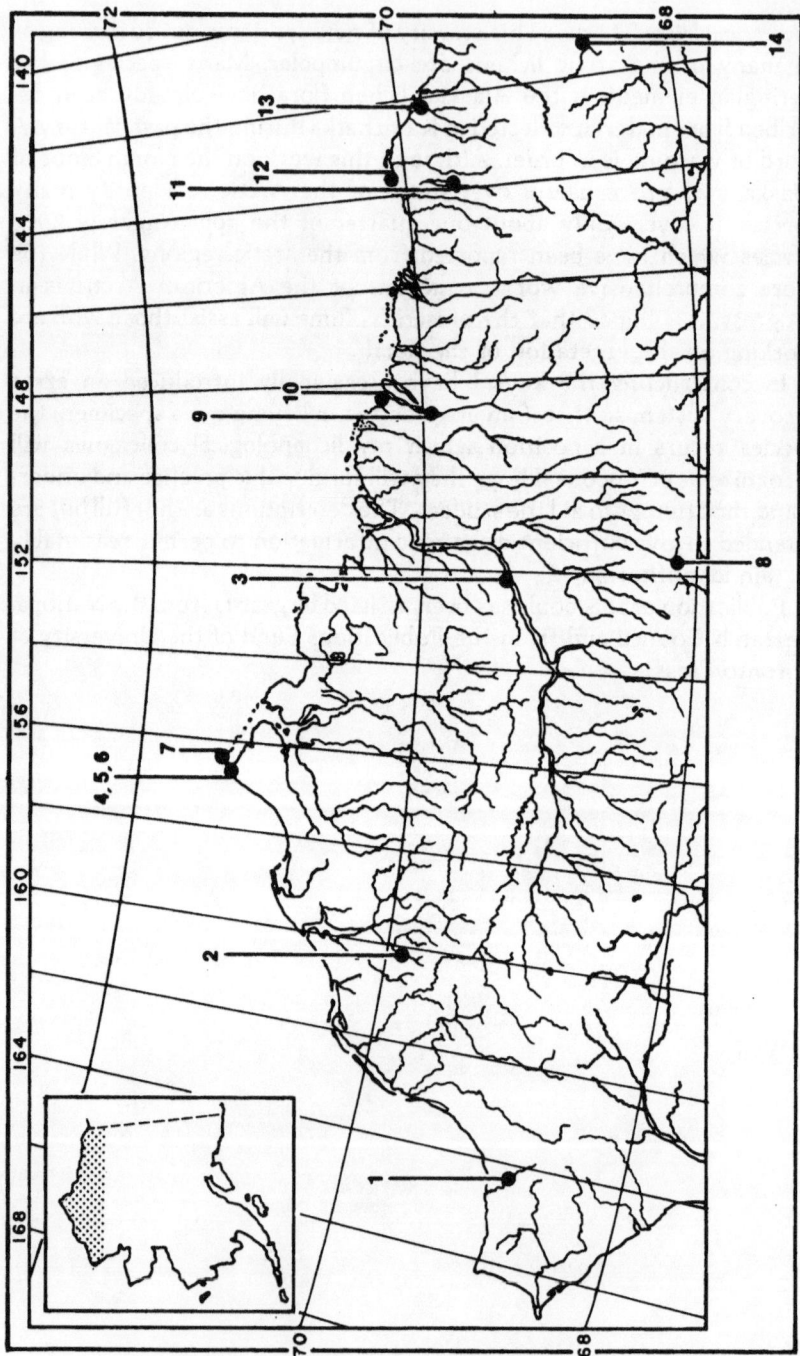

The North Slope of Alaska (after I.L. Wiggins and J.H. Thomas, *A Flora of the Alaskan Arctic Slope*).

THE LOCALITIES STUDIED

Note that the following numbers are used to indicate the localities for the individual species in the text.

1   Along the Pitmegea River, 15 miles upstream from Cape Sabine, 68° 48′ N, 164° 20′ W, 10–17 July 1958 by J.W. Thomson.

2   Along the Kaolak River, south of Wainright, 69° 56′ N, 159° 57′ W, 9–18 July 1958 by Sam Shushan and W.J. Maher.

3   Along the north ridge of the Colville River near Umiat, 69° 25′ N, 152° 10′ W, 22–25 July 1958 by J.W. Thomson and Sam Shushan.

4   On old beach ridge at Circular Marsh near Barrow, 71° 18′ N, 156° 11′ W, 9 and 21 July 1958 by J.W. Thomson and Sam Shushan.

5   Vicinity of Imiksonn Lake, 6 miles south-southwest of Barrow base, 71° 14′ N, 156° 45′ W, 27 July 1958 by J.W. Thomson and Sam Shushan.

6   Near Loon Lake, 6½ miles south-southwest of Barrow Base, 71° 12′ N, 156° 30′ W, 20 July 1958 by J.W. Thomson, Sam Shushan, and A.J. Sharp.

7   On the point of Point Barrow, 71° 23′ N, 156° 27′ W, 18 July 1958 by J.W. Thomson, Sam Shushan, and A.J. Sharp.

8   Anaktuvuk Pass, 68° 10′ N, 152° 00′ W, summer 1976 by Thomas H. Nash III.

9   Franklin Bluffs on the Sagavanirktok River, 68° 50′ N, 148° 15′ W, 29 July to 2 August 1958 by J.W. Thomson, Sam Shushan, and J. Koranda.

10 Prudhoe Bay, 70° 17′ N, 148° 20′ W, 1970, 1971, and 1974 by Barbara M. Murray.

11 Valley and ridges of the Okpilak River near Okpilak Lake and Mt Michelson, 69° 34′ N, 144° 5′ W, 2–7 August 1958 by J.W. Thomson and Sam Shushan.

12 Barter Island, 70° 8′ N, 143° 35′ W, 7 August 1958 by J.W. Thomson, 1970 by Barbara M. Murray.

13 Beaufort Lagoon, Nuvagapak Point, 69° 40′ N, 141° 10′ W, 1970 and 1971 by Barbara M. Murray.

14 Firth River at the junction with Mancha Creek, 68° 40′ N, 141° W, August 1958 by A.J. Sharp.

TOP Willow (*Salix alaxensis*) thicket along the Pitmegea River. Bark at base with *Caloplaca pinicola, C. ferruginea, C. cerina, Physcia adscendens, P. aipolia*. Locality 1.

BOTTOM Whale bones from 'ice-houses' on Point Barrow, which were excellent sources of lichens. S. Shushan and A.J. Sharp in photo. Locality 7.

TOP Franklin Bluffs on the east side of the Sagavanirktok River. Limestone barrens with typical calciphile lichens on the bluffs. Locality 9.

BOTTOM Raised polygon area near Barrow with heath-type tundra with lichens on tops, sedges between. Locality 4.

TOP The valley of the Okpilak River above Okpilak Lakes. Lichen collections made along the slopes. Locality 11.

BOTTOM Collecting in the talus on the slopes of the Okpilak River valley. Heath type of tundra with lichens (light) interspersed and boulders covered with lichens. Locality 11.

TOP The low tundra largely dominated by *Betula* in the valley of the Sagavanirktok River. Locality 9.

BOTTOM *Thamnolia subuliformis*, which is abundant in many types of tundra vegetation.

Lichens of the Alaskan Arctic Slope

Introduction

The geographical area known as the North Slope of Alaska, in which these lichen studies were made, includes the Arctic Coastal Plain from the Pitmegea River near Cape Lisburne in the west to the Beaufort Lagoon near the Canadian boundary in the east. It also includes the Arctic Foothills Province described by Britton (1957, 1973) and reaches into the Brooks Range at Anaktuvuk Pass, the Okpilak River, and the upper drainage of the Firth River to the east.

The landforms which support and influence the vegetation of the region are strongly connected with its geological history. During much of geological time the area which is now the North Slope was under water. To the north, a landmass which was probably part of the Canadian Shield contributed reworked (clastic) sediments to the shelf and to a geosyncline which extended south to central Alaska during the middle Paleozoic and into the middle Mesozoic times. Most of the rocks are middle Paleozoic limestones, sandstones, shales, and their metamorphic equivalents. These rocks, of course, bear characteristic lichen floras dependent upon the acidity or alkalinity of the rocks as well as their porosity and other characteristics. In the foothills occur shales, sandstones, conglomerate, and cherts.

During the late Mesozoic, tectonic action brought about uplift of the Brooks Range. In portions of the range, igneous rocks were added to the materials available for colonization by plants. For example, Mt Michelson in the region of the upper Okpilak River is on a granitic batholith. Through the Brooks Range there was intense folding and faulting along with some of the igneous intrusion. At this time, in the late Mesozoic and Cenozoic, the direction from which materials were deposited onto the area which became the North Slope was reversed. The northern source of materials became eliminated by rifting and drifting of the continental plates and formation of the Arctic Ocean. The main source of sedimentary rocks then became the Brooks Range to the south, which continues to be a source today. Over much of the Arctic Coastal Plain the underlying sediments consist of undeformed Mesozoic sediments which underlie a mantle of unconsolidated silt, sand, and gravel of the Quaternary age known as the Gubik formation.

The advent of the Pleistocene brought changes in climate which strongly influence the present landforms and vegetation. Several glaciations occurred in the Brooks Range. During the last 6,000 to 10,000 years several advances and retreats of glaciers have occurred in the range. Still today there are remnant glaciers in cirques in the eastern portion of the Brooks Range. Such remnants are not present in the De Long Mountains in the west. The absence of sufficient precipitation, however, strongly limited the extent to which the glaciers were able to flow down from the Brooks Range and onto the North Slope. Even during the maximum advances of the ice, the area covered was never more than fifteen per cent of the North Slope, and the glaciers did not advance more than seventy kilometers north of the Brooks Range. Most of the valleys north of the Brooks Range have a series of recessional moraines up-valley from terminal moraines.

The features which dominate the North Slope derive from the presence of permafrost, which underlies the slope to a depth of 700 meters and which thaws a few millimeters each year in the active layer just under the surface. The freezing and thawing give rise to such phenomena as stone nets, strips, circles, and solifluction lobes, all of these prominent in the Foothills Province and influencing the succession of plant species on the soils. On the Arctic Coastal Plain the development of the famous ice wedge polygons and oriented lakes as well as such structural phenomena as pingoes also influences the vegetational cover.

The freezing and thawing over the permafrost influence the development of the arctic soils so that movement of materials is characteristic, and displacement and disruption are common. Most of the poorly drained Arctic Coastal Plain is underlain by Tundra and Bog soils, the former being the most common soils of northern Alaska. Tundra soils are silty and usually acid soils, although in limestone areas they may be alkaline. They thaw to a depth of about thirty centimeters in the active season of about three months. They have four horizons: an upper organic horizon, an upper mineral soil, a buried organic horizon, and the permanently frozen substrate. The Bog soils, confined to low-lying areas, are composed mainly of peats with permafrost close to the surface. Arctic Brown soil consisting of dark brown to sandy loam grading to a dark yellow-brown in the B-horizon and a dark gray or gray-brown silt loam in the C-horizon is more common in the alpine region and in the forested regions southward. Frost boils are also a very characteristic feature of the arctic soils and create microhabitats important for some species of lichens.

The climate of the North Slope is severe. It is cold and dry. It cannot quite be called a cold desert, however, as the relative humidity remains high and the subsurface melting of the permafrost contributes a source of moisture for the abundant plants which cover the low-lying soils. Rock barrens are clothed with lichens where the vascular plants thin out over the deeper thaw layers.

Few data appear to be available on climatic conditions on the North Slope except at Barrow where a weather station has been maintained. There the winter temperature may drop to -48°C, and the temperature may only rarely rise above 21°C in July and August. Only June, July, and August have average temperatures above freezing, and freezing temperatures can occur or snow fall on any day during the summer. The average monthly temperatures at Barrow for the period 1921–56 were: January -26.1°C, February -27.7°C, March -26.3°C, April -17.9°C, May -7.3°C, June 1°C, July 4.3°C, August 3°C, September -0.9°C, October -8.3°C, November -16.1°C, and December -23.5°C. The temperatures at Prudhoe Bay during the summer average about two Celsius degrees above those at Barrow.

The precipitation averages about 103 millimeters annually at Barrow, most of which falls during the summer months. In the mountains the total annual precipitation has been estimated at 450 millimeters. The mean annual snowfall at Barrow is 355 millimeters, the greatest fall occurring in October. In the Brooks Range the snowfall has been estimated at 850 millimeters. The relative humidity remains over 90% during the growing season, June to September, a circumstance which is all-important for plant growth in view of the low precipitation. The mean monthly wind velocities range between 11 and 14 miles per hour, the higher figures occurring in August, September, and October. Winds of highest velocities occur, however, in November, December, and January, when they can cause ice crystal abrasion, an important factor in plant stresses.

GENERAL REFERENCES

Britton, M.E. 1957. Vegetation of the Arctic Tundra. From *18th Biology Colloquium*, 26–61. Oregon State College, Corvallis, Ore.
– 1973. *Alaskan Arctic Tundra*. Arctic Inst. N. Amer. Tech. Paper 25
Brown, Jerry. 1966. *Soils of the Okpilak River region*. Research Report 188, Cold Regions Research and Engineering Laboratory, Hanover, NH
Wiggins, I.L., and J.H. Thomas. 1962. *A Flora of the Alaskan Arctic Slope*. Toronto: University of Toronto Press

Ecological considerations

EPIGAEIC LICHENS

A very large number of species of lichens in the arctic occur on humus-rich soils. These soils vary immensely in their moisture-holding capacity and texture and must be studied carefully before we can describe the lichens and their roles in the arctic ecosystems. Ordinarily the *Eriophorum* tussock tundras are low in numbers of lichens, but some *Cladonia stellaris*, *Asahinea chrysantha*, *Peltigera aphthosa*, *P. malacea*, *Parmelia sulcata* or *P. omphalodes*, and other such foliose and fruticose lichens may occur in the tops of the tussocks and between the sedge stems. Yet at the type locality of *Cetrelia alaskana* in the Pitmegea River valley there were considerable mats of this lichen both on the tussocks and on the low, wet soils between them. The dry humus soils of the heath types of tundra are very rich in numbers of species, particularly those which are circumpolar in distribution. Here gradations in quantities of a species are gradual, dependent perhaps upon the slope, the exposure to wind or sunlight, the amount of snow present and its persistence, as well as even less visible parameters characterizing the substratum. The tundras are full of surprises. One may spend a long time gathering fragment after fragment of a species for study and then over the next ridge come upon a nearly solid stand of the same species. Sometimes the humus soil may take a form quite unexpected by the student accustomed to the lichens of temperate zones. The dry dung of arctic hares and the droppings of lemmings, long preserved in the arctic climate, are a frequent habitat for *Caloplaca tiroliensis*, *C. stillicidiorum*, *C. cinnamomea*, *Lecanora epibryon*, *L. behringii*, *Lecidea assimilata*, *Rinodina turfacea*, and *R. mniaraea*.

The species found on the North Slope growing on humus soils include: *Alectoria ochroleuca*, *A. nigricans*, *Asahinea chrysantha*, *Bacidia alpina*, *B. bagliettoana*, *B. melaena*, *B. obscurata*, *B. sphaeroides*, *B. trisepta*, *Biatorella fossarum*, *Bryoria nitidula*, *Buellia papillata*, *Caloplaca cinnamomea*, *C. jungermanniae*, *C. stillicidiorum*, *C. tetraspora*, *C. tiroliensis*, *Candelariella terrigena*, *Catillaria muscicola*, *Cetraria andrejevii* (in low, wet areas), *C. cucullata*, *C. delisei* (in low, wet areas), *C. elenkienii, C. ericetorum, C. islandica, C. kamczatica, C. nigricans, C. nivalis, Cetrelia alaskana, Cladonia aberrans, C. acuminata, C. alaskana, C. amaurocraea, C. arbuscula, C. bacilliformis, C. bellidiflora, C. cariosa, C. carneola, C. cenotea, C. chlorophaea, C. coccifera, C. cornuta, C. crispata, C. cyanipes, C. decorticata, C. deformis, C. ecmocyna, C. fimbriata, C. glauca, C. gonecha, C. gracilis, C. lepidota, C. metacorallifera, C. mitis, C. norrlini, C. phyllophora, C. pleurota, C. pseudostellata, C. pyxidata, C. rangiferina, C. scabriuscula, S. stellaris, C. subfurcata, C. subulata, C. tenuis, C. uncialis, C. verticillata, Coccocarpia pellita, Collema ceraniscum, Coriscium viride, Cornicularia divergens, Dactylina arctica, D. beringica, D. madreporiformis, D. ramulosa, Hypogymnia austerodes, H. subobscura, Icmadophila ericetorum, Lecanora behringii, L. castanea, L. epibryon, L. leptacina, L. urceolaria, Lecidea assimilata, L. berengeriana, L. fusca, L. granulosa, L. tornoensis, L. uliginosa, L. vernalis, Lecidella wulfenii, Leciophysma finmarkicum, Leptogium palmatum, L. sinuatum, L. tenuissimum, Lobaria linita, L. pseudopulmonaria, Lopadium coralloideum, L. fecundum, L. pezizoideum, Massalongia carnosa, Microglaena muscorum, Nephroma arcticum, N. expallidum, N. isidiosum, Ochrolechia androgyna, O. frigida, O. grimmiae, O. inaequatula, O. upsaliensis, Pannaria pityrea, Parmelia disjuncta, P. omphalodes, P. saxatilis, P. substygia, P. sulcata, Parmeliella praetermissa, Peltigera aphthosa, P. canina, P. malacea, P. polydactyla, P. pulverulenta (P. scabrosa), Pertusaria bryontha, P. coriacea, P. dactylina, P. glomerata, P. octomela, P. oculata, P. subdactylina, P. subobducens, P. trochisea, Physcia muscigena, Platismatia glauca, Polyblastia gelatinosa, P. gothica, P. sendtneri, Polychidium muscicola, P. umhausense, Psoroma hypnorum, Rinodina mniaraea, R. nimbosa, R. roscida, R. turfacea, Sphaerophorus globosus, Stereocaulon paschale, Thamnolia subuliformis,* and *T. vermicularis.*

Soils in the very low, wet seepages below late snowbanks bear a very characteristic flora. In the wettest, coldest parts, underwater for long periods, may be found the blue gray papillate crust of *Lecidea ramulosa* or the erect white thalli of *Siphula ceratites.* Farther down the seepages, where it is warmer and slightly drier, will be the habitat of *Cetraria delisei, C. andrejevii, Cladonia lepidota,* and sometimes *Solorina crocea.*

The lichens of mineral soils differ according as the soils are calcareous or acidic. Calcareous soils are especially well represented along the North Slope, particularly along the De Long Mountains and the west end of the Brooks Range. Characteristic lichens on such soils included *Cetraria tilesii, Collema bachmanianum, C. glebulentum, Dactylina madreporiformis, D. ramulosa, Dermatocarpon cinereum, D. hepaticum, Endocarpon pusillum, Evernia perfragilis, Fulgensia bracteata, Gyalecta foveolaris, Heppia lutosa, Lecanora lentigera, Lecidea decipiens, L. rubiformis, L. subcandida, Leptogium lichenoides, Pannaria pezizoides, Peltigera lepidophora, P. venosa, Protoblastenia terricola, Solorina crocea, S. bispora, S. octospora, S. saccata, S. spongiosa, Thamnolia subuliformis, T. vermicularis, Toninia caeruleonigricans, T. lobulata,* and *T. tristis.* Many of these pioneer in colonizing and stabilizing the surface of old frost boils along with the primary thalli of *Stereocaulon* sp., the beginnings of *Pertusaria* sp., and *Lecanora epibryon.*

The non-calcareous soils seemed to be dominated by a substantially different assemblage of species: *Alectoria nigricans, Baeomyces carneus, B. placophyllus, B. roseus, B. rufus* (especially on moist seepages), *Bryoria nitidula, B. tenuis, Buellia elegans, Cladonia cariosa, C. decorticata, C. thomsonii, C. verticillata, Cornicularia divergens, C. muricata, Geisleria sychnogonoides, Lecidea cuprea, L. demissa, L. globifera, L. ramulosa, Mycoblastus alpinus, M. sanguinarius, Pseudephebe minuscula, P. pubescens, Ramalina almquistii, R. pollinaria, Stereocaulon alpinum, S. glareosum, S. incrustatum, S. paschale, S. rivulorum, S. saxatile, S. tomentosum, Thamnolia subuliformis,* and *T. vermicularis.*

EPILITHIC LICHENS

The calcareous rocks, like the calcareous soils, were best represented in the western portions of the North Slope, although excellent lichens of calcareous rocks were also obtained on the Lisborne limestone outcrop located among the predominantly acid rocks of the Okpilak River valley. The list of species found on the calcareous rocks is large: *Acarospora badiofusca, A. glaucocarpa, A. scabrida, A. schleicheri, A. smaragdula, A. veronensis, Agyrophora rigida, Bacidia coprodes, Blastenia exsecuta, Buellia alboatra, B. immersa, B. notabilis, B. vilis, Caloplaca cirrochroa, C. festiva, C. fraudans, C. granulosa, C. murorum, Candelariella athallina, C. aurella, Collema multipartitum, C. polycarpon, C. tuniforme, C. undulatum, Glypholecia scabra, Ionaspis euplotica, I. heteromorpha, I. ochraceella, I. reducta, I. schismatopsis, Lecania alpivega, Lecanora annulata, L. anseris, L. atra, L. caesiocinerea, L. candida, L. cinerea, L. cinereorufescens, L. crenulata, L. dispersa, L. disserpens, L. fimbriata, L. flavida, L. heteroplaca, L. intricata, L. lesleyana, L. microfusca, L. muralis, L. nikrapensis, L. nordenskioeldii, L. pergibbosa, L. polychroma, L. polytropa, L. rupicola, L. ryrkapiae, L. sublapponica, L. torrida, Lecidea atromarginata, L. hypocrita, L. jurana, L. lapicida, L. lithophila, L. marginata, L. melaphanoides, L. speirea, L. umbonata, L. vorticosa, Lecidella carpathica, L. goniophila, L. stigmatea, Microglaena muscorum, Pertusaria subplicans, Placopsis gelida, Placynthium aspratile, P. nigrum, Polyblastia hyperborea, P. integrascens, P. sommerfeltii, P. theleodes, Porocyphus coccodes, Protoblastenia rupestris, Pyrenopsis pulvinata, Rhizocarpon alaxensis, R. chioneum, R. disporum, R. geographicum, R. inarense, R. intermediellum, R. umbilicatum, Rinodina bischoffii, R. occidentalis, Sarcogyne regularis, Staurothele clopima, S. perradiata, Thelidium acrotellum, T. aenovinosum, T. pyrenophorum, Verrucaria aethiobola, V. devergens, V. muralis, V. nigrescens, V. obnigrescens, V. rupestris, Xanthoria candelaria, X. elegans,* and *X. sorediata.*

The old whale bones around Point Barrow and the caribou bones scattered over the tundra yielded such lichens as *Bacidia beckhausii, B. microcarpa, Buellia punctata, Candelariella aurella, Lecanora behringii, Physcia caesia,* and *P. dubia.*

The acid rocks, especially well seen at the east end of the Brooks Range in the Okpilak River valley, proved to have a very rich cover of lichens: *Acarospora chlorophana, A. fuscata, A. veronensis, Agyrophora rigida, A. scholanderi, Asahinea scholanderi, Bacidia lugubris, Bryoria chalybeiformis, Buellia malmei, B. moriopsis, Caloplaca festiva, Candelariella aurella, Catillaria chalybeia, Cetraria commixta, C. hepatizon, Diploschistes scruposus, Haematomma lapponicum, Hypogymnia bitteri, H. oroarctica, H. subobscura, Ionaspis suaveolens, Lasallia pensylvanica, Lecanora aquatica, L. atrosulphurea, L. badia, L. caesiosulphurea, L. chrysoleuca, L. cinerea, L. cingulata, L. composita, L. concinnum, L. dispersa, L. elevata, L. frustulosa, L. granatina, L. intricata, L. lacustris, L. melanaspis, L. melanopthalma, L. pelobotryon, L. peltata, L. perradiata, L. plicigera, L. polytropa, L. rosulata, L. rupicola, L. stygioplaca, L. supertegens, Lecidea aglaea, L. armeniaca, L. atrata, L. atrobrunnea, L. auriculata, L. brachyspora, L. carbonoidea, L. cinereorufa, L. confluens, L. crustulata, L. diducens, L. elegantior, L. flavocaerulescens, L. glaucophaea, L. instrata, L. lapicida, L. leucophaea, L. lucida, L. macrocarpa, L. melinodes, L. pallida, L. panaeola, L. pantherina, L. paupercula, L. picea, L. shushanii, L. tessellata, Lecidella stigmatea, Nephroma isidiosum, N. parile, Omphalodiscus decussatus, O. krascheninnikovii, O. virginis, Parmelia almquistii, P. alpicola, P. centrifuga, P. exasperatula, P. fraudans, P. incurva, P. infumata, P. panniformis, P. separata, P. stygia, P. sulcata, P. tasmanica, Pertusaria amara var. flotowiana, P. excludens, Physcia caesia, P. dubia, P. endococcinea, P. grisea, P. intermedia, P. sciastra, Pilophorus robustus, Platismatia glauca, Pseudephebe minuscula, P. pubescens, Rhizocarpon badioatrum, R. copelandii, R. crystalligenum, R. disporum, R. eupetraeoides, R. eupetraeum, R. geographicum, R. grande, R. hochstetteri, R. polycarpum, R. riparium, R. superficiale, Rinodina milvina, R. oreina, Sphaerophorus fragilis, Sporastatia testudinea, Staurothele fissa, Stereocaulon botryosum, S. dactylophyllum, S. subcoralloides, S. symphycheilum, Umbilicaria angulata, U. arctica, U. caroliniana, U. cinereorufescens, U. cylindrica, U. deusta, U. hirsuta, U. hyperborea, U. proboscidea, U. torrefacta, U. vellea, Verrucaria deverges-cens, V. margacea,* and *Vestergrenopsis isidiata.*

EPIPHYTIC LICHENS

The epiphytic lichens were most often seen on the barks of the willows (*Salix*), birch (*Betula*), and alder (*Alnus*), but occasionally also on other slightly woody plants such as *Cassiope*. At the extreme eastern edge of Alaska where Dr Sharp collected on the conifers (*Picea mariana*) at Mancha Creek, a considerable number of lichens of the boreal forest accompanied the spruce on its bark and old wood. The lichens recorded on the barks included *Bacidia subincompta*, *Bryoria simplicior*, *Buellia disciformis*, *B. punctata*, *Caloplaca cerina*, *C. discolor*, *C. ferruginea*, *C. holocarpa*, *C. pinicola*, *Candelariella lutella*, *C. xanthostigma*, *Cetraria pinastri*, *C. sepincola*, *C. sibirica*, *Evernia mesomorpha*, *Hypogymnia austerodes*, *H. bitteri*, *H. physodes*, *Lecanora chlarona*, *L. coilocarpa*, *L. distans*, *L. expallens*, *L. hageni*, *L. rugosella*, *Lecidea vernalis*, *Lecidella elaeochroma*, *L. glomerulosa*, *Leptogium saturninum*, *Lobaria scrobiculata*, *Lopadium pezizoideum*, *Ochrolechia androgyna*, *Parmelia exasperatula*, *P. septentrionalis*, *P. sulcata*, *Parmeliopsis ambigua*, *P. hyperopta*, *Pertusaria alpina*, *P. dactylina*, *P. sommerfeltii*, *P. velata*, *Physcia adscendens*, *P. aipolia*, *P. orbicularis*, *Ramalina pollinaria*, *Rinodina archaea*, *R. hyperborea*, *R. laevigata*, *R. septentrionalis*, *Sticta weigelii*, *Usnea glabrescens*, *U. lapponica*, and *Xanthoria fallax*.

On the rotting woods, especially of the spruces, there grew *Buellia disciformis*, *B. punctata*, *Caloplaca discolor*, *Candelariella vitellina*, *Cetraria pinastri*, *Cladonia cenotea*, *C. cyanipes*, *C. deformis*, *C. fimbriata*, *C. gonecha*, *Cyphelium inquinans*, *C. tigillare*, *Lecanora vegae*, *Lecidea granulosa*, *L. plebeja*, *L. symmicta*, *Mycocalicium parietinum*, and *Rinodina lecideoides*.

Of unusual interest are the lichens which grew over other lichens: *Caloplaca invadens* on the thalli of *Lecanora* and *Placynthium*, *Buellia nivalis* on *Xanthoria* and *Caloplaca*, *Buellia scabrosa* on *Baeomyces*, and *Rhizocarpon pusillum* on *Sporastatia testudinea*.

A MIGRATORY LICHEN

The Amphi-Beringian species *Cetraria richardsonii* appears to be unattached to a substratum, blowing with the winds and rolling curled-up when dry, coming to rest in hollows and spreading open when moist.

Artificial key to the lichen genera of the North Slope

1		Thallus fruticose	2
		Thallus foliose, squamulose, or crustose	18
2	(1)	Thallus hollow	3
		Thallus with solid or at most partly fistulose center	5
3	(2)	Thallus bone white, pointed	*Thamnolia*, p. 248
		Thallus greenish yellow, yellow, or straw colored	4
4	(3)	Podetia with inner cartilaginous layer	*Cladonia*, p. 119
		Podetia lacking inner cartilaginous layer	*Dactylina*, p. 249
5	(2)	Thallus pendent with a solid central axis within a cottony medulla and an outer cortex	*Usnea*, p. 236
		Thallus lacking the solid central axis strand	6
6	(5)	Thallus all white, tips rounded, growing in moist seepages	*Siphula*, p. 238
		Thallus greenish yellow, straw colored, blue gray, or brown	7
7	(6)	Thallus greenish yellow, straw colored, or brown	8
		Thallus blue gray (sometimes with a brownish tinge)	16
8	(7)	Thallus with a continuous primary crustose or squamulose thallus on which are scattered short, erect stipes bearing large capitate apothecia	*Baeomyces*, p. 116
		Primary thallus lacking, thallus composed mainly of erect or pendent, branching podetia	9
9	(8)	Longitudinal sections showing cartilaginous strands interspersed in a cottony medulla	10
		Longitudinal sections showing only a uniform medulla	11
10	(9)	Cross section of branches flattened	*Ramalina*, p. 246
		Cross section of branches round or angular	*Evernia*, p. 247
11	(9)	Thallus cylindrical in cross section	12
		Thallus flattened or channeled in cross section	*Cetraria*, p. 213

12 (11) Cortex of longitudinally oriented hyphae   13
Cortex of irregularly or radially oriented hyphae
Cornicularia, p. 239

13 (12) Algae of thallus bright green; thallus over 4 mm tall   14
Algae of thallus blue green; thallus less than 4 mm tall
Polychidium, p. 32

14 (13) Thallus lacking pseudocyphellae, dark brown to black, appressed and attached to rock or gravel; cortex with superficial cells differentiated; spores frequently present, 8 per ascus, hyaline, simple   *Pseudephebe*, p. 240
Thallus often with pseudocyphellae, yellow green or brown to black, erect or pendent, loose on the substratum if on soil or attached if on trees; cortical cells not differentiated; spores rare, 8 or 2–4 per ascus, hyaline or brown, simple
15

15 (14) Thallus on ground, rarely on lower tree branches, yellow green or brown to black with paler base, pseudocyphellae always present, conspicuously raised; medullary hyphae knobby ornamented; containing either usnic acid or alectoralic acid; if spores present, 2–4 per ascus and brown
Alectoria, p. 241

Thallus epiphytic on trees, brown, pseudocyphellae present or not, adpressed to only slightly raised; medullary hyphae not ornamented; lacking usnic acid, species in this area lacking alectoralic acid; spores rare to unknown, 8 per ascus, hyaline   *Bryoria*, p. 243

16 (7) Thallus covered with granular to squamulose or coralloid phyllocladia, often with a tomentum between, and with scattered cephalodia containing blue green algae, the center solid   *Stereocaulon*, p. 108
Lacking phyllocladia, cephalodia, or tomentum   17

17 (16) Thallus surface smooth, usually shining, the center cottony; spores spherical   *Sphaerophorus*, p. 294
Thallus surface decorticate and warty, the center composed of longitudinal loose hyphae surrounded by dense mechanical tissue; spores ellipsoid   *Pilophorus*, p. 107

18 (1) Thallus foliose   19
Thallus squamulose or crustose   51

19 (18) Thallus umbilicate   20
Thallus not umbilicate   26

20 (19) Fruiting structure's flask-like perithecia imbedded in the thallus, only the mouths showing on the upper surface
Dermatocarpon, p. 289

Fruiting structure's apothecia usually superficial to sessile on the thallus 21

21 (20) Apothecia lecanorine *Lecanora*, p. 174
Apothecia lecideine 22

22 (21) Apothecia with flat disk, surface smooth and with proper margin 23

Apothecia with flat or convex disk, the surface with a central sterile button or fissure, or with gyrose secondary furrows 24

23 (22) Thallus pustulate *Lasallia*, p. 146
Thallus non-pustulate *Agyrophora*, p. 147

24 (22) Apothecia with flat disk with a sterile secondary button or fissure *Omphalodiscus*, p. 148
Apothecia convex, the disk furrowed 25

25 (24) Furrows of the disk concentric, a proper margin present
Umbilicaria, p. 150

Furrows of the disk radial, proper margin absent
Actinogyra, p. 155

26 (19) Thallus of small cochleate, blue green lobes with paler inrolled margins, upper cortex paraplectenchymatous, algae *Polycoccus* (primary squamules of *Cladonia* have been mistaken for this but lack inrolled margins and their algae are *Trebouxia*) *Coriscium*, p. 295

Thallus lacking the above criteria 27

27 (26) Thallus gelatinous when wet, homiomerous, with nostocoid algae 28

Thallus nongelatinous, heteromerous 31

28 (27) Cross section showing cortical layer 1 or more cells thick
Leptogium, p. 22

Cross section lacking cortex 29

29 (28) Spores with several cells, apothecia with thalloid margin
Collema, p. 25

Spores single-celled 30

30 (29) Apothecia with only proper margin *Leciophysma*, p. 29
Apothecia with thalloid margin (*Lempholemma*)

31 (27) Algae of thallus blue green   32
          Algae of thallus green   38

32 (31) Thallus lobes very small, less than 2 mm broad   33
          Thallus lobes more than 2 mm broad   34

33 (32) Marginal lobes longitudinally striated; spores 12–16 in ascus
                                                          Vestergrenopsis, p. 31
          Marginal lobes flattened, not longitudinally striated; spores
          8 in ascus   *Massalongia*, p. 32

34 (32) Underside cyphellate   *Sticta*, p. 53
          Underside lacking cyphellae   35

35 (34) Underside densely clothed with bluish black rhizoids
                                                            Coccocarpia, p. 39
          Underside with rhizoids but these not dense or bluish black
                                                                                 36

36 (35) Apothecia borne on the upper surface or margins, lower
          surface usually veined   37
          Apothecia borne on the lower surface which is then reflexed,
          lower surface pubescent or glabrous, sometimes with
          tubercles or papillae   *Nephroma*, p. 49

37 (36) Apothecia borne on margins, lower surface lacking cortex;
          spores 4–8-celled, thin walled, hyaline, pointed
                                                             Peltigera, p. 40
          Apothecia borne on upper surface, lower surface with
          cortex; spores 2-celled, dark brown, tips round
                                                             Solorina, p. 46

38 (31) Thallus with external or internal cephalodia containing blue
          green algae   39
          Thallus without such cephalodia   40

39 (38) Thallus with warty cephalodia external on the upper or
          lower surface of the thallus   *Peltigera*, p. 40
          Thallus with the cephalodia dispersed within the medulla
                                                            Nephroma, p. 49

40 (38) Thallus with pit-like cyphellae below   *Sticta*, p. 53
          Thallus without pit-like cyphellae   41

41 (40) Thallus bright yellow   *Cetraria*, p. 213
          Thallus gray, yellow green, brown, or black   42

42 (41) Lobes small, narrower than 1 mm   43
          Lobes large, more than 1 mm broad   44

43 (42) Tips shiny, medulla UV+ (divaricatic acid), spores simple,
          hyaline   *Parmeliopsis*, p. 209
          Tips dull, medulla UV- (no divaricatic acid), spores 2-celled,
          brown   *Physcia*, p. 263

44 (42) Upper side with network of raised ridging around depres-
          sions (scrobiculate), underside with fine pale tomentum
          interspersed by bald spots   *Lobaria*, p. 51
          Upper side lacking network (or if with network then lacking
          the fine tomentum below), underside bare or else with
          rhizinae but not fine tomentum   45

45 (44) Medulla hollow   *Hypogymnia*, p. 210
          Medulla well developed   46

46 (45) Apothecia almost always present, disk dark brown; spores
          dark brown, 2-celled   *Physcia*, p. 263
          Apothecia often absent; spores if present simple, hyaline (if
          upper side is white pruinose check *Physcia*)   47

47 (46) Pycnidia internal, the mouths appearing as black spots dis-
          tributed over the surface of the thallus *Parmelia*, p. 226
          Pycnidia superficial and mainly along the margins or lacking
                                                              48

48 (47) Cortex paraplectenchymatous (with the hyphae densely
          coherent but with the lumina large and a cellular appear-
          ance)   *Cetraria*, p. 213
          Cortex prosoplectenchymatous (the hyphae very thick
          walled and with minute lumina)   49

49 (48) Rhizines absent, purple pigment usually produced in lower
          part of medulla, especially dying parts *Asahinea*, p. 223
          Rhizines present, no purple pigment produced in medulla
                                                              50

50 (49) Caperatic acid present (feathery crystals in GE); upper cortex
          I-, medulla I+ blue; upper surface usually pseudocyphel-
          late; soredia common; pycnoconidia without inflated ends;
          spores small, subspherical   *Platismatia*, p. 225
          Caperatic acid absent; upper cortex often I+ blue, always
          pseudocyphellate; soredia rare; pycnoconidia with inflated
          ends; spores larger, ellipsoid   *Cetrelia*, p. 225

51 (18) Fruit a slender stalked or sessile structure tipped by an
          apothecium containing a mazaedium of spores   52
          Fruiting structure an apothecium or a perithecium   54

52 (51) Fruit a slender stalked structure   53
        Fruiting structure sessile   *Cyphelium*, p. 292

53 (52) Spores brown, ovoid to ellipsoid   *Mycocalicium*, p. 292
        Spores hyaline, spherical   (*Coniocybe*)

54 (51) Fruiting structure a perithecium   55
        Fruiting structure an apothecium (in *Pertusaria* it often
        opens by small pores but the spores are relatively very large
        and thick walled as compared to the Verrucariaceae) 62

55 (54) Thallus crustose, closely applied to or immersed in the sub-
        stratum   56
        Thallus squamulose   61

56 (55) Paraphyses soon gelatinizing or disappearing   57
        Paraphyses persistent   60

57 (56) Spores simple   *Verrucaria*, p. 280
        Spores several-celled to muriform   58

58 (57) Perithecia containing algae in the hymenial layer
                                               Staurothele, p. 287
        Perithecia without algae in the hymenial layer   59

59 (58) Spores 2–4(–6)-celled with only transverse septa
                                                Thelidium, p. 283
        Spores muriform, with both transverse and longitudinal
        septa   *Polyblastia*, p. 285

60 (56) Spores 3-septate   *Geisleria*, p. 289
        Spores muriform   *Microglaena*, p. 291

61 (55) Spores simple, hyaline, hymenial layer without algae
                                              Dermatocarpon, p. 289
        Spores muriform, brown, hymenial layer containing algae
                                               Endocarpon, p. 290

62 (54) Apothecia with double margins, an inner proper exciple and
        an outer thalloid one   63
        Apothecia with either a proper exciple or a thalloid margin
        but not both   64

63 (62) Inner exciple dark, fimbriate, spores dark, muriform
                                              Diploschistes, p. 279
        Inner exciple pale waxy, entire, spores hyaline, 3-septate
                                                  Gyalecta, p. 54

64 (62) Apothecia with proper exciple   65
        Apothecia with thalloid margin (margin containing algae)
                                                             83

65 (64) Algae of thallus blue green   66
          Algae of thallus green   70

66 (65) Spores simple, hyaline   67
          Spores fusiform, 2-celled   69

67 (66) Apothecia immersed in the thallus   68
          Apothecia sessile on the thallus, thallus heteromerous with upper cortex paraplectenchymatous, alga *Nostoc*, spores ellipsoid, thallus squamulose   *Parmeliella*, p. 37

68 (67) Thallus coralloid or granular, on rocks, medulla with hyphal cells in fountain-like arrangement, alga *Calothrix*, spores ellipsoid   *Porocyphus*, p. 21
          Thallus squamulose-areolate, on soil, medulla paraplectenchymatous, alga *Scytonema*, spores fusiform
                                          Heppia, p. 30

69 (66) Spores 2-celled, alga *Nostoc* in glomerules
                                     Massalongia, p. 32
          Spores 2-3-4-celled, alga *Rivularia* or *Scytonema* in chains
                                     Placynthium, p. 34

70 (65) Apothecium surface turning purple with KOH   71
          Apothecium surface KOH-   72

71 (70) Spores simple   *Protoblastenia*, p. 268
          Spores polaribilocular   *Blastenia*, p. 269

72 (70) Spores simple   73
          Spores 2- or more-celled   78

73 (72) Spores 1-8, rarely 32 per ascus   74
          Spores many per ascus   76

74 (73) Spores small, lees than 40 μ long, thin walled   75
          Spores large, 70-100 × 35-45 μ, thick walled
                                     Mycoblastus, p. 88

75 (74) Asci with conspicuous 'tholus' (upper part thick gelatinous and staining dark blue with iodine); paraphyses free or only slightly coherent; medulla I-, containing xanthones
                                     Lecidella, p. 56
          Asci without 'tholus' (ascus I- or the entire ascus only slightly blue); paraphyses always gelatinized and coherent; medulla I+ blue or I-, lacking xanthones   *Lecidea*, p. 59

76 (73) Apothecia immersed in the thallus which is radiate-areolate and C+ red   *Sporastatia*, p. 170
          Apothecia sessile, thallus not radiate-areolate, C-   77

77 (76) Apothecia lecideine (upper part of hymenium and/or exciple black), on rocks *Sarcogyne*, p. 168
Apothecia biatorine (upper part of hymenium and exciple pale, not black), on soil or bark *Biatorella*, p. 169

78 (72) Spores without gelatinous epispore, paraphyses unbranched, conglutinate or free 79
Spores with gelatinous epispore, 2-celled to muriform, usually soon brown but hyaline in some species, paraphyses anastomosing branched *Rhizocarpon*, p. 98

79 (78) Spores brown, 2–4-celled or muriform *Buellia*, p. 252
Spores hyaline to pale brownish 80

80 (79) Spores 2-celled, hyaline, ellipsoid; thallus granulose, not squamulose *Catillaria*, p. 89
Spores 2-celled fusiform and thallus squamulose, or spores parallel 4- or more-celled or muriform 81

81 (80) Spores parallel-celled, not muriform 82
Spores muriform *Lopadium*, p. 97

82 (81) Spores 2-celled fusiform to parallel 8-celled; thallus pseudocorticate, squamulose *Toninia*, p. 90
Spores 3-many-celled, cells parallel; thallus ecorticate, granulose crustose but not squamulose *Bacidia*, p. 92

83 (64) Algae of thallus blue green 84
Algae of thallus green 88

84 (83) Thallus tiny lobate-crustose and closely applied to the rock 85
Thallus squamulose or granulose 87

85 (84) Thallus margins striate or grooved lobate, thallus heteromerous 86
Thallus margins not striated or grooved but lobate, thallus homiomerous *Pyrenopsis*, p. 31

86 (85) Thallus lobes longitudinally striated; spores 12–16 per ascus, ellipsoid or globose *Vestergrenopsis*, p. 31
Thallus lobes at margin with at most 1 groove; spores 8 per ascus, ellipsoid *Placynthium*, p. 34

87 (84) Thallus granulose to coralloid crustose; apothecia immersed; spores ellipsoid, walls even *Porocyphus*, p. 21
Thallus small squamulose; apothecia superficial; spores ellipsoid with rough to tuberculate walls *Pannaria*, p. 38

88 (83) Thalli forming rosettes on rocks with the center covered by a rosette-like cephalodium containing blue green algae
Placopsis, p. 171

Thallus without central rosette-like cephalodia 89

89 (88) Spores many per ascus 90
Spores 1–8 per ascus 95

90 (89) Apothecia with completely dark proper exciple, immersed in a radiate areolate thallus which is C+ red; hymenium blue green above *Sporastatia*, p. 170

Apothecia with a proper exciple dark only on the outside or lacking and immersed in the thallus; hymenium pale or brownish above 91

91 (90) Apothecia immersed in the thallus 92
Apothecia not immersed in the thallus, thallus often poorly developed 93

92 (91) Apothecia simple; thallus attached along most of the underside by the medullary layer *Acarospora*, p. 164

Apothecia compound; thallus nearly umbillicate, only the central portion attached, the rest free *Glypholecia*, p. 168

93 (91) Thallus and apothecia golden yellow, K-, spores simple or 2-celled *Candelariella*, p. 205

Thallus (if present) not yellow, apothecia not golden yellow, spores simple 94

94 (93) Apothecia lecideine (upper part of hymenium and/or the exciple black), on rocks *Sarcogyne*, p. 168

Apothecia biatorine (upper part of the hymenium and the exciple pale, not black), on soil or bark *Biatorella*, p. 169

95 (89) Spores large, over 30 μ long, paraphyses anastomosing 96
Spores less than 30 μ, thin walled, paraphyses unbranched or only slightly branched, not anastomosing 97

96 (95) Spore walls thick *Pertusaria*, p. 156
Spore walls thin *Ochrolechia*, p. 172

97 (95) Spores with tuberculate walls, on humus and over mosses, apothecia with granulose-squamulose margins
Psoroma, p. 36

Spores with evenly thickened walls or polaribilocular, not tuberculate, apothecial margins not granulose-squamulose 98

98 (97) Apothecial disk turning purple with KOH   99
           Apothecial disk not turning purple with KOH   100

99 (97) Spores unicellular   *Fulgensia*, p. 270
           Spores polarbilocular   *Caloplaca*, p. 270

100 (98) Thallus and apothecia bright yellow or orange yellow; spores simple to 2-celled, hyaline, partly bean-shaped
                                      Candelariella, p. 205
           Thallus and apothecia not bright yellow   101

101 (100) Spores simple   102
           Spores septate   103

102 (101) Algae *Trebouxia*; margin distinctly lecanorine *Lecanora*, p. 174
           Algae *Trentepolia*; with a distinct proper margin covered more or less by a thalloid margin   *Ionaspis*, p. 200

103 (101) Spores hyaline   104
           Spores brown, 2- or 4-celled   *Rinodina*, p. 256

104 (103) Spores 1(–3)-septate, ellipsoid; paraphyses free
                                      Lecania, p. 203
           Spores (1–)3-septate or more, fusiform or acicular; paraphyses conglutinate   105

105 (104) Disk pale yellowish to rosy flesh colored; spores 1–3-septate, fusiform; on humus or moss   *Icmadophila*, p. 203
           Disk scarlet; spores 3–7-septate, acicular; on rocks
                                      Haematomma, p. 204

Lichens of the North Slope

LICHINACEAE

Gelatinous lichens with crustose, squamulose, or thread-like fruticose thalli, with or without rhizinae, the interior hyphae in parallel or fan- or fountain-like arrangements in erect portions of the thallus. Apothecia immersed or sessile, closed or becoming open, with thalloid exciple and/or proper exciple, the proper exciple black green or brown; asci cylindrical or clavate; spores 8 to 48, single-celled, hyaline. Pycnidia on thallus margins, pycnoconidia terminal, round or cylindrical. Algae: various blue-green genera.

1. POROCYPHUS Körb.

Thallus tiny, squamulose, granular, coralloid, or fruticose, lacking rhizinae, interior hyphae with rounded cells in loose fan-like or fountain-like arrangement or pseudoparenchymatous toward the base of thallus. Apothecia on the surface and immersed or sessile or terminal at tips of erect lobes, with thalloid exciple or later also with proper exciple; asci cylindrical to clavate; spores 8 or 16, single-celled, hyaline; paraphyses net-like anastomosing. Algae: *Calothrix*.

1. **Porocyphus coccodes** (Flot.) Körb. Thallus squamulose, granular or coralloid, forming areoles to 2.5 mm broad, 1 mm high, squamules of flat thallus thick, smooth or slightly effigurate, the erect thallus more or less developed, black, rarely green. Apothecia to 0.3 mm broad, when young with thalloid exciple and blackish, later with brown proper exciple, superficial on the flat thallus or at tips of erect lobes, the disk red brown; hymenium 110–120 μ; hypothecium 45–55 μ; asci cylindrical, 55–80 × 9–11 μ, paraphyses net-like anastomosing or unbranched with thicker tips; spores 8, 1-celled, oval, hyaline, 11–17.5 × 7–13 μ. The hymenial gelatin is greenish blue, becoming wine red with I.

This is a species of moist rocks. It is reported from Europe, North Africa, and North America.

Locality: 1.

COLLEMATACEAE

Foliose homiomerous lichens with or without a cortical layer. Apothecia lecanorine, with a thalline exciple as well as a proper exciple, or with only a proper exciple; spores non-septate or from 2-celled to muriform. Alga: *Nostoc*.

1   Spores several- to many-celled; apothecia lecanorine with thalloid margin

              2

    Spores single-celled   3

2 (1) Thallus with a single cortical layer one or more cells thick   1. *Leptogium*

       Thallus lacking layer of cortex   2. *Collema*

3 (1) Apothecia with only proper exciple   3. *Leciophysma*

       Apothecia with thalloid margin   (*Lempholemma*)

1. LEPTOGIUM S. Gray

Thallus foliose, rarely crustose or fruticose, with upper and lower cortices of single layers of isodiametric cells in the arctic species, the internal structure homiomerous with strands of algal cells scattered among the loosely interwoven hyphae in some species or distributed among the paraplectenchymatous cells in others. Apothecia usually adnate to sessile, laminal or marginal, discoid, the disk brown to black, both thalloid and proper exciples present; spores usually 8, 2-celled to several-celled or muriform, hyaline.

1   Thallus paraplectenchymatous throughout, lobes isidiate on margins and surface   1. *L. tenuissimum*

    Thallus with upper and lower cortices, hyphae loose within   2

2 (1) Underside tomentose, upper side with black granules   2. *L. saturninum*

       Underside lacking tomentum   3

3 (2) Surface of thallus longitudinally wrinkled   4

       Surface of thallus not wrinkled or at most irregularly wrinkled, lobe tips entire or irregularly cut, not isidiate-fringed   5. *L. sinuatum*

4 (3) Lobes with curled-in antler-like tips   3. *L. palmatum*

       Lobes with margins and tips finely isidiate-fringed   4. *L. lichenoides*

1. **Leptogium tenuissimum** (Dicks.) Fr. Foliose thalli very small, to 2 mm broad, the lobes to 0.2 mm broad, flattened to almost round, sometimes short and crowded appearing as coralloid outgrowths, surface smooth or granular, lead gray to brown or blackish, paraplectenchymatous throughout, the algal cells in chains or clusters throughout the thallus. Apothecia common, adnate to sessile on upper surface of thallus; disk concave to flat, light to dark brown or black; thalloid exciple entire to lobulate; proper exciple euparaplectenchymatous; hymenium 105–180 μ thick, hyaline with yellowish to brownish epithecium; paraphyses unbranched, 1 μ with slightly thickened apices; asci cylindrico-clavate, 90–150 × 12–14 μ; spores 8, uniseriate to irregular, ovoid to ellipsoid or subfusiform, apices rounded or pointed, 17–37 × 9–14 μ, 3–7-septate transversely, 0–2-septate longitudinally.

On sandy or clay soil, also reported on sandstone and tree barks. This species is known from Europe and Asia, and in North America from Quebec to Alaska, south to South Carolina, Minnesota, Colorado, and California.

Localities: 1, 2, 10, 13.

2. **Leptogium saturninum** (Dicks.) Nyl. Thallus foliose, of broad lobes 3–10 mm broad, spreading, the margins entire to occasionally irregularly cut or isidiate, sometimes curling under, lead gray to usually black, upper surface with abundant granular isidia which are occasionally slightly coralloid, upper surface smooth, lower surface with a dense tomentum; upper cortex of one or sometimes partly two cells, lower cortex of larger cells than the upper, hyphae of interior loosely interwoven, algal cells and chains distributed throughout the thallus. Apothecia rare, sessile on the upper surface, to 2.5 mm broad, disk flat to convex, brown to red brown, thalloid exciple entire or granulose, of the same color as the thallus or cream colored, proper exciple euparaplectenchymatous; hymenium 100–125 μ, hyaline with thin brown epithecium; paraphyses unbranched, 1–2 μ thick, slightly thicker at the apices; asci cylindrico-clavate, 85–115 × 12–16 μ; spores 8, irregular in the ascus, 20–25 × 7–10 μ ellipsoid, the apices rounded to pointed, 3–4-septate transversely, 1-septate longitudinally.

This species is commonest on the bases of willows and birch in the arctic, occasionally on rocks, usually in very damp low sites. It is circumboreal and low arctic as well as temperate. It is known from Europe, Asia, and in North America from Baffin Island to Alaska and south to New England, Iowa, New Mexico, Arizona, and California.

Localities: 1, 2, 3, 9, 11, 14.

3. **Leptogium palmatum** (Huds.) Mont. *in* Webb. & Berth. Thallus foliose, of tufted mainly erect lobes to 20 mm long, 6 mm broad with the margins curling inward especially at the apices to form antler-like tips, surface wrinkled, the heaviest wrinkles running the length of the lobes, becoming shiny at the apices, tufts of hairs sometimes developed; cortex of one layer of cells on each side of thallus, interior with interwoven somewhat compact hyphae with the algal chains distributed through the thallus. Apothecia frequent, adnate to sessile to 0.6 mm broad; disk concave to flat, brown to red brown; thalloid exciple entire, of same color as thallus or cream colored; proper exciple euparaplectenchymatous; hymenium 125–300 μ thick, hyaline with thin brown epithecium; paraphyses unbranched, ca. 1 μ thick, tips slightly thickened; asci cylindrico-clavate, 130–185 × 16–25 μ; spores 8, biseriate or irregular, ellipsoid to subfusiform, apices rounded or pointed, 30–56 × 10–20 μ, 5–9-septate transversely, 1–2-septate longitudinally.

This species grows on soil and rocks among mosses. It is an oceanic species in Europe and Japan, and in North America grows along the west coast from California to Alaska.

Locality: 1.

4. **Leptogium lichenoides** (L.) Zahlbr. Foliose, of erect to semi-erect lobes to 4 mm broad, rounded to elongate with the margins finely lobulate to fimbriate, the surfaces distinctly wrinkled, the heaviest wrinkles running the length of the lobes, the underside with scattered hapters; cortex of single layer of cells, interior of loosely interwoven hyphae and the algal chains scattered through the thallus. Apothecia common, sessile on the upper surface of the thallus, to 1 mm broad; disk concave, brown to dark red brown; thalloid exciple with numerous lobulate to coralloid outgrowths of same color as thallus; proper exciple euparaplectenchymatous; hymenium 130–200 μ thick, hyaline with thin yellow to brown epithecium; paraphyses unbranched, about 1.5 μ thick, the apices little thickened; asci cylindrico-clavate, 125–185 × 16–21 μ; spores 8, biseriate to irregular, ellipsoid to subfusiform, apices rounded to pointed, 18–45 × 11–16 μ, 5–9-septate transversely, 1–3-septate longitudinally.

This species grows on calcareous rocks and soils among mosses, occasionally forming tufts or cushion-like masses (f. *pulvinatum*). It is a circumpolar temperate and arctic species with a wide range in North America, lacking only in the southeast coastal plains, the central plains, and the Great Basin.

Localities: 1, 9, 14.

5. Leptogium sinuatum (Huds.) Mass. Foliose with lobes which may become erect or semi-erect, forming small cushions, the lobes rounded, to 4 mm broad, the margins entire to irregularly cut, surface roughened to distinctly and irregularly wrinkled, dull or shiny especially near margins, cortex a single layer of cells, the interior loosely interwoven with the algae in chains through the thallus. Apothecia common and numerous, adnate to sessile on the upper surface; disk concave to flat, light to dark brown; thalloid exciple entire, of same color as thallus, sometimes with wrinkles following around the edge; proper exciple euparaplectenchymatous; hymenium 115–170 μ thick; paraphyses slightly thicker at apices; asci cylindrico-clavate, 95–150 × 14–23 μ; spores 8, uniseriate to irregular, ellipsoid, the apices rounded to pointed, 25–35 × 12–14 μ, 7-septate transversely, 1–2-septate longitudinally.

Growing on soil, bark, and rocks. A circumpolar species little collected in North America but known from the eastern states, the Rocky Mountains, and the west coast from California to Alaska.

Locality: 2.

2. COLLEMA G.H. Web.

Foliose, rarely subfruticose or crustose lichens, gelatinous when wet, lobed and often also lobulate, olive green or blackish, underside attached by hapters, homiomerous (the algae distributed throughout the thallus), the hyphal context lax, cortex lacking except in apothecial tissues. Apothecia round, sessile, flat, concave or convex, lecanorine or with proper margin; disk reddish or darker; the thalline exciple lacking cortex; proper exciple with the cells more or less parallel to the surfaces (euthyplectenchymatous) or distinctly irregularly cellular (paraplectenchymatous); paraphyses simple or branched, septate, the apices thickened; asci subcylindrical to clavate, the apical membrane thick; spores 4 or 8, variable in shape, 1- to several-septate to usually muriform, hyaline or pale colored. Pycnidia immersed, globose, fulcra endobasidial, pycnoconidia scarce, oblong, usually straight, the ends usually slightly swollen. Some species with internal conidia (*C. bachmanianum* in arctic).

1   Thallus with swollen lobules over the surface   2
     Thallus lacking swollen lobules   3

2 (1) On rocks; lobules raised on edge; margins of apothecia entire; pycnidia present   1. *C. polycarpon*
     On soil and humus; with flattened or globular lobules; margins of apothecia crenulate; pycnidia lacking
                    2. *C. bachmanianum* var. *millegranum*

3 (1) Thallus foliose to subfruticose with terete, papilliform nodules, forming
      cushions; spores 4 per ascus   3. *C. ceraniscum*
      Thallus distinctly foliose; spores 8 per ascus   4

4 (3) Isidia lacking   5
      Isidia present   6

5 (4) Lobes broad; spores straight, 3-celled, 26–43 μ
                                    4. *C. undulatum* var. *undulatum*
      Lobes narrow; spores curved, 3-celled, 17–30 μ 5. *C. multipartitum*

6 (4) Isidia more or less globular   7
      Isidia coralloid and abundant   7. *C. glebulentum*

7 (6) Thallus very blackish, lobules with strongly undulating margins,
      100–300 μ thick; spores 4-celled with rounded ends
                                 4. *C. undulatum* var. *granulosum*
      Thallus olive green, 100–200 μ thick, somewhat pustulate; spores
      muriform   6. *C. tuniforme*

1. **Collema polycarpon** Hoffm. Thallus foliose, of numerous radiating
lobes covered with flattened lobules which are raised on edge, the tips
usually swollen, with usually dense branching to give a cushion-like
appearance; dark olive green to blackish. Apothecia numerous, sessile on
the margins or tips of the lobes, circular, small, to 1.5 mm in diameter;
the disk red brown to blackish, more red when moist; the margin thin,
entire, smooth, of the same color as the thallus; exciple euthyparaplec-
tenchymatous (of cells running parallel to the outer surface and somewhat elongate); hymenium 65–106 μ high, I+ blue; paraphyses 2–3 μ,
simple or branched toward the 4–6.5 μ apices; asci subcylindrical to
clavate, 65–80 × 13–20 μ; spores 8, more or less biseriate, straight, with
acute or sometimes partly with rounded tips, usually 4-celled, hyaline,
13–30 × 5–8.5 μ. Pycnidia frequent, immersed, globose; pycnoconidia
straight or slightly curved, swollen at ends, 5–6.5 × 1–1.5 μ.

On limestone rocks, usually on the bare surface, frequently in seepages and fissures of the rocks. Circumpolar, arctic-alpine as well as
temperate, known from Europe, Asia, North Africa, Greenland, Spitzbergen, and in North America south to Wisconsin and Colorado.
Locality: 11.

2. **Collema bachmanianum** (Fink) Degel. Thallus foliose, rigid, thick,
strongly swollen when moist, deeply lobed, the lobes radiating, with
flattened lobules, often coarsely verrucose with globular lobules on the
upper surface, dark olive green, occasionally yellowish, bluish, or
grayish. Apothecia usually common, on the upper surface or margins,
sessile with constricted base, to 3 mm diameter; disk plane to very

concave, light to dark red, smooth; the margin becoming thick, coarsely crenulate to lobulate; exciple of a few layers of hyphae parallel to the surface of the apothecium; hymenium 100–135 μ thick, I+ blue, paraphyses 2 μ, simple or branched at the apices; asci subcylindrical to clavate, 65–110 × 13–26 μ; spores 8, uniseriate to biseriate, ellipsoid with pointed or blunted ends, muriform with 3 transverse, 1 longitudinal septa, yellowish brown, 20–36 × 8.5–15 μ. Pycnidia lacking; conidia internal, 6–16 × 2–3 μ.

A species of bare soil containing lime. Known from Europe, Siberia, Alaska, Canada, and in the United States south to Tennessee, Illinois, Iowa, and Colorado.

Localities: 1, 2, 9, 10.

The specimens are all of var. *millegranum* Degel. which is distinguished from the species by the globular isidia on the thallus, the usually non-crenulate apothecial margin, and thinner, broader lobes. In the typical var. *bachmanianum* the upper surface lacks the globular isidia (but has terminal lobules) and the apothecial margin is markedly crenulate. When sterile it is indistinguishable from C. *tenax*.

3. **Collema ceraniscum** Nyl. Thallus very small, foliose to subfruticose, in small cushions, deeply lobate, the lobes ascendent, the lower parts flattened, to 2 mm broad, smooth to nodulose, repeatedly branched, the end branches terete to papilliform or nodulose, sometimes flattened, brown or blackish brown. Apothecia usually numerous and covering the thallus, sometimes sparse, sessile with constricted base in or near apices of lobules, subglobose; disk strongly concave to flat, to 0.8 mm in diameter, red to red brown or blackish, paler when wet, glossy, smooth; the margin thick, entire or papillose or lobulate; exciple subparaplecten-chymatous; hymenium 150–190 μ, I+ blue; paraphyses 1.5–3 μ, simple or branched, the tips darkened and thicker, 4.5–8.5 μ; asci subcylindrical to clavate 78–130 × 13–34 μ; spores 4 (rarely 2), broadly oblong or oval or subcubic, muriform with 3–4 transverse, 1–3 longitudinal septa, the cells more or less globose, 20–40 × 13–22 μ. Pycnidia not seen.

This arctic species grows mainly over mosses near the ground, rarely on base soil, in limestone districts. It is known from Europe, Siberia, Iceland, Greenland, Bear Island, Spitzbergen, Novaya Zemlya, in Canada from Ellesmere Island, Baffin Island, Fort Reliance, and Coppermine, and from Alaska.

Localities: 7, 8, 9.

4. **Collema undulatum** Flot. Thallus foliose, deeply lobed, thin to fairly thick, repeatedly branched, concave with ascending undulate margins,

usually entire, neither swollen nor glossy, smooth or in var. *granulosum* Degel. with globular isidia, underside with hapters or dense rhizinae, dark olive green to blackish. Apothecia usually numerous, often lacking in var. *granulosum*, on the margins or upper surface, sessile, becoming constricted at base; disk flat to convex, dark red to red brown, slightly glossy, epruinose; margin thin to thick, entire or crenulate, of same level as disk or lower; exciple euparaplectenchymatous; hymenium 85–105 μ, I+ blue; paraphyses simple or abundantly branched, 2–4 μ; apices clavate, 4.5–6.5 μ; asci clavate, 60–70 × 13–20 μ; spores 8, biseriate, linear oblong with rounded ends, 3-septate, hyaline, 17–30 × 6.5–9 μ. Pycnidia common, marginal, globose; pycnoconidia straight, 4–4.5 × 1–1.5 μ.

On calcareous rocks or in var. *granulosum* also on earth. Known from Europe, Iceland, Greenland, Spitzbergen, Novaya Zemlya, and Alaska.

Localities: 1, 8, 14. Latter two of var. *granulosum* Degel.

The var. *granulosum* differs from typical var. *undulatum* in having globular isidia 0.1–0.2 μ in diameter and in being more often without apothecia. It grows on both rocks and soil, usually calcareous.

5. **Collema multipartitum** Sm. Thallus deeply lobate, more or less adnate, fragile, dark olive green to brownish or blackish, dull or slightly glossy, epruinose, lacking isidia, smooth or usually minutely longitudinally striate, the lobes quite convex but not swollen, the margin entire. Apothecia 1–2 mm diameter, sessile, flat, the margin entire or lobulate; disk flat to slightly convex, dark red to blackish, smooth, epruinose; proper exciple euparaplectenchymatous (cells ± isodiametric globose to somewhat polygonal); hymenium 90–110 μ thick, upper part brownish, I+ blue; paraphyses simple or upper part irregularly branched, 2–4.5 μ thick; asci clavate, 60–80 × 17–22 μ; spores 8, hyaline, linear oblong, straight or curved, 3-septate, 20–60 × 4.5–6.5 μ.

This species grows on open calcareous rocks. It is known from Europe and North Africa, and Degelius (1954, 1974) has reported it from British Columbia, Alberta, and Alaska (central Yukon River valley, White Mountains).

Locality: 8.

6. **Collema tuniforme** (Ach.) Ach. em Degel. Thallus foliose, deeply and broadly lobed, fairly thin, irregularly minutely pustulate, isidiate with globular isidia to 0.3 mm diameter, dull, dark olive green to blackish. Apothecia numerous or lacking, on the surface, sessile to with constricted base; disk flat, occasionally convex or concave, dark red to red brown, occasionally glossy; margin thin to thickish, smooth or isidiate; exciple euparaplectenchymatous; hymenium 85–130 μ, I+ blue; para-

physes simple or branched, 2–3 μ, tips globose or clavate 4.5–5 μ; asci clavate, 80–90 × 15–20 μ; spores 8, biseriate, broadly oval, sometimes ovoid or globose, tips rounded, muriform, 3 transverse, 1–3 longitudinal septa, 15–28 × 6.5–15 μ. Pycnoconidia straight 4.5–6 × 1.5 μ.

On calcareous rocks, rarely in calcareous areas on wood or roots, among mosses and on calcareous soils. It is very widely distributed in the temperate as well as arctic zones and is known from Europe, Siberia, North Africa, Iceland, Greenland, Canada, Iowa, Utah, and Alaska.

Localities: 2, 8, 13, 14.

The isidiate forms of C. *tenax*, which also occur in Alaska (Port Clarence), may be distinguished by their swollen and plicate lobules and differing exciple structure (of cells parallel to the surface) as well as the habitat on soil.

7. **Collema glebulentum** (Cromb.) Degel. Thallus foliose, thin, rounded, deeply lobate and also lobulate with superficial and marginal terete papilliform to coralloid isidia; upper surface dull or slightly glossy, dark olive green to blackish; lower surface with hapters. Apothecia unknown, pycnidia very rare and undeveloped.

Growing on siliceous as well as calcareous rocks, this is an arctic-alpine species of northern Europe and the Alps, Novaya Zemlya, Greenland, in Canada from Baffin Island, and Fort Reliance, NWT, and Alaska.

Locality: 11.

3. LECIOPHYSMA Th. Fr.

Thallus very short, granulose to subfruticose, lobes globose to cylindrical, lacking cortex, the algae in glomerules distributed in the medulla. Apothecia lecideine, lacking a thalline margin; the proper margin distinct or not; the proper exciple with radially arranged hyphae, hymenium I+ blue; paraphyses sparingly branched, tips dark; asci cylindrical; spores 8, hyaline, 1-celled, spherical or ellipsoidal. Pycnidia in the thallus, pycnoconidia short, cylindrical. Algae: *Nostoc*.

1. **Leciophysma finmarkicum** Th. Fr. Thallus caespitose in small rosettes to 15 mm broad, lobes erect, branching, to 0.8 mm tall attached by tufts of rhizinae, verruculose, blackish brown, lacking cortex, the medulla with a loose network of hyphae forming broad lacunae, the *Nostoc* in chains and distributed through the thallus. Apothecia black, rarely brown, to 1.0 mm broad; the proper margin indistinct; subhymenium hyaline or brownish 110–170 μ high; hymenium 120–190 μ, upper part brown; paraphyses filiform, 1–1.5 μ, lax, gelatin abundant; asci

subclavate 85–115 × 12.5–17.5 μ; spores 8, hyaline, globose to ovoid 16–23 × 11–14 μ.

Growing on mosses and other lichens over limestone soils. This species is known from Scandinavia, Novaya Zemlya, Bear Island, Iceland, Greenland, and in North America from Ellesmere Island, Axel Heiberg Island, Coppermine, NWT, and Alaska.

Localities: 1, 9.

HEPPIACEAE

Squamulose, rarely crustose-areolate or subfruticose lichens with a lower paraplectenchymatous thallus usually lacking an upper cortex but covered by a necral layer, a paraplectenchymatous or loose medullary layer, immersed or sessile apothecia; 8 or numerous spores which are single-celled and hyaline. Pycnidia on upper surface or tips of lobes, conidia exobasal, fusiform to bacilliform. Algae: *Scytonema* or *Anacystis* in Cyanophyceae.

One genus, *Heppia*, is in the arctic and has 8 spores; the more southerly genus *Peltula* has numerous spores.

1. HEPPIA Naeg. in Hepp

Thallus squamulose, broadly attached by hapters, more or less paraplectenchymatous throughout but heteromerous with medulla, asci with 8 spores, without gelatinous sheath, spores hyaline 1-celled. Algae: *Scytonema*.

1. **Heppia lutosa** (Ach.) Nyl. Thallus of squamules 1–7 mm broad, concave or flat, rounded to irregular or elongate but not deeply lobed, margins sometimes granular, upper surface rough and cracked, tan to dark olive, most of lower surface attached by hapters; algal cells in columns, clumped in pairs or single. Apothecia usually 1 per squamule, immersed in the upper surface; disk to 1.5 mm broad, lacking thalloid exciple; epithecium yellowish brown; hymenium 103–120 μ I+ wine red; paraphyses 3–4.6 μ thick, septate; asci clavate, 68 × 23 μ; spores 8, irregular, hyaline, fusiform, 15–24 × 6.8–10.3 μ.

A species of calcareous soil. It is temperate circumpolar in distribution, in North America from New England to Florida and westward to California and Saskatchewan, the Alaska report seemingly disjunct. It is reported from Greenland (Bocher 1954).

Locality: 9.

PYRENOPSIDACEAE

Thallus crustose, tiny, squamulose or fruticose, attached to the substratum with rhizoids, homiomerous, the interior hyphae net-like, anastomosing. Apothecia immersed or adnate with thalloid exciple, paraphyses unbranched, spores 8(–32) spherical to ovoid, 1-celled, hyaline. Pycnidia immersed, pycnoconidia elongate, straight or curved. Algae: *Nostoc*.

1. PYRENOPSIS Nyl.

Crustose thallus of small granular to coralloid lobes which may form areolate masses, the lobes homiomerous without layers, the inner part with loosely interwoven hyphae and glomerules of the alga. Apothecia immersed or sessile, discoid, closed to open, concave to flat; the disk brown to blackish; the exciple thalloid; paraphyses more or less clear, unbranched, septate; asci clavate; spores 8, ellipsoid to subspherical, 1-celled, hyaline.

1. **Pyrenopsis pulvinata** (Schaer.) Th. Fr. Crustose red brown thallus, warty to coralloid, lacking layers, attached to the substratum by rhizoids. Apothecia adnate or immersed; with thalloid exciple, on the margins of the lobes; the disk almost closed to open, red brown to black; hymenium 150 μ, brownish; asci clavate; spores ellipsoid, hyaline, 9–12 × 5–6 μ.

A species of calcareous rocks, sometimes among mosses. Reported from Europe, Greenland, Iceland, Spitzbergen, and in North America from New England, Devon Island, west to Minnesota and Alberta, south to Arizona.

Locality: 1.

PLACYNTHIACEAE

Thallus small foliose to crustose. Apothecia hemiangiocarpic. Algae: *Nostoc*, *Rivularia*, or *Scytonema*.

1   Asci with 12–16 spores, spores 1-celled, ellipsoid or globose
                                                        1. *Vestergrenopsis*
      Asci with 8 or less spores, spores 2–4-celled, ellipsoid to fusiform   2

2 (1) Alga *Nostoc*, cells 2–5 μ, in rounded clumps, the cells round   3
      Alga *Rivularia* or *Scytonema*, cells over 5 μ, in straight or curved chains, the cells oval   4. *Placynthium*

3 (2) Thallus foliose, flattened, isidiate, dull   2. *Massalongia*
     Thallus fruticose, terete, non-isidiate, but coralloid, tips shining or dull
                 3. *Polychidium*

1. VESTERGRENOPSIS Gyel.

Thallus of radiating small lobes closely attached to each other and to the substratum, isidiose or not, esorediate; interior hyphae mainly longitudinally extended. Apothecia hemiangiocarpic with thalloid exciple. Hymenium brown above, I+ blue; paraphyses septate, simple or sparingly branched, tips thickened and moniliform; asci cylindrical or clavate with apically thickened walls; spores 12-16, hyaline, 1-celled ellipsoid. Algae: *Scytonema*.

1. **Vestergrenopsis isidiata** (Degel.) Dahl Thallus of tiny radiating lobes, flat, longitudinally grooved-striated or rarely smooth, 2-6 mm long, 0.2-0.4 mm broad, tips broadened to 0.8 mm, isidia on the upper surface, globose or cylindrical, becoming flattened with pale underside, upper side dark olive brown, lower side pale, rhizinae pale; interior hyphae longitudinally oriented, those in the center cylindrical, those toward the surfaces ellipsoid, loosely arranged. Apothecia to 1 mm diameter with thalloid exciple; disk brown; hymenium 80-125 μ thick, hyaline, upper part brown; subhymenium hyaline, 40-70 μ, of ellipsoid cells in radiating rows; pseudoexcipulum with large, broadly ellipsoid cells; paraphyses septate, simple or sparingly branched, 2 μ thick, the tips broadened to 7 μ, often moniliform; asci 55-58 × 11.5-13 μ; spores (8-)12-16, hyaline, 1-celled but frequently with cytoplasmic bridges and appearing 2-celled, 7-10 × 4-6 μ.

This species may grow on vertical cliffs or on rock faces which have temporarily running water over them. It is known from Sweden, Greenland, Baffin Island, Devon Island, and Alaska, where it has been reported from Mt McKinley National Park and Mendenhall Glacier near Juneau.

Localities: 8, 11.

2. MASSALONGIA Körb.

Thallus small foliose, lobes squamulose or elongated, more or less isidiose, esorediate, underside pale with brown or blackish rhizinae; upper cortex paraplectenchymatous, medulla with loosely interwoven hyphae, filled with globose *Nostoc* colonies, lower surface of densely interwoven hyphae. Apothecia hemiangiocarpic, laminal or marginal, with proper exciple; hymenium brown above, hyaline below, I+ blue; paraphyses septate, simple or branched; asci cylindrical with apically thickened walls; spores 8, hyaline, ellipsoid to fusiform, 2-4-celled, pycnidia brown, conidiophores short-celled, branched, producing the pycnoconidia terminally and laterally, pycnoconidia short, cylindrical.

1. **Massalongia carnosa** (Dicks.) Körb. Prothallus lacking. Thallus tiny foliose, of flattened lobes 0.5-2.0 mm broad, 0.5-3(-10) mm long, irregularly branched and overlapping, upper side brown, smooth or with round or globose isidia, the isidial tips sometimes breaking and simulating soredia, dull, underside whitish to brown with brown rhizinae; hyphae forming an upper cortex of 3-8 cells covered by a gelatinous layer, pseudoparenchymatous, medulla of loosely interwoven, mainly vertical, hyphae, with numerous *Nostoc* colonies, the lower cortex of longitudinal hyphae more or less closely interwoven. Apothecia often stipitate, brown; with proper margin composed of exciple plus pseudoexcipulum; hymenium 70-130 μ, upper part brown; paraphyses septate, unbranched, 2 μ thick, the tips tickened to 5-6 μ; subhymenial layer hyaline, of interwoven hyphae, 45-80 μ thick; exciple pseudoparenchymatous, pale; asci clavate, 57-79 × 11.5-14.5 μ; spores 8, hyaline, 2-3-celled, ellipsoid to fusiform, 11-27 × 4.5-8.5 μ. Pycnoconidia slightly dumbbell-shaped, 4-6 × 1 μ.

This species grows over mosses and humus, on boulders and rockfaces, rarely on earth. It is arctic-alpine in Europe, Greenland, and in North America from Baffin Island to Alaska, south to Nova Scotia in the east and Colorado and California in the west.

Localities: 3, 11.

Henssen (1963) commented that *M. carnosa* and *Parmeliella praetermissa* are easily mistaken. The latter becomes sorediose and the upper cortex has angular cells with strongly gelatinous walls and connecting pores; the former lacks soredia and the upper and the upper cortex has rounded cells with thin walls.

3. POLYCHIDIUM (Ach.) S.F. Gray

Thallus fruticose, dichotomously branched, the branching progressively smaller toward the tips, forming small dendroid masses; interior hyphae longitudinally oriented in outer parts, paraplectenchymatous in the older, basal parts; thallus with a single cortical layer of cells. Apothecia lateral; disk brown; lacking thalloid margin; with thick proper exciple; hymenium brown above; subhymenium of loose hyphae; exciple paraplectenchymatous; paraphyses septate, tips thickened; asci cylindrical, apical wall thickened; spores 8, 1- or 2-celled, hyaline, ellipsoid or spindle-shaped, thin or thick walled. Algae: *Nostoc* in arctic species.

1 Thallus tips shining, dichotomous, not coralloid, usually with apothecia
                                                    1. *P. muscicola*
  Thallus tips dull, coralloid branched, apothecia unknown   2. *P. umhausense*

1. **Polychidium muscicola** (Sw.) S.F. Gray Thallus fruticose, chestnut brown or blackish, tips shining, interior of masses dull, forming small cushion-like masses, lobes dichotomous or palmately divided, the lateral axes of similar size or with a main axis and smaller side branches, to 4 mm tall, lobes 0.2 mm thick, terete at the tips; cortex of 1–2(–3) cells thick with short cilia which toward the base become rhizinae; interior hyphae loosely interwoven, more or less parallel to the surface, toward the base forming a paraplectenchyma. Apothecia to 2 mm broad; disk red brown; exciple of same color as thallus; subhymenium 70–110 μ, hymenium 90–100 μ; paraphyses septate, unbranched, 1–2 μ thick; asci clavate, 45–62 × 9–22 μ; spores 8, 2-celled, uniseriate or biseriate, spindle-shaped, 19–25.5 × 4.5–6.5 μ.

This species grows among mosses usually in moist situations and over acidic rocks. It is known from Europe, Iceland, Greenland, and North America, where it ranges from Ellesmere Island to Alaska, south to New England, Colorado, and California.

Locality: 11.

2. **Polychidium umhausense** (Auersw.) Henss. *Dendriscocaulon umhausense* (Auersw.) Degel. Thallus small fruticose, brown, paler in shaded interior of the cushion-like mass, the tips coralloid, the branching irregular, sides with many fine white hairs; cortex 3–4 cells thick, interior hyphae longitudinally oriented and radiating out into the branches. Apothecia unknown.

Growing among mosses in moist situations. This species is known from Europe, and from the United States in North Carolina, Massachusetts, and Alaska.

Localities: 3, 11. Also known from Wonder Lake, Mt McKinley National Park, Weber & Viereck 7231 (WIS).

4. PLACYNTHIUM (Ach.) S.F. Gray

Thallus small squamulose or stellate radiating, flattened or filiform, a dark prothallus present in some species, with or without isidia, attached by rhizinae; internal anatomy radial or dorsiventral, the hyphae reticulate or pseudoparenchymatous. Apothecia hemiangiocarpic with thalloid or proper exciples; hymenium brown or green in upper part, I+ blue; asci cylindrical with apically thickened wall; spores 8, hyaline, 2-, 3-, or 4-celled, ellipsoid or fusiform; paraphyses septate, ends pointed, septate or not. Pycnidia with dark or brown ostiole; pycnidial paraphyses dark brown; pycnoconidia rod- or dumbell-shaped. Algae: Rivulariaceae or Scytonemaceae.

1 Black protothallus usually present, lobes papillose-granulose, not effigurate
1. *P. nigrum*
   Black prothallus lacking, lobes filiform or canaliculate, effigurate at the margin
2. *P. aspratile*

1. **Placynthium nigrum** (Huds.) S.F. Gray Prothallus usually present, blue black, sometimes quite fimbriate; thallus papillate-granulose, not effigurate, lobes flat, 0.4–1.5 mm broad with crenate or digitate margin, the squamules sometimes scattered over the prothallus, crowded toward the center and forming areoles 1–4 mm broad, cylindrical isidia sometimes present and mainly horizontal; upper side olive or blackish; lower side blue green and with dark rhizinae; hyphae forming a reticulum in most of the thallus with a few strands in the lower part of the thallus being oriented parallel to the surface and forming a lower cortex. Apothecia to 1 mm broad with dark proper exciple; disk brown or black, urceolate, becoming flat or convex; hymenium 70–175 μ, the upper part violet brown or dark green; paraphyses septate, branched, 2–5 μ thick, end cells pointed or thickened; subhymenium brown, 90–280 μ, of interwoven or pseudoparenchymatic hyphae; cells in lower part of dark violet or green exciple producing hairs, the cells of the exciple radiating; asci cylindrical, 39–54 × 8–15.5 μ; spores 8, hyaline, narrowly ellipsoid, 2-, 3- or 4-celled, 7–17 × 3.5–5.5 μ. Algae: *Dichothrix* in Rivulariaceae.

This species grows on calcareous rocks and is widely distributed in temperate and boreal parts of the northern hemisphere. It is known from Novaya Zemlya and Greenland, and in North America from Baffin Island to Alaska and southward through the United States. In addition to the specimens cited below I have seen one from the Bering Straits, Ukinyik Creek, *Viereck & Bucknell* 4384 (COLO).

Localities: 1, 2.

2. **Placynthium aspratile** (Ach.) Henss. Prothallus seldom present; thallus of more or less aggregated canaliculated lobes, effigurate at the border, the center becoming granulose or isidiose and breaking up into areoles 1–2 mm broad; upper side shining, olive or blackish; underside dark blue green, rhizinose; isidia erect, cylindrical, to 3 mm tall; interior hyphae mainly parallel, more or less connected to an upper pseudopa-

renchyma; lower layers blue green. Apothecia to 1 mm, with dark proper exciple, the exciple radiate; disk black, urceolate, becoming flat or convex; hymenium 80–100 μ, upper part blue green; paraphyses septate, branched, to 2.5 μ thick, end cells pointed or thickened; subhymenium brown, 100–130 μ thick, the hyphae interwoven; asci cylindrical, 44–45 × 12.5–18 μ; spores 4 or 8, broad or narrowly ellipsoid, 11.5–18 × 4.5–7 μ. Algae: a filiform blue-green alga in Rivulariaceae or Scytonemataceae.

This is a very common lichen on both calcareous and acidic rocks. It is known from northern Europe and northern North America, from Novaya Zemlya, Spitzbergen, Bear Island, Greenland, and across Canada and Alaska, south to Lake Superior, Colorado, and Wyoming.

Localities: 1, 2, 3, 8, 9, 11, 14.

The canaliculate shining marginal lobes and the granulose or isidiose center aggregated into areolate masses are characteristic of this species. A dendroid prothallus is mentioned by Henssen (1963c) as sometimes being developed in arctic specimens. 'Pannularia nigra' of Nylander from Konyam Bay, Siberia side of Bering Sea, No. 9827 Nyl. (H) is this species.

PANNARIACEAE

Thallus squamulose, sometimes becoming a granulose crust; heteromerous, with upper cortex, lower cortex present or absent, attached to substratum by rhizinae. Apothecia with thalloid or proper exciple; paraphyses simple or branched; spores 8, 1-celled, hyaline, thin-walled or with warty wall. Pycnidia immersed, fulcra endobasidial, pycnoconidia short, straight.

1	Thallus with bright green algae; apothecia with thalloid margin	
		1. *Psoroma*
	Thallus with blue-green algae	2
2 (1)	Apothecia with proper exciple	2. *Parmeliella*
	Apothecia with thalloid exciple	3. *Pannaria*

1. PSOROMA

Thallus squamulose, granulose-squamulose, close to substratum or raised; upper cortex paraplectenchymatous of several cells thickness, algal and medullary layers indistinct, lower cortex of interwoven hyphae; attached by rhizinae. Apothecia adnate to sessile; disk concave or flat, red to red brown; thalloid exciple colored like thallus, crenate to granulose; subhymenium yellowish brown; hymenium hyaline; paraphyses simple, coherent; asci clavate; spores 8, hyaline, 1-celled with a thin but warty wall, ellipsoid to ovoid. Algae: *Myrmecia*.

1. **Psoroma hypnorum** (Vahl) S.F. Gray Thallus of coarsely granular squamules to 1 mm long and broad, smooth yellowish brown to brown, dull, lacking isidia, upper cortex paraplectenchymatous, lower of interwoven hyphae. Apothecia sessile, to slightly immersed among squamules, to 6 mm broad; disk concave to flat, red brown, dull; margin of same color as thallus, granular to squamulose; subhymenium yellowish brown; hymenium 100–120 μ, upper part yellow brown; paraphyses thin, 1–2 μ; spores 8, hyaline, simple, ellipsoid to spindle-shaped or ovoid, rough walled, 19–26 × 8–12 μ.

Usually growing on earth containing humus or over mosses in moist habitats. A circumpolar arctic and boreal species common in arctic and alpine tundras, and at the edge of the boreal forest across Canada and Alaska and south in New England and the Rocky Mountains to New Mexico.

Localities: 1, 2, 3, 4, 5, 6, 7, 8, 9, 11, 14.

2. PARMELIELLA Müll. Arg.

Thallus squamulose or crustose, usually over a black prothallus, attached by rhizines; upper cortex paraplectenchymatous, lower cortex lacking; algae in glomerules in the medulla. Apothecia sessile, with proper exciple; disk flat to convex, red brown to blackish; hypothecium pale to brownish; hymenium pale with yellow or brown epithecium; spores 8, hyaline, single-celled. Pycnidia immersed, pycnoconidia short, straight. Algae: *Nostoc*.

1. **Parmeliella praetermissa** (Nyl.) P. James [P. *lepidiota* (Sommerf.) Vain.] Thallus of small squamules to 3 mm long, 1.5 mm broad, the margins becoming bluish sorediate, granular divided and dactyliform isidiate, becoming erect, much imbricated, sometimes so thickly covered with the smaller marginal soredia as to form a dense crustose appearance; white pruinose at edges of lobes and granules; upper side of thallus reddish brown, lead gray, or blackish; underside pale brownish or white with pale rhizinae, a black prothallus sometimes apparent. Apothecia infrequent in arctic material, adnate, to 2 mm, with proper exciple, the exciple pale; the disk flat or convex, reddish brown to brownish black or black, epruinose; subhymenium hyaline or yellowish; hymenium hyaline, the upper part yellowish brown, 110–120 μ; paraphyses unbranched, septate, 2 μ, little thickened at the tips; asci clavate 18–88 μ

thick; spores 8, hyaline, 1-celled, ellipsoid, the tips rounded, 13–23 × 8–11.5 μ.

Growing on humus and mosses and over rocks. An arctic-alpine circumpolar species which ranges across northern Canada and Alaska south to New England, South Dakota, Colorado, and California.

Localities: 1, 2, 3, 9, 11.

There is a close resemblance of this species to *Massalongia carnosa* but the latter lacks the soredia of this species and the cells of the upper cortex are roundish with thin walls instead of angular with strongly gelatinous walls and connecting pores as in *P. praetermissa* according to Henssen (1963c). The dactyliform isidia are covered at the tips with a thin bluish white pruina which gives them a very characteristic appearance.

3. PANNARIA Del.

Thallus foliose or squamulose, usually with dark green to black prothallus present; upper cortex paraplectenchymatous, lower cortex lacking; attached to substratum by small rhizinae; heteromerous, algal colonies glomerulate in the medulla. Apothecia sessile on the upper surface, with thalloid exciple; subhymenium hyaline; paraphyses unbranched, septate, not thickened at apices; spores 8, 1-celled, hyaline, the outer wall often tuberculate. Pycnidia immersed, hyaline, pycnoconidia straight, cylindrical. Algae: *Nostoc*.

1 Thallus coarsely blue gray sorediate-isidiate, lobate toward the margins; apothecia rare   1. *P. pityrea*
  Thallus forming a nearly continuous crust of small gray to brown squamules, esorediate; apothecia common   2. *P. pezizoides*

1. **Pannaria pityrea** (DC.) Degel. Thallus squamulose, blue gray, coarsely sorediate-isidiate, the center sometimes becoming a mass of soredia, lobed toward the margins, P+ red (pannarin), upper side brown, smooth, dull, pruinose toward the tips, underside black, dull, with numerous black rhizinae. Apothecia rare, to 1.5 mm diameter, margin pale brown, dull, usually becoming sorediate; disk chestnut brown, smooth, dull, concave or flat; epithecium granular, red brown; hymenium 75–95 μ; paraphyses slender, 1.5 μ, unbranched, the tips not capitate; asci cylindrical clavate, spores 8, smooth walled, 15–16 × 9–10 μ.

Growing over mosses and over rocks. A temperate to arctic circumpolar species preferring oceanic habitats.

Locality: 8.

2. **Pannaria pezizoides** (Web.) Trev. Thallus squamulose, the squamules flattened against the substrate, tiny, to 0.75 mm broad, neither sorediate nor isidiate, but the center of the thallus becoming coarsely granular, pale gray to brown, underside white or pale brown over a black prothallus; upper cortex paraplectenchymatous, 35–45 μ thick, the cells rounded or angular with 2 μ thick walls, the algae glomerulate in a compact medullary tissue, medulla sometimes nearly lacking below the algal layer; lower cortex lacking. Apothecia sessile or immersed in the squamules of the thallus, to 2 mm broad, the disk flat to slightly convex, bright red brown to black in exposed specimens, the thalloid exciple conspicuously crenate-lobulate and gray contrasting with the reddish disk; subhymenium hyaline to dusky; hymenium 95–135 μ, the upper part yellowish brown; paraphyses unbranched, non-septate, 105 μ thick, the tips not thickened; asci cylindrico-clavate, 15–16 μ thick; spores 8, hyaline, ovoid with tuberculate episporium, tips somewhat pointed, 1-celled, 26–31 × 9.5–12 μ.

This species grows on soil, sometimes on the edge of frost boils, less often on soils with humus or over mosses, usually in somewhat moist habitats. It is circumpolar, arctic-alpine and boreal, in Greenland and in North America from Baffin Island to Alaska and south to New England, Wisconsin, New Mexico, and California.

Localities: 2, 3, 4, 8, 9, 11, 14.

COCCOCARPIACEAE

Thallus fruticose or foliose, attached to the substratum by hyaline or greenish rhizinae. Apothecia adnate, flat to convex; proper exciple more or less developed, cellular; hymenium gelatinous, the upper part brown or violet black; subhymenium pseudoparenchymatous; hypothecium not distinct; paraphyses branched, 1-septate; asci clavate with I+ blue walls; spores 8, hyaline, 1-celled, thin walled, round or oval. Pycnidia adnate or immersed, pycnoconidia short. Of the two genera, *Spilonema* and *Coccocarpia*, the latter reaches the North Slope.

1. COCCOCARPIA Pers.

Thallus foliose squamulose, corticate above and below, the upper cortex with longitudinally oriented hyphae, the medulla and lower cortex not clearly distinguished, the underside with rhizinae and tomentum. Apothecia with pale proper margin, lacking thalloid margin, disk black, spores 8, hyaline, elliptical. Pycnidia immersed. Algae: *Scytonema*.

1. **Coccoparpia pellita** (Ach.) Müll. Arg. Thallus small foliose, the lobes entire margined, lead colored or greenish gray, upper side smooth, lower side densely covered with blue black rhizinae which may project beyond the margins. Apothecia red brown to black, adnate on the upper surface; exciple of radiating paraplectenchymatous hyphae; hymenium 50–70 μ; paraphyses coherent, the apices scarcely thickened; asci clavate; spores 7–13 × 2–5 μ.

Growing among mosses over soil. A tropical to temperate species known from the southeastern United States. Its disjunct occurrence at Anaktuvuk Pass correlates with those of *Umbilicaria caroliniana* and *Baeomyces roseus*.

Locality: 8.

PELTIGERACEAE Zahlbr.

Thallus foliose, large lobed or reduced to small single lobes, the margins usually raised, the underside usually more or less veined, attached by rhizinae; upper cortex of several layers of paraplectenchyma, heteromerous, the algal layer in upper part of medulla, lower part loosely interwoven, lower cortex present or lacking. Apothecia hemiangiocarpic, immersed in upper surface or borne at margins of lobes on upper surface; subhymenium hyaline; hymenium hyaline; paraphyses slender, unbranched, septate; asci clavate, 2–8-spored; spores hyaline or brown, ellipsoid, spindle-shaped, or acicular, 2-celled to many cells in a row, thin walled. Cephalodia often present. Algae: *Nostoc* or *Coccomyxa*.

1 Thallus lacking lower cortex; apothecia borne at the edge of thallus, the disk not concave; spores 4–8-celled, thin walled, hyaline or pale brownish
1. *Peltigera*
Thallus with lower cortex; apothecia borne immersed in middle of thallus, the disk concave; spores 2-celled, dark brown when ripe, with thick wall
2. *Solorina*

1. PELTIGERA Willd.

Thallus foliose, more or less lobed, attached to substratum by rhizinae; upper cortex paraplectenchymatous, algae in continuous upper layer of medulla, lower cortex lacking, the lower surface more or less definitely veined; upper surface smooth, scabrid, or with a tomentum of superficial hyphae, especially near the margins. Apothecia round, marginal or on marginal lobules, horizontal or vertical, often with the sides reflexed; proper exciple lacking, a thalloid rim around the apothecium which is sunken in the thallus, the margins usually becoming crenate; subhymenium hyaline to brown; hymenium hyaline to brownish; paraphyses simple, septate, upper part red brown; asci cylindrico-clavate, upper part considerably gelatinized; spores 8, fusiform or acicular, sometimes slightly curved, 3-9-septate. Algae: *Nostoc* and/or *Coccomyxa*; in species in which the latter is the main alga, the *Nostoc* is in cephalodia on the upper or lower surface of the thallus.

1   Algae of thallus green, cephalodia containing *Nostoc* on the upper or lower
        surface of the thallus   2
     Algae of the thallus blue green (*Nostoc*), lacking cephalodia   3

2 (1) Upper side of the thallus with warty cephalodia and with erect tomentum
        towards margins; under surface with scattered fasciculate rhizinae;
        thallus and lobes large, to 6 cm across; apothecia vertical
                                                                    1. *P. aphthosa*
     Upper side of thallus lacking cephalodia, etomentose, underside with
        small black cephalodia on the veins; attached centrally or at one side by
        a group of rhizinae; thallus small, to 2 cm broad; apothecia horizontal
                                                                     2. *P. venosa*

3 (1) Upper side of thallus lacking tomentum, smooth to verruculose-scabrid,
        veins of underside not well developed, broad and with small interspaces
                                                                                   4
     Thallus with erect tomentum toward the margins; veins lacking on
        underside or else with very well developed veins and broad interspaces
                                                                                   5

4 (3) Upper side verruculose-scabrid   3. *P. pulverulenta*
     Upper side smooth   4. *P. polydactyla*

5 (3) Tomentum at margins erect, veins lacking below   5. *P. malacea*
     Tomentum more or less appressed to upper surface; lower surface with
        very definite narrow veins   6

6 (5) Upper surface with scattered peltate isidia   6. *P. lepidophora*
     Upper surface lacking peltate isidia but sometimes with small marginally
        attached lobules associated with cracks and injuries   7. *P. canina*

1. **Peltigera aphthosa** (L.) Willd. Thallus large, sometimes over 1 m broad, the lobes to 10 cm long and 6 cm broad, usually much less in arctic conditions; upper surface with smooth center and erect tomentum toward margins, warty cephalodia scattered over the surface; under surface pale at the margins blackening toward the center, either veinless and covered with an even nap of tomentum or becoming veined with white or dark brown to black veins, rhizinae scattered, fasciculate. Apothecia large, to 16 mm broad, on marginal lobules, the sides reflexed; the margins becoming crenate, the reverse with a cortex at first, this

breaking up leaving scaly patches of cortex on the back of larger apothecia; disk reddish brown to blackish brown; subhymenium brown; hymenium hyaline to pale brown; paraphyses simple, thickened and red brown at tips; asci cylindrico-clavate 80 × 12 μ; spores 8, acicular, 3–9-septate, 46–63 × 4–5 μ.

This species occupies a wide variety of habitats. In willow thickets and by waterfalls it may grow to very large size, in the drier heaths and limestone barrens it tends to be very small. It is usually among mosses and on earth, occasionally on rotting logs and stumps. It is circumpolar low arctic and alpine and is common in the boreal forest. It also occurs in the southern hemisphere. It grows from as far north as Thule in Greenland, Devon Island, and Baffin Island to Alaska, south to North Carolina, Wisconsin, New Mexico, and California.

Localities: for var. *aphthosa* without definite veins on the underside, 1, 2, 3, 8, 11, 13, 14; for f. *complicata* (Th. Fr.) Zahlbr. with the margins markedly crisped and auriculate, 4, 6; for var. *leucophlebia* with definite veins on the underside, 2, 4, 8, 9.

There exist considerable differences of opinion concerning the variability of the morphology of this species, particularly as to the significance of the venation pattern on the underside. European lichenologists (and the 4th edition of the North American Checklist) recognize the strongly veined plants as a species, *Peltigera leucophlebia*, separate from those which lack veins and which are recognized as *Peltigera aphthosa*. Examination of long series of specimens convinces me that these are only extremes of a long series which has ecotypic but not taxonomic significance. The varieties (or 'species') are also often separated on a correlation of the vein pattern with the smoothness or verruculose condition of the reverse of the apothecium. Examination of those specimens in WIS which possess apothecia showed that of those specimens with the var. *aphthosa* type of underside (broad veined or no veins), 27 had the reverse of the apothecium smooth and 31 verruculose; of those with the var. *leucophlebia* underside (narrow, raised veins), 12 had the reverse of the apothecium smooth and 35 verruculose. These data do not indicate a satisfactory correlation between the vein type and apothecium reverse sufficient to warrant separating them on this correlation. During the examination it appeared that the ontogeny of the reverse of the apothecium would be to commence with a continuous cortex, usually somewhat coarsely wrinkled. This then appears to break up into coarse verrucules which appear to be shed, the reverse of the apothecium then becoming decorticate, sometimes with a few verrucules persisting along the rim of the back. As Krog says in her treatment of these species in her Alaska paper

(Krog 1968), they 'may at times be difficult to separate.' This is exactly as one would expect of transitional specimens in a series. Needed to clarify this situation are transplant and other experiments in an arctic environment where the species is readily available as at Barrow.

2. **Peltigera venosa** (L.) Baumg. Thallus small, of single rounded, fan-shaped lobes to 2 cm broad, flattened; upper surface smooth, shining, apple green when moist, grayish- or brownish green when dry; under surface white or pale brown with dark brown or black veins with small dark green or brown to black cephalodia on the veins and containing the *Nostoc* algae; a single group of rhizinae at one side attach the thallus to the substrate; upper cortex paraplectenchymatous, the algal layer below with green *Coccomyxa* cells in continuous layer; medulla of interwoven hyphae; lower cortex lacking. Apothecia small, to 5 mm broad, horizontal, borne at the margin of the thallus, the margin crenate; disk reddish to blackish brown, the reverse with a thick, dark paraplectenchymatous cortex; subhymenium pale to dark brown with layer of hyphae inspersed with air immediately below; hymenium hyaline to pale brown; paraphyses unbranched, coherent, thickened and dark red brown at tips; asci cylindrico-clavate, 75 × 11 μ; spores 8, fusiform, (1-)3(-5)-septate, 24–40 × 6–8 μ.

On moist calcareous soils, especially where some seepage is present. Often on moist calcareous rock outcrops. A circumpolar arctic and boreal species which in North America ranges from Ellesmere Island to Alaska south to New England, Wisconsin, New Mexico, and California.

Localities: 2, 8, 9, 11, 14.

3. **Peltigera pulverulenta** (Tayl.) Kremph. (*Peltigera scabrosa* Th. Fr.) Thallus medium sized to large, lobes to 5 cm broad but usually smaller; upper surface greenish gray to yellowish brown, dull, scabrid; underside with broad flat veins like *P. polydactyla*, the veins brown to blackish brown, the interspaces white to pale brown; rhizinae fasciculate, scattered to the margins, thicker and sometimes forming a mat to the center of the thallus; cortex paraplectenchymatous with groups of cells forming the scabridity. Alga: *Nostoc*. Apothecia large, to 6 mm broad, borne on extended lobules, vertical, often round or with the sides reflexed; disk dark reddish brown, the margin entire to slightly irregular, the reverse with a cortex; subhymenium hyaline to pale brown; hymenium hyaline below, deep brown above; paraphyses unbranched, thickened and red brown at tips; asci cylindrico-clavate, 10 × 105 μ; spores 8, acicular, 5-7-septate, 63–95 × 3–4 μ.

This species grows on soil and among mosses, preferring the more moist types of tundras and growing larger in such habitats. It is circumpolar, arctic-alpine, in North America from Baffin Island to Alaska, south to New York, Colorado, and Washington.

Localities: 1, 2, 3, 4, 6, 9.

4. **Peltigera polydactyla** (Neck.) Hoffm. Thallus varying from very small, single lobes to very large, the lobes up to 4 cm broad but much smaller in arctic material; margins ascending, some forms crisped or proliferate; upper surface dark greenish blue when moist, slaty or bluish or greenish gray or brownish when dry, smooth, shining; underside with broad white, brown, or black veins with very small white interspaces; rhizinae short and fasciculate or very long and simple; cortex paraplectenchymatous. Algae: *Nostoc.* Apothecia 2–5 mm broad, on erect narrow lobules at the ascendent tips of the lobes, the sides reflexed, margin crenate; disk reddish to chestnut brown or black, the reverse corticate; subhymenium brown; hymenium hyaline, 100–110 μ, upper part brown; paraphyses unbranched, tips thickened; asci cylindrico-clavate, 80–105 μ; spores 8, acicular, slightly curved, 3–7-septate, 48–105 × 3–4 μ.

This species grows on soil and humus among mosses, occasionally at base of willows and trees, in moist places. It is circumpolar and ranges in North America south of a line from Baffin Island to Alaska and throughout the United States. It is uncommon on the North Slope.

Localities: 1, 3, 8. One specimen at locality 1 is of var. *crassoides* Gyel. with small, dark, rigid thallus and lack of interspaces between veins on the underside.

5. **Peltigera malacea** (Ach.) Funck. Thallus foliose, the lobes ascending, to 4 cm long, 2 cm broad; upper surface brown to brownish green or apple green, shining toward the center, with a sparse erect tomentum and sometimes also whitish pulverulent toward the apices of the lobes; margins inrolled and form a light border to the lobes; underside brownish at the margins, blackish brown to the center, lacking veins and covered with a fine, even tomentum, sometimes with a few whitish interspaces; rhizinae sparse, black, fasciculate. Apothecia vertical, borne on extended lobules, the sides reflexed; disk chestnut to blackish brown, 4–8 mm broad, margin crenate; subhymenium pale to dark brown; hymenium pale to dark brown, 110 μ thick; paraphyses unbranched, thickened and darkened to red brown at tips; asci cylindrico-clavate, 80–112 μ; spores 8, acicular, 3–5-septate, 46–72 × 3.5–6 μ.

Growing on soil and among mosses. A circumpolar arctic and boreal species known from Europe, Asia, North Africa, Novaya Zemlya, Spitzbergen, Iceland, Greenland, and in North America from Ellesmere Island to Alaska, south to New Hampshire, Minnesota, Colorado, and Washington.

Localities: 1, 3, 10, 11, 13.

6. **Peltigera lepidophora** (Nyl.) Vain. Thallus small, of small lobes to 2 cm broad, the edges ascending, sometimes making the lobes cochleate; upper surface tomentose and with peltate isidia scattered over the surface and containing the same alga as the thallus, brown to grayish blue; underside pale to brown with irregular raised venation; rhizinae simple to fibrillose, pale to brown. Apothecia rare, small, to 4.5 mm broad, on extended lobules, margin crenate; disk brown; hymenium hyaline to pale brown; paraphyses thickened and red brown at tips; asci sparse; spores 8, acicular, 55–85 × 4–4.5 μ.

Growing on soil and among mosses, the lobes usually very tiny under arctic conditions. Reported from Europe, Iceland, Greenland, and in North America from Baffin Island to Alaska and south to Connecticut, Wisconsin, Colorado, and Washington.

Localities: 2, 3, 8, 9, 11.

7. **Peltigera canina** (L.) Willd. Thallus exceedingly variable, small or very large, the lobes to 8 cm broad but usually much smaller, particularly in dry situations, upper surface slate gray, greenish brown or brown with arachnoid to dense tomentum toward margins, the center dull or shining, in one form with coarse gray soredia in rounded spots on the upper surface, sometimes with flat platelets attached by their margins along cracks or spots of injury and representing regeneration lobules; underside with a strong network of raised white to brown veins; rhizinae simple, fasciculate, or fibrillose, sometimes forming a confluent spongy mat below, white to brown. Apothecia vertical on extended lobules, to 10 mm broad, the sides reflexed, the margins crenate; disk light brown, reddish brown, chestnut brown to black; subhymenium brown; hymenium hyaline to brown, 70–110 μ; paraphyses thickened and red brown at tips; asci cylindrico-clavate, 90 × 10 μ; spores 8, acicular, hyaline to yellowish, 3–7-septate, 23–67 × 3–6.5 μ.

A species which grows on soil and also among mosses. It is cosmopolitan, reported over much of the world from the Arctic to Antarctica, and over most of North America.

Localities: Typical var. *canina* specimens: 2, 3, 8, 10, 11; var. *rufescens* (Weis.) Mudd. with short, narrow, brittle lobes: 1, 2, 3, 7, 9, 11, 13, 14; var. *spongiosa* Tuck. with spongy mat of rhizinae on underside: 2, 8, 9; var. spuria (Ach.) Schaer. with small ascending lobes: 2, 7, 8; f. *sorediata* Schaer. with soredia above: 1, 2, 3, 6, 7.

2. SOLORINA Ach.

Thallus foliose to much reduced in one species, elongate lobes or forming small thalli around the apothecia; upper cortex paraplectenchymatous, lower cortex paraplectenchymatous in vicinity of apothecia; heteromerous. Algae: *Nostoc* and *Coccomyxa*, the two in separated layers in the same thallus in *S. crocea*. Apothecia round, immersed in the upper surface, slightly to very deeply concave, red brown; lacking proper exciple; subhymenium narrow; hymenium hyaline to brownish; paraphyses unbranched, the tips red brown; asci clavate; spores 2–4–8, brown, ellipsoid with constriction between the 2 cells.

1	Thallus yellow to reddish orange below	1. *S. crocea*
	Thallus gray to brown below	2
2 (1)	Ascus containing 4 spores	3
	Ascus containing 2 or 8 spores, thallus sometimes a small squamule with the deep disk of apothecium in center	4
3 (2)	Thallus broad, foliose lobed	2. *S. saccata*
	Thallus reduced, consisting of granules, squamules, and cephalodia around the apothecium	3. *S. spongiosa*
4 (2)	Ascus containing 2 large spores 60–104 × 34–40 μ, thallus usually pruinose	4. *S. bispora*
	Ascus containing 8 smaller spores, 35–40 × 18–21 μ, thallus usually epruinose	5. *S. octospora*

1. **Solorina crocea** (L.) Ach. Ashy to reddish brown when dry, dark green when moist, large lobed, to 3 cm broad in large examples, flat to slightly concave above; upper surface smooth to roughened; lower side with conspicuous orange flabellate veins and yellow to orange interspaces; rhizinae black; upper cortex paraplectenchymatous, thick; algae of two types in two layers, *Coccomyxa* above, *Nostoc* below; lower cortex under apothecia only, medullary hyphae pigmented orange. Apothecia round to slightly angular, immersed in the upper surface, to 1 cm broad, red to chestnut colored; subhymenium hyaline, 40–70 μ; hymenium 140–180 μ, hyaline below, yellow above; paraphyses coherent, unbranched, septate, the tips thickened and red brown; asci 110–130 ×

22-24 μ; spores 6 or 8, brown, 2-celled, not constricted between cells, 34-42 × 10-14 μ.

A species of very moist places, below late snowbanks, and in seepages but also on drier limestone 'barrens' in the arctic, usually on soil high in clay. This is a circumpolar arctic-alpine species. In North America it grows from Ellesmere Island to Alaska, and south to Labrador, Quebec, New Mexico, and Washington.

Localities: 8, 11.

2. **Solorina saccata** (L.) Ach. Thallus foliaceous, membranaceous thin, to 2 cm broad, seldom broader, forming larger groupings of lobes, loosely attached to substratum, flat or slightly convex, sometimes slightly scrobiculate, apple green to slaty when fresh, becoming reddish in the herbarium, sometimes slightly pulverulent above; underside pale tan; rhizinae scattered, pale; upper cortex paraplectenchymatous, algal layer with only 1 alga, Coccomyxa, lower side lacking cortex. Apothecia rounded, 2-5 mm broad, immersed in the upper side of the thallus, sometimes several per lobe, lacking proper exciple, breaking through the upper cortex and with remnants of this sometimes around the concave disk which is deep in the thallus, dark brown red; hymenium 120-280 μ, hyaline to reddish; paraphyses 5-6 μ, unbranched, septate, the tips red brown, moniliform; asci 180-210 × 24-32 μ; spores 4, brown, 2-celled, 34-60 × 16-26 μ.

This is a calciphile species which grows best in very moist places such as spray from waterfalls but in open tundras growing in moist micro-habitats as sides of hummocks, sides of animal burrow entrances, edges of solifluction lobes, seepages and moist cliff sides. It is circumpolar and arctic and boreal, occurring in calcareous districts across arctic America south to the northern states, Vermont, Wisconsin, and South Dakota, and in Alberta and British Columbia. Washington reports are incorrect *fide* Imshaug (1957).

Localities: 1, 2, 4, 5, 8, 9, 10, 13, 14.

3. **Solorina spongiosa** (Sm.) Anzi Thallus scant and fragile, of small squamules and granules around the very urceolate apothecium, bluish gray to reddish gray, mixed with Nostoc colonies forming cephalodia; upper cortex paraplectenchymatous, the algal layer containing Coccomyxa, no lower cortex, the lower side of the apothecium and squamules attached to substratum by rhizines. Apothecia deeply urceolate, the disk dark brownish red to blackening; hymenium hyaline, 160-200 μ; paraphyses unbranched, hyaline with the tips red brown, 6-8 μ, coherent,

little thickened; asci 90–120 μ; spores 4, becoming brown, 2-celled ellipsoid, 30–50 × 18–22 μ.

A species of moist calcareous habitats, reported from Europe, Novaya Zemlya, Iceland, Bear Island, Greenland, and in North America from Baffin Island to Alaska south to Newfoundland, Ontario, Colorado, and Washington.

Localities: 3, 8, 10, 11, 13.

The spongy reverse of the apothecium is an easily recognized character for this species.

4. **Solorina bispora** Nyl. Thallus foliose, sometimes reduced to a small area around an apothecium, sometimes slightly lobate, slaty or reddish brown, white pruinose to becoming scabrid above, the margins slightly raised; underside pale; attached centrally by rhizinae, cortex paraplectenchymatous. Algae: *Coccomyxa* (green). Apothecia deeply sunken in the upper surface of the thallus, red to red brown or blackish, to 2 mm broad, sometimes surrounded by remnants of the cortex through which the apothecium broke; subhymenium hyaline; hymenium 220–260 μ, hyaline below, reddish in upper part; paraphyses slightly branched, septate, upper part little thickened, tips sometimes darkened; asci clavate, 180 × 54 μ; spores 2, becoming red brown, 1-septate, ellipsoid, walls sometimes slightly roughened, 60–104 × 34–40 μ.

Growing on calcareous soils, sometimes on humus, at edges of frost boils, on earth edges, usually somewhat moist. An arctic-alpine species known from Siberia, Europe, Novaya Zemlya, Spitzbergen, Bear Island, Iceland, Greenland, and in North America from Ellesmere Island to Alaska and south to Colorado in the Rocky Mountains. Its range in eastern North America is unknown south of the arctic islands, specimens from the Belcher Islands, Southampton Island, and Churchill are in WIS.

Localities: 1, 2, 3, 9, 11, 13, 14.

5. **Solorina octospora** (Arn.) Arn. Thallus foliose, single lobed to several lobed, thin, reddish brown, epruinose, somewhat concave with raised margins, upper surface roughened, underside pale to reddish brown, with pale to brown rhizinae; upper cortex paraplectenchymatous; algae: *Coccomyxa* (green); KC+ red, containing methylgyrophorate. Apothecia reddish brown, in the upper side of the thallus, usually not quite as sunken as the preceeding species but concave, often surrounded by cortical remnants; subhymenium pale brownish; hymenium hyaline below, yellowing above, 210–260 μ; paraphyses coherent, septate, scarcely branched, the tips darkening, not thickened; asci clavate, 180 ×

26 μ; spores 8, 2-celled, becoming red brown, ellipsoid, 1-septate, 35–40 × 18–21 μ.

A species of soils and humus, sometimes on soil over rocks. Reported from Europe, Greenland, Spitzbergen, and in North America from Ellesmere Island to Alaska, south to Colorado and New Mexico in the Rocky Mountains.

Localities: 1, 2.

This species may resemble *S. saccata* but has 8 instead of 4 spores. It may also be mistaken for *S. bispora* but is usually not pruinose, has shallower apothecia, and 8 instead of 2 spores.

NEPHROMATACEAE

Thallus foliose, large, lobed, attached by rhizinae; upper and lower cortex present, paraplectenchymatous, heteromerous, internal cephalodia present. Apothecia borne on the lower surface, immersed, the tips usually reflexed to bring the disks up; hymenium coherent, 60–90 μ thick; paraphyses unbranched, septate, the tips darkened and enlarged; asci clavate, without apparatus at apex which is present in Peltigeraceae; spores 8, 3-septate, pale brown. Algae: *Coccomyxa* in some species, *Nostoc* in cephalodia and the only component in the thallus in other species.

1. NEPHROMA Nyl.

Thallus foliose, usually quite thin, greenish yellow or brown; upper and lower cortices paraplectenchymatous, algal layer with either *Coccomyxa* (green) or *Nostoc* (blue green), medulla of loosely interwoven hyphae; lower surface glabrous, pubescent or tomentose, sometimes with tubercles or papillae. Apothecia immersed in the lower surface at tips of lobes; disks light to dark brown with narrow proper exciple; hymenium 60–90 μ thick; paraphyses unbranched; spores 8, 3-septate, light brown. The green species containing *Coccomyxa* with internal cephalodia containing *Nostoc*.

1   Algal layer bright green   2
     Algal layer blue green   3

2 (1) Thallus yellow green, cephalodia containing *Nostoc* visible as bluer warty areas on upper surface   1. *N. arcticum*
        Thallus dull slaty green or brownish, cephalodia visible on lower surface, upper surface scaly or slightly pubescent   2. *N. expallidum*

3 (1) Gray soredia in patches on upper surface or on margins   3. *N. parile*
        Brown coralloid isidia from ridges on upper side   4. *N. isidiosum*

1. **Nephroma arcticum** (L.) Torss. Thallus foliose with broad lobes, bright yellow green above, smooth to slightly undulating, the cephalodia showing as bluish green warts on upper surface, regeneration squamules common; lower surface pale at margins, blackening toward the center, corticate toward the margins and becoming increasingly tomentose toward the center, the tomentum blackening. Apothecia large, to 30 mm broad; disks light brown; subhymenium thin; hymenium hyaline below, pale brown above; spores 8, 3-septate, subfusiform, light brown, 23–27 × 4–6 μ.

This species grows in protected places in the tundras, in mossy, wet meadows, willow thickets, spruce outliers, etc. It is low arctic and alpine. This is a circumpolar species in the northern part of the boreal forest and arctic tundras. In North America it grows from Baffin Island to Alaska, south to the New England mountains in the east, Alberta in the west.

Localities: 1, 2, 8, 9, 13, 14.

2. **Nephroma expallidum** Nyl. Foliose, darker brownish green or slaty green, uneven in color, lobes middle sized, upper surface scaly to finely pubescent, sometimes slightly pruinose, underside tan at the margins, blackening toward the center, becoming closely tomentose; upper cortex paraplectenchymatous, algal layer green, cephalodia internal and near lower surface. Apothecia uncommon, to 10 mm broad, disk light brown; spores 8, light brown, 3-celled or occasionally abnormal and misshapen, otherwise subfusiform, 17–21 × 5–6 μ.

A species of moist meadows and tundras. Subarctic to arctic and circumpolar. In North America ranging from Ellesmere Island to Alaska, south to Southampton Island, northern Ontario (Fort Severn), and Alberta. It is more northern than the preceding species.

Localities: 1, 2, 3, 4, 5, 6, 8, 9, 11, 13.

3. **Nephroma parile** (Ach.) Vain. Thallus foliose, brown, middle sized, with bluish gray soredia on upper surface in patches or on margins, upper surface otherwise smooth, underside brown, darkening to center, variably pubescent, rugulose. Algae: *Nostoc*. Apothecia rare, specimens usually sterile, disks light brown; spores 8, ellipsoid, pale brown, 16–18 × 6–7 μ.

This commonly grows on rocks and tree bases in moist places. It is circumpolar, arctic to temperate, in North America occurring from Quebec to Alaska, south to North Carolina, Wisconsin, Arizona, and Oregon.

Localities: 8, 11.

4. **Nephroma isidiosum** (Nyl.) Gyel. Thallus foliose, middle sized, lobes to about 1 cm broad, upper surface somewhat lacunose and with many clusters of coralloid isidia arising from the ridges; lower side brown with dense short tomentum to margins and scattered rhizinae. Algae in thallus: *Nostoc*, algal layer blue green. Apothecia not seen.

This species grows on mosses, bases of shrubs and on rocks. It appears to be rare and with a disjunct distribution, being reported from Scandinavia, western Siberia, and Alaska. it is quite possible that the Blomberg and Forssell (1880) report from Europe is incorrect and that this is an amphi-Beringian species. Wetmore reported these Okpilak collections from Alaska and an additional one from Valdez, Alaska. Krog reported additional localities in Alaska from near Juneau and from the Seward Peninsula area.

Locality: 11.

STICTACEAE

Foliose, large, lichens with scrobiculate upperside or smooth upperside, with upper and lower cortex, the lower cortex covered with tomentum or bald, with or without cyphellae on the underside, the cortices paraplectenchymatous. Apothecia marginal or on surface, adnate or sessile, with or without thalloid margin, with thick paraplectenchymatous proper exciple; hymenium pale below, darkening above; paraphyses unbranched, septate, not thickening toward tips; asci clavate, spores 8, hyaline or brown, spindle-shaped to acicular, 1–5-septate. Pycnidia in small warts with pale conceptacle, pycnoconidia short, straight, cylindrical. Algae: *Myrmecia* and *Nostoc*, the latter in cephalodia.

1 Lower surface lacking cyphellae	1. *Lobaria*
Lower surface with cyphellae	2. *Sticta*

1. LOBARIA Schreb.

Large foliose, scrobiculate (in arctic species), lichens with upper and lower cortex paraplectenchymatous; the medulla loosely interwoven; the algae green or blue green, with internal cephalodia containing *Nostoc*; underside with or without tomentum, attached to substratum by rhizinae. Apothecia with thalloid margin; hymenium hyaline or yellowish; paraphyses unbranched, septate, not thickened at tips; asci clavate; spores 8, spindle-shaped to acicular, 1–5-septate, hyaline to pale brown.

1   Algae of thallus green; underside with tan network; thallus K-,
        containing tenuorin   1. *L. linita*
     Algae of thallus blue green   2
2 (1) Isidia present; underside with black network which gives a K+ purple dif-
        fusion; thallus K+, P+, containing stictic acid   2. *L. pseudopulmonaria*
     Soredia present, underside without black network   3. *L. scrobiculata*

1. **Lobaria linita** (Ach.) Rabenh. Thallus large foliose, scrobiculate
above, upper side smooth, yellow green to brownish; undersurface with
bald spots among a fine, even pale to tan tomentum, cephalodia showing
as swellings on both surfaces; upper cortex paraplectenchymatous;
algae: *Myrmecia* with *Nostoc* in the cephalodia; lower cortex paraplectenchymatous. Apothecia not formed in tundra material but abundant on
material in forested areas, adnate on the upper surface; disk red brown,
margin crenated, the algae often confined to below the subhymenium;
subhymenium pale yellowish; hymenium 75–113 μ, hyaline below, yellowish brown above; paraphyses unbranched, not coherent, septate, not
thickened above; asci clavate, 90–100 × 15–21 μ; spores 8, hyaline to pale
brownish, 1–2-septate, fusiform, 16–34 × 5–10 μ. Reactions K-, C-, P-;
contains tenuorin.

This species grows among mosses and on hummocks over humus
soils in the tundras. Farther south in the forests it may also grow on tree
trunks and even branches in the rain forests. This species is circumpolar
arctic-alpine in the tundra material which is non-fertile, smaller, and less
reticulated. Such material is known as var. *linita* and occurs in Scandinavia and the alps in Europe; in Spitzbergen, Novaya Zemlya, and Siberia;
in North America from Alaska east to Chesterfield Inlet. The fertile
material with abundant pycnidia and apothecia, more reticulated upper
surface and growing on trees as well as mosses is found in eastern Asia
and along the coastal forests from Alaska to Washington in North
America. It is known as var. *tenuior* (Hue) Asah. Krog (1968) suggests
that the latter is temperate amphi-Pacific.

Localities: 1, 2, 3, 6, 8, 9, 11, 13.

2. **Lobaria pseudopulmonaria** Gyel. Thallus large foliose, scrobiculate
above, upper side smooth, yellow green to brown, lacking soredia or
isidia; underside with bald spots among a very black network of tomentum, lacking cephalodia; upper and lower cortices paraplectenchymatous. Algae: *Nostoc*. Apothecia not seen in North American material. In an
Asiatic specimen they were adnate on the upper surface, to 2 mm broad,
the margin entire to crenulate and in some with tomentum on the
reverse, the exciple paraplectenchymatous and lacking algae; the sub-

hymenium brownish; hymenium 110 μ thick; hyaline below, yellow brown above; paraphyses unbranched, not coherent, septate, the tips not thickened; spores 8, hyaline, 1–3-septate, 25 × 7–8 μ. Containing thelephorin ± stictic acid, traces of norstictic acid and constictic acid.

Like *L. linita* this species grows on moss and humus in the tundra and on tree trunks in more southerly climates. It is an amphi-Pacific species, growing in eastern Asia and in Alaska. From the latter state it has already been reported by Weber and Viereck (1967) and Krog (1968), under the name of *L. retigera* which, however, is an isidiate species.

Localities: 1, 11.

3. **Lobaria scrobiculata** (Scop.) DC. in Lam. & DC. Thallus large foliose, very scrobiculate, slaty green or yellowish green or olive green when dry, becoming blackish when wet, ascending, the lobes rounded, the tips strongly pulverulent, soredia on the upper surface and margins in rounded or elongate masses, slaty gray to brown; underside smooth in parts heavily tomentose in others, pale brown to blackening, the medulla K+ orange red, KC+ red, C-, P+ orange, containing usnic acid, scrobiculin, norstictic, stictic and constictic acids. Apothecia rare, not present in Alaskan material.

This species is commonly on trees and soil in most of its range, rarely on rocks. It is circumpolar, arctic to temperate in range.

Locality: 8.

2. STICTA Schreb.

Thallus foliose, dorsiventral, erect in some non arctic species; upper side smooth to tomentose (in non-arctic species); underside with a fine nape of tomentum, cyphellate below; upper and lower cortex paraplectenchymatous; algae either green (*Palmella*) or blue green (*Nostoc*); attached by rhizinae. Apothecia adnate or marginal, with or without algae in the margin; the margin paraplectenchymatous; subhymenium hyaline or brownish; upper part of the hymenium colored; paraphyses unbranched, septate, coherent; spores 8, hyaline to brown, spindle-shaped to acicular, 1–7-septate. Pycnidia immersed, marginal or in the surface; pycnoconidia short, straight, cylindrical or with the ends thickened.

1 Thallus lacking soredia or isidia	1. *S. arctica*
Thallus, especially on the margins, with soredia or isidia	2. *S. weigelii*

1. **Sticta arctica** Degel. Thallus foliose, dark brown, the lobes small, to 30 mm long, 12 mm broad, the edges somewhat crisped, upper side

smooth; underside pale at the edges, dark toward the center, covered with a fine tomentum with scattered cyphellae; attached by rhizinae; upper and lower cortex paraplectenchymatous, medulla lax. Apothecia unknown. Algae: *Nostoc*. Reactions all negative with usual reagents and no substances known from this species.

Growing over mosses and on hummocks in tundras, including moist and also the dry *Dryas* types of tundra. An amphi-Beringian species known from central northern Siberia, Kamtchatka, and North America from Baffin Island in the east to Alaska.

Localities: 1, 11.

2. **Sticta weigelii** (Isert ex Ach.) Vain. Thallus foliose, brown, the lobes narrow and elongate to 1.2 mm broad, the edges crisped and with an abundance of black soredia and isidia, the isidia may also line the edges of cuts and holes in the upper side; underside covered with a mat of fine tomentum, lighter brown toward the lobe tips, blackening centrally, with an abundance of scattered impressed cyphellae. Apothecia not seen.

This species grows on tree bases and over rocks in more southerly parts of its range. It is tropical and temperate, Asian, and North and South American, in North America being distributed in the southeast and in the western states. It has previously been reported from Alaska by Krog (1968) and McCullough (1965). The occurrence in Anaktuvuk Pass fits well with the occurrence of such southeastern lichens as *Umbilicaria caroliniana*.

Locality: 8.

GYALECTACEAE

Crustose lichens, homiomerous or heteromerous, lacking cortex, lacking rhizinae. Apothecia disk-shaped, immersed or sessile; proper exciple pale and hyaline or carbonaceous, covered by thallus or naked; paraphyses unbranched, loose; asci 6-many-spored; spores hyaline, 2-many-septate or muriform, ovate or acicular, the cells cylindrical and thin walled. Algae: *Scytonema* or *Trentepolia*.

1. GYALECTA Ach.

Thallus crustose with little differentiation, lacking cortices, with loosely interwoven hyphae of medulla mixed with threads and glomerules of the *Trentepolia* algae forming a loose layer over the substrate. Apothecia discoid, sunken or sessile, the margin waxy, naked or surrounded by thallus, the disk usually deeply concave and pale waxy; subhymenium pale; hymenium hyaline; paraphyses unbranched; asci clavate; spores 8, hyaline, ellipsoid or elongate, 3-septate to muriform.

1. **Gyalecta foveolaris** (Ach.) Schaer. Thallus very thin, membranace-ous to coarsely granular, pale olive yellowish, or ashy gray to whitish or darkening to brownish, with a translucent texture. Apothecia deeply urceolate usually with a thalloid covering, to 2 mm broad, the disk an orange reddish pale color or hyaline whitish; subhymenium hyaline; hymenium hyaline below, pale brownish orange above, 90–140 μ thick; paraphyses unbranched, septate indistinctly, 3–5 μ thick, the tips scarcely thickened; asci cylindrico-clavate, 75–110 × 6–11 μ; spores 8 hyaline, long-ellipsoid, 3-septate, 12–24 × 5–8 μ.

This species grows on mineral soil at the edge of frost boils and also on humus soils. It is calciphile, arctic-alpine and circumpolar. It is known from Greenland to Alaska in the high arctic and south to Colorado in the Rocky Mountains in North America.

Localities; 8, 9, 10, 11, 13.

LECIDEACEAE

Thallus crustose, sometimes with lobate edges, chinky, areolate, granu-lar or squamulose, attached directly by the lower hyphae to the substra-tum, lacking clearly developed cortex but sometimes with layer of vertical hyphae to upper side. Apothecia discoid, adnate, sometimes immersed or stipitate; with proper exciple which is hyaline to carbon-aceous; no thalloid margin; subhymenium hyaline to carbonaceous; paraphyses rarely branched, septate, loose or conglutinate; asci cylindri-cal to clavate; spores 8 (rarely 16), hyaline to darker, 1-celled, 2-many-celled, or muriform, rarely the wall thick, halonate with gelatinous epispore in *Rhizocarpon*. Pycnidia immersed in thallus; pycnoconidia ellip-tical to cylindrical. Algae: *Trebouxia*.

1	Paraphyses unbranched, conglutinate or free; spores lacking gelatinous epispore (or if with one then 1-celled)	2
	Paraphyses anastomosing branched; spores with gelatinous epispore, 2-celled to muriform, brown	8. *Rhizocarpon*
2 (1)	Spores 1-celled	3
	Spores 2- or more-celled	5
3 (2)	Spores small, less than 40 μ, thin walled	4
	Spores large, 70–100 × 35–45 μ, thick walled	3. *Mycoblastus*

4 (3) Asci with a conspicuous I+ dark blue 'tholus' at the tip; paraphyses free or only slightly coherent; medulla always I-; spores medium-sized, broadly ellipsoid; nearly all species containing xanthones   1. *Lecidella*

Asci without an amyloid 'tholus' (asci I- or entirely pale blue); paraphyses always coherent; medulla I+ blue or I-; spores very small to large; no species contain xanthones   2. *Lecidea*

5 (2) Spores 2-celled, hyaline, ellipsoid; thallus granulose, not squamulose
   4. *Catillaria*

Spores 2-celled fusiform and thallus squamulose, or spores parallel 4- or more-celled to muriform   6

6 (5) Spores parallel-celled, not muriform   7
Spores muriform   7. *Lopadium*

7 (6) Spores 2-celled fusiform to parallel 8-celled; thallus pseudocorticate, squamulose   5. *Toninia*

Spores 3-many-celled, cells parallel; thallus ecorticate, granular crustose but not squamulose   6. *Bacidia*

1. LECIDELLA Körb. em. Hertel & Leuckert

Thallus crustose, medullary hyphae I-. Apothecia commonly adnate with narrow base, black (rarely blackish brown), occasionally pruinose; exciple well developed, radiate, pale, not at all carbonized except for a narrow marginal zone; paraphyses 1.7–2.5 μ thick, free or easily so, simple (the bases rarely connected, the apices occasionally branching or bicapitate), the apices thickened to 5 μ; asci clavate, the tip with a broad I+ dark blue upper portion called the 'tholus'; spores 8, ellipsoid, middle sized, hyaline, without halo of gelatin. Pycnidia immersed; fulcra exobasidial; pycnoconidiospores filiform. Thallus usually containing xanthones.

1   On rocks   2
   On mosses, humus, soil, or bark   5

2 (1) Thallus K+ yellow (atranorin)   3
Thallus K-, lacking atranorin (thallus sometimes lacking)   4

3 (2) Exciple dark; thallus bullate-areolate; hypothecium yellow brown (K+ bright orange); epithecium blue green; exciple subparaplectenchymatus
   1. *L. carpathica*

Exciple hyaline within, the edge green or violet; thallus granular warty or irregularly areolate to commonly lacking (f. *egena*); hypothecium hyaline or in age becoming reddish brown; upper part of hymenium blue, blue green, gray, violet, or reddish black, the lower part hyaline to pale rosy brownish; exciple appearing columnar   3. *L. stigmatea*

4 (2) Exciple green black externally, dark brown internally   2. *L. goniophila*
Exciple green black to brownish externally, hyaline internally
                                                             3. *L. stigmatea*

5 (1) Thallus C-, on bark and old wood   4. *L. glomerulosa*
Thallus C+ red (xanthones)   6

6 (5) On bark and old wood   5. *L. elaeochroma*
On humus and soil   6. *L. wulfenii*

1. **Lecidella carpathica** Körb. (*Lecidea latypiza* Nyl.) Thallus granulose areolate or granular in thin to thick crust, whitish, yellowish or grayish; granules to 0.1 mm, areoles of granules to 1 mm; K+ yellow (atranorin), I-. Apothecia small, to 0.6 mm, adnate on areolae and between the granules; disk flat, black, dull; margin prominent, shining, black; exciple blue brown at upper edge, dark brown at sides of apothecium and continuous with dark brown hypothecium, the hypothecium and sub-hymenium bright or orange in KOH; epithecium blue black; upper hymenium blue, lower, brownish; hymenium 65 μ; paraphyses unbranched, septate, tips slightly wider; asci clavate; spores 8, biseriate, hyaline, ellipsoid, 12–13 × 7.5–8.5 μ.

On rocks, both calcareous and acid, in open places. Circumpolar, boreal, and arctic; in North America from southern Greenland and Quebec, west to Alaska, south to Quebec, Ontario, Wisconsin, Alberta, Washington.

Locality: 1.

2. **Lecidella goniophila** (Flörke) Körb. (*Lecidea goniophila* Flörke) Thallus absent or very thin of verrucules, about 0.1–0.2 mm broad, sometimes becoming verrucose-areolate, I-, K-. Apothecia solitary or conglomerate, to 1 mm broad; disk flat to slightly convex, black or brown black, dull, bare; margin thin, black; exciple from under the edges of the apothecium brown on interior, blue black or green black externally; hypothecium hyaline; epithecium blue green or brown green; hymenium 60–80 μ, olive brown to bluish above, pale brownish below; paraphyses lax, unbranched, septate, 2 μ, tips slightly thickened and darkened; asci clavate; spores 8, biseriate, hyaline, ellipsoid, 11–14 × 7–10 μ.

On rocks, usually acid but also calcareous, especially on cliffs. Known from Europe, Spitzbergen, Greenland, and North America where it appears to be distributed as an arctic-alpine species from Ellesmere Island to Alaska, south to New York, Michigan, South Dakota, New Mexico.

Localities: 3, 9.

3. **Lecidella stigmatea** (Ach.) Hert. & Leuck. (*Lecidea stigmatea* Ach.) Thallus thin or disappearing, verrucose with large and small verrucules intermixed, white or ashy white, K- or K+ yellowish brown, lacking soredia, hypothallus indistinct. Apothecia to 0.8 mm, adnate, base constricted or not; disk flat to becoming convex, black, bare; exciple black, shining, persistent to disappearing, the outer part dark green, more or less violet, inner part pale; hypothecium hyaline to pale brownish; epithecium brownish black to blue green; hymenium 70 μ, upper part blue green; paraphyses lax, the tips slightly thickened; spores 8, biseriate, ellipsoid, simple, 8–14 × 5–10 μ.

Growing on acid rocks, occasionally also on calcareous rocks. Circumpolar, arctic-alpine. In North America south to Quebec, Wisconsin, Colorado, and California.

Localities: 1, 2, 8, 9, 11, 14.

4. **Lecidella glomerulosa** (DC.) Choisy (*Lecidea glomerulosa* (DC.) Steud., *Lecidea euphorea* (Flörke) Nyl.) Thallus thin to moderately thick, verruculose, white or ashy greenish to subolivaceous, medulla K- to K+ yellowish (traces of atranorin), C-. Black hypothallus usually indistinct. Apothecia small, to 0.6 mm, adnate, not constricted; disk flat or becoming convex, black, bare, dull or slightly shining; margin thin, black, disappearing; exciple exterior blue, interior golden brown to violet or reddish; epithecium blue or bluish brown; hypothecium dark brown or reddish brown; hymenium 70–110 μ, upper part bluish, I+ blue becoming wine red; paraphyses lax, slender, the tips slightly thickened, darkened, slightly gelatinized; asci clavate; spores 8, biseriate, simple, hyaline, ellipsoid, 10–16 × 6–9 μ.

On bark and rotting wood, occasionally on old caribou antlers. Circumpolar, boreal and temperate, in North America from Baffin Island to Alaska south to New York, Wisconsin, Colorado, California.

Localities: 1, 8, 14.

5. **Lecidella elaeochroma** (Ach.) Haszl. (*Lecidea elaeochroma* (Ach.) Ach.; *Lecidea olivacea* (Hoffm.) Mass.) Thallus thin, to moderately thick, smooth to verruculose, bluish-, yellowish- to whitish-ashy, esorediate, medulla K+ yellow, KC+ red (gyrophoric acid), a black hypothallus sometimes present and limiting the margins. Apothecia small, to 1 mm, adnate, not constricted; disk flat or becoming convex, very black, bare, dull or slightly shining; margin thin, black, becoming excluded; exciple with exterior bluish brown to violet brown, interior violet brown; epithecium very blue; hypothecium reddish, brownish or violet, upper part pale; hymenium 70 μ, I+ blue becoming wine red; upper part blue; paraphyses

lax, unbranched, slender with tips slightly thickened, slightly gelatinized; asci clavate; spores 8, biseriate, hyaline, ellipsoid to subglobose 7–16 × 6–8 μ.

On barks of many kinds of trees and on wood. European and western Asiatic, and in western North America, Alaska to Alberta, Saskatchewan, California, and west of this line.

Localities: 1, 2, 14.

6. **Lecidella wulfenii** (Hepp) Körb. (*Lecidea wulfenii* (Hepp) Arn. *Lecidea heppii* Anderson & Weber) Thallus crustose, verrucose, dispersed or contiguous, esorediate, white or ashy white, K+ yellow, KC+ orange red, hypothallus indistinct. Apothecia often becoming confluent and tuberculate, becoming convex and immarginate, to 1 mm broad; disk flat to convex, black, dull, bare, rarely shining; exciple brownish purple below, or pale, upper part bluish brown, hyphae radiate; epithecium bluish green, rarely olivaceous; hypothecium brown red or brown, the upper part of vertical hyphae; hymenium 80–100 μ, upper part bluish, I+ blue becoming wine red; paraphyses lax, unbranched, the tips slightly thickened; asci clavate; spores 8, biseriate, hyaline, simple, ellipsoid, 7–16 × 7–8 μ.

Growing on mosses and earth as well as humus. Circumpolar, arctic-alpine, in North America from Ellesmere Island to Alaska, south to New Hampshire, and Churchill, Manitoba. A scantily known species.

This species which grows on soil and humus is certainly exceedingly closely related to *L. elaeochroma* which grows on barks.

2. LECIDEA Ach.

Thallus crustose, granulose, verrucose, areolate, or squamulose, lacking cortex in most species, heteromerous, attached to the substratum without rhizinae, hypothallus sometimes present. Apothecia with proper exciple only, immersed to adnate or sessile, pale to dark or black; the exciple radiate or with the hyphae irregularly cellular, usually of same color as the disk; hypothecium hyaline to dark brown or black; epithecium pale to brown, blue black, olivaceous or greenish; hymenium hyaline to colored, brownish or bluish; paraphyses simple or branched toward the apices or branched and anastomosing; asci clavate or cylindrico-clavate; spores 8–12 hyaline, simple or partly 1-septate, oblong, ellipsoid, fusiform, or globose, usually thin walled, rarely halonate. Algae: *Trebouxia, Jaagia, Chlorella, Coccobotrys.*

1	On rocks	2
	On humus, mosses, plant remains, bark, wood, or mineral soil	3

2 (1) Thallus either sorediate or cephalodiate   3
        Thallus with neither soredia nor cephalodia   10

3 (2) Thallus sorediate, lacking cephalodia   4
        Thallus with red brown cephalodia, gyrophoric acid present, soredia
            present or absent   6

4 (3) Thallus composed entirely of greenish yellow soredia, on shaded sides
            of boulders; apothecia rare, yellow, convex, margin disappearing,
            epithecium yellowish, hypothecium hyaline, paraphyses lax, spores
            $3-7 \times 1-3 \ \mu$ 1. *L. lucida*
        Thallus of gray or orange areolae with craterform soralia bearing blue
            gray soredia; apothecia rare   5

5 (4) Thallus orange to partly gray   2. *L. melinodes*
        Thallus bluish gray or white   3. *L. glaucophaea*

6 (3) Cortex C+ red, spores large, $20-30 \times 10-16 \ \mu$ 7
        Cortex C-, medulla C-, spores small, $9-13 \times 5-7 \ \mu$ 6. *L. shushanii*

7 (6) Both cortex and medulla C+ red, margins of thallus effigurate,
        cephalodia centered on thallus, soredia present or absent
                               Placopsis gelida

        Cortex C+ red, medulla C-, margins not effigurate, cephalodia
            scattered over thallus area   8

8 (7) Usually sorediate, apothecia between the squamules, areolae high
            bullate   9
        Esorediate, apothecia immersed in the areolae, areolae bullate if
            fruiting, only slightly convex if not   *Lecanora pelobotrya*

9 (8) Thallus thin, apothecia rare, epruinose, 0.5–1.0 mm diam.
                                 4. *L. panaeola*

        Thallus thick (–2 mm), apothecia abundant, pruinose, over 1 mm diam.
                              5. *L. elegantior*

10 (2) Thallus lacking except beneath apothecium   11
        Thallus present, well developed, areolate or verrucose   19

11 (10) Hypothecium or exciple changed by reagents K, C, or I   12
        Hypothecium or exciple not changing with reagents   15

12 (11) Hypothecium K+ violet   7. *L. hypocrita*
        Hypothecium K-   13

13 (12) Thallus disappearing; spores narrow, $6-12 \times 2.5-3.5 \ \mu$, exciple K-, I+
            blue   8. *L. auriculata*
        Thallus white to blue gray, disappearing; spores broader, $8-13 \times 5-6 \ \mu$
                                14

14 (13) Thallus K-   9. *L. lapicida*
        Thallus K+ red (norstictic acid)   10. *L. pantherina*

15 (11) Hypothecium and exciple dark brown to blackish   16
Hypothecium hyaline, pale or brownish; spores 10–15 × 7–10 μ; exciple margin dark green ± violet; hymenium 70 μ; epithecium blue green to brown; paraphyses lax   *Lecidella stigmatea*

16 (15) Spores less than 16 μ long   17
Spores more than 16 μ long   18

17 (16) Spores 11–13 × 4–5 μ, upper part of hymenium emerald, epithecium dark green, apothecia small (–1 mm)   11. *L. vorticosa*
Spores 12–16 × 5–7 μ, upper part of hymenium and epithecium olive to black, apothecium to 1 mm   12. *L. crustulata*

18 (16) Apothecia 0.5–0.8 mm flat, often pruinose; hymenium 90 μ, epithecium dark brown ± violet, mucilagenous; asci transversely wrinkled when old; spores 16–29 × 9–12 μ 13. *L. jurana*
Apothecia to 3 mm, flat and marginate or convex and immarginate; hymenium 100–120 μ; epithecium brown green; asci not wrinkled; spores 15–30 × 7–11 μ 14. *L. macrocarpa*

19 (10) Thallus red orange or orange   20
Thallus red brown, coppery brown to black, white or ashy   22

20 (19) Hypothecium and exciple dark brown to black   21
Hypothecium pale to brownish; epithecium blue black; exciple dark, inner part pale, the hyphae erect   9. *L. lapicida* f. *ochracea*

21 (20) Thallus red orange; apothecia very small, 0.2–0.5 mm, concave to flat, epruinose; paraphyses anastomosing; hymenium I+ blue; spores 9–16 × 5–9 μ 15. *L. atrata*
Thallus orange; apothecia large, 1–1.5 mm, flat to convex, pruinose; paraphyses unbranched, clavate; hymenium I–; spores 15–23 × 7.5–12 μ 16. *L. flavocaerulescens*

22 (19) Thallus red brown, coppery brown, or black   23
Thallus gray, white, ashy, or yellowish   29

23 (22) Thallus I+ blue   24
Thallus I–   26

24 (23) Apothecia adnate, convex, thallus red brown   25
Apothecia immersed in thallus, concave; thallus sooty brown, blackish, with both ashy and black borders   19. *L. carbonoidea*

25 (24) Areolae flat, scattered, K+ yellowish; hypothecium brown black; exciple brown black; hymenium 65–85 μ; spores 10–12 × 5–6 μ
     17. *L. paupercula*
Areolae verrucose, thick, coppery or gray brown, K–; hypothecium very pale; exciple bluish black; hymenium 40–70 μ; spores 7–9 (–12) × 4–6 μ
     18. *L. atrobrunnea*

26 (23) Thallus brown   27
Thallus black, thin, diffract areolate, hypothecium and hymenium violet black, the upper part of the hymenium blue black; spores 7–8 × 5 μ   20. *L. picea*

27 (26) Thallus thick, verrucose areolate to globulose or squamulose; hypothecium and exciple dark brown, hymenium 100–200 μ; spores 7–10 μ, globose   21. *L. cinereorufa*
Thallus thin, effuse, granular, reddish brown   28

28 (27) Proper exciple present, hymenium I+ blue at tips of paraphyses, paraphyses free; epithecium blue green to brown; exciple dark green ± violet; hymenium 70 μ; spores ellipsoid, 10–15 × 7–10 μ
       Lecidella stigmatea
Proper exciple disappearing; hymenium I+ entirely blue; paraphyses coherent; epithecium brown; hymenium 65 μ; spores 10–12 × 5–7 μ
       22. *L. instrata*

29 (22) Medulla I+ blue   30
Medulla I–   36

30 (29) Hypothecium black to dark brown   31
Hypothecium hyaline to pale   33

31 (30) Exciple well developed; apothecia adnate   32
Exciple poorly developed; apothecia usually immersed; hypothallus white; calcicolous; spores 12–18 × 5.5–9 μ   24. *L. speirea*

32 (31) Thallus chinky to areolate, bluish, smoky gray or whitish gray, on black hypothallus; apothecia to 2 mm, flat to convex, immersed or projecting; spores 7.5–10 × 5–6.5 μ   23. *L. confluens*
Thallus white to gray, verrucose to disappearing; apothecia adnate, marginate; exciple pale, I+ blue; epithecium blue brown; spores narrow, 6–12 × 2.5–3.5 μ   8. *L. auriculata*

33 (30) Thallus K–   34
Thallus K+ red (norstictic acid), ashy gray, apothecia from innate to adnate; exciple hyaline with brown margin; epithecium black; spores 9–15 × 4–7 μ   10. *L. pantherina*

34 (33) Thallus white to gray, areolate, with distinct black hypothallus, marginate; exciple green brown; epithecium blue black; spores 8–13 × 5–6 μ   9. *L. lapicida*
Thallus white to gray, flat areolate; apothecia between the areolae and level with thallus; exciple hyaline with narrow brown margin   35

35 (34) Thallus chalky white; apothecia often umbonate; subhymenium coppery green; hypothecium ocher brown; spores pseudo 2-celled, 10–13 × 5–7 μ   25. *L. umbonata*
Thallus blue gray; apothecia not umbonate; subhymenium colorless; spores 1-celled, 8–11 × 4.5–6 μ   26. *L. tessellata*

36 (29) Paraphyses lax   see *Lecidella*
        Paraphyses conglutinate   37

37 (36) Spores large, average over 17 μ long   38
        Spores less than 17 μ average   39

38 (37) Epithecium olive or brown, ± violet; apothecia usually pruinose; thallus not well developed   13. *L. jurana*
        Epithecium blue green, apothecia conglomerate, shining; thallus of glomerate verrucules   27. *L. pallida*

39 (37) Spores less than 10 μ long   40
        Spores 10–15 (–17) μ long   41

40 (39) Spores globose, 7–10 μ; thallus gray to red brown, verrucose to globulose; hypothecium and exciple dark brown; epithecium blue green   21. *L. cinereorufa*
        Spores ellipsoid, 9–14 × 4–6 μ; thallus bluish white, areolate, areolae farinose; hypothecium pale; exciple brown green; epithecium olive brown   28. *L. lithophila*

41 (39) Hypothecium and exciple dark, black brown   42
        Hypothecium yellow brown or paler to hyaline   43

42 (41) Base of exciple and hypothecium K+ violet; thallus yellow gray, chinky areolate or poorly developed; epithecium olive or brown, I± violet   13. *L. jurana*
        Hypothecium and exciple K–; thallus thin, gray white or orange; epithecium olive to black   12. *L. crustulata*

43 (41) Hypothecium yellow brown or brownish   44
        Hypothecium hyaline   47

44 (43) Apothecia large, over 0.5 mm; epithecium blue green; hymenium 50–55 μ   45
        Apothecia tiny, less than 0.4 mm; epithecium brown; hymenium 40–50 μ; spores 9–14 × 5–6 μ; thallus brown or olive   31. *L. melaphanoides*

45 (44) Exciple C–   46
        Exciple C+ red; spores narrowly ellipsoid, 6.5–10 × 2.1–3.6 μ   30. *L. diducens*

46 (45) Spores narrow, subcylindrical to elongate-oblong, 6–11 × 2.5–3.5 μ   8. *L. auriculata*
        Spores subglobose, 4–5 μ diam. or 6–7 × 4.5 μ   29. *L. brachyspora*

47 (43) Epithecium brown to green brown   48
        Epithecium green or black   51

48 (47) Thallus yellowish, thick, chinky areolate   49
        Thallus ashy gray or orange, disk black, reddening when moist   50

49 (48) Thallus P+ yellow, areolae convex; tips of paraphyses thickened, over 4 μ; spores broadly ellipsoid, 8–15 × 5–8 μ   32. *L. aglaea*

Thallus P-, areolae flat; tips of paraphyses slender (-1.7 μ); spores 8–14 × 4.5–7.5 μ 33. *L. atromarginata*

50 (48) Thallus ashy, thin, chinky areolate, the areolae flat; apothecia immersed, brown black, pruinose; spores 9–14 × 4–6 μ

28. *L. lithophila*

Thallus dark ashy, granulate, the margins effigurate, areolae convex; apothecia adnate, disk epruinose; spores 8–14 × 4–7 μ

34. *L. leucophaea*

51 (47) Thallus yellow, brown yellow, rarely white, P+ orange (psoromic acid), K+ orange; spores 10–12 × 4–5 μ 35. *L. armeniaca*

Thallus white, K+ yellow; apothecia immersed to adnate, margin dissappearing; spores 8–14 × 4.5–7.5 μ 36. *L. marginata*

52 (1) On bark or old wood   53
        On humus, mosses, plant remains or mineral soil   56

53 (52) Paraphyses lax   see *Lecidella*
        Paraphyses conglutinate   54

54 (53) Hypothecium red brown to black; spores thin walled, 6–8 × 2.5–3 μ; thallus verruculose, disappearing; apothecia black, flat to convex, margin very thin; epithecium brown; exciple edge brown, inside pale

37. *L. plebeja*

        Hypothecium pale, hyaline to yellow   55

55 (54) Spores thick walled, 15–23 × 8–13 μ; thallus thin, gray white; apothecia red brown to black, hemispherical; epithecium red brown; exciple reddish brown   38. *L. tornoënsis*

        Spores thin-walled, 7–16 × 4–5 μ; thallus yellowish, sometimes sorediate, KC+ orange, smooth to verruculose; apothecia yellowish; epithecium yellowish; exciple yellowish   39. *L. symmicta*

56 (52) Thallus squamulose   57
        Thallus not squamulose, granular to verruculose   60

57 (56) Edge or underside of squamules white   58
        Edge or underside of squamules black   59

58 (57) Squamules brown red, brick colored, dark or white when dying, concave, crenulate; spores 9–15 × 7–9 μ 40. *L. decipiens*

        Squamules white or brownish, overlapping, ascending, concave; spores 10–17 × 4–7 μ 41. *L. rubiformis*

59 (57) Squamules appressed to substratum, gray to sordid gray, or brownish, dull   42. *L. demissa*

        Squamules overlapping, ascending, greenish chestnut to red brown, shining   43. *L. globifera*

60 (56) Thallus dark red brown to black or dark olive brown   61
        Thallus whitish, grayish, brownish gray, varnish like, or disappearing

62

61 (60) Spores thick walled, large, 15–23 × 8–13 μ; hypothecium pale; epithecium red brown; apothecia hemispherical   38. *L. tornoënsis*
Spores thin walled, small, 11–16 × 5–7 μ; hypothecium dark yellow brown; epithecium brown   44. *L. uliginosa*

62 (60) Thallus with soredia, verruculose granular, gray or green gray, C+ red, K+ yellow (gyrophoric acid); apothecia brick colored to black
       45. *L. granulosa*
Thallus esorediate   63

63 (62) Paraphyses lax; thallus white verruculose, K+ yellow, KC+ red; apothecia black, shining, convex; hypothecium yellow brown; epithecium blue green; spores 7–16 × 7–8 μ   *Lecidella wulfenii*
Paraphyses conglutinate   64

64 (63) Hypothecium dark brown, red brown, or black   65
Hypothecium hyaline to very pale brownish   68

65 (64) Apothecia brown   66
Apothecia black   67

66 (65) Exciple of coarse hyphae with many cross walls forming paraplectenchyma; thallus thick, verruculose, white to green gray
       46. *L. berengeriana*
Exciple of thin walled slender hyphae, radiate; thallus thin, grayish, disappearing   47. *L. fusca*

67 (65) Thallus of small white granules; apothecia lacking exciple, almost spherical, pruinose; hypothecium very uneven, of brown prongs going up into hymenium; paraphyses very gelatinous; spores uniseriate, 8–12 × 3–5 μ   48. *L. subcandida*

Thallus of coarse verrucules; apothecia with radiate exciple, adnate, base not constricted, flat to convex or conglomerate, bare; hypothecium even across upper side; paraphyses only slightly gelatinized; spores biseriate, 8–19 × 3–6 μ   49. *L. assimilata*

68 (64) Apothecia dark yellow to red brown   69
Apothecia black   70

69 (68) Apothecia dark yellow to red brown; epithecium and exciple pale; thallus thin, glaucous, P–; spores 10–20 × 3–7 μ   50. *L. vernalis*
Apothecia red brown to orange; epithecium and edge of exciple red brown; thallus whitish, P+ red (baeomycesic acid); spores 10–22 × 3–6 μ   51. *L. cuprea*

70 (68) Thallus of rounded isidia-like glomerules, thick, blue gray; apothecia flat; hymenium 45 μ, red brown, K+ violet; spores 9–18 × 3–6 μ
       52. *L. ramulosa*
Thallus thin, of small, contiguous or dispersed verrucules; apothecia convex; hymenium 100 μ, bluish; spores 10–16 × 3–5 μ
       53. *L. limosa*

1. **Lecidea lucida** (Ach.) Ach. Thallus entirely farinose-sorediate to partly verruculose, thin or forming a thick covering on the substrate, lemon yellow; hypothallus indistinct; K-, C-. Apothecia rare, 0.2–0.5–1 mm, adnate, convex, soon immarginate, lemon yellow to yolk colored; disk smooth to tuberculate; hypothecium hyaline, slightly granular, of loose thin walled cells; exciple of radiating hyphae, hyaline to yellowish in outer part; hymenium 30 μ, I+ blue, becoming wine red; paraphyses unbranched, little-septate, slender, hyaline, the tips slightly thickened, lax in KOH; epithecium yellow granular; spores 8, biseriate, oblong ovate to clavate, hyaline, 3–7 × 1–3 μ.

In shaded moist microhabitats as the underside of rocks, underhangs of cliffs, on acid rocks, rarely on earth and humus. A circumpolar arctic and temperate species, New England, Northwest Territories to Alaska, south to Virginia, Oklahoma, Washington.

Localities: 3, 8.

2. **Lecidea melinodes** (Körb.) Magn. Thallus crustose, thin to moderately thick, chinky to areolate, areolae to 1 mm, very orange colored, the centers of the areolae with crateriform soralia producing blue gray soredia, I-, K-. Apothecia (not seen, description after Lynge) rare, large, to 1.5 mm, adnate; disk black, epruinose, slightly convex, margin indistinct; exciple black, hyaline in the upper part and verging into the hymenium; hymenium 100 μ, inspersed, upper part olive brown; epithecium more or less granulose; paraphyses conglutinate, the tips not much thickened or not at all; spores poorly developed, 16 × 10–11.

A species of acid rocks. Probably circumpolar arctic. Reported from Jan Mayen Island, Novaya Zemlya, northern Scandinavia, and under the name *L. soredizodes* f. *ochracea* Lynge from Greenland and Baffin Island. Specimens from Iceland, Devon Island, Labrador, Northwest Territories, and Alaska are in WIS.

Locality: 3.

3. **Lecidea glaucophaea** Körb. (*L. soredizodes* (Nyl.) Vain.) Thallus crustose, moderately to quite thick, bluish ashy to ashy, smooth, K-, I-, soredia in crateriform soralia over the surface; a black hypothallus usually visible. Apothecia adnate, base constricted, to 2.5 mm broad; disk bluish or whitish pruinose; margin thick, black, bare; hypothecium pale above, brownish black below, of vertical hyphae, the lower part irregular; outer part of exciple brownish black; epithecium olive to brownish; paraphyses coherent but separable under pressure, unbranched, septate, the tips thickened; asci clavate; spores 8, biseriate, commonly with gelatinous epispore, simple, hyaline, elongate, 17–30 × 9–12 μ.

A species of acid rocks. Known from Scandinavia, Greenland, Jan Mayen Island. In North America I have seen specimens from Baffin Island, Devon Island, Axel Heiberg Island, the Adirondack Mts, and Wisconsin.

Localities: 8, 11.

This species also includes arctic reports under *L. soredizodes* (Nyl.) Vain. and *L. albuginosa* Nyl. It closely resembles *L. subsorediza* Nyl. which is K+ red, I+ blue instead of the K-, I- reactions of this species.

4. Lecidea panaeola Ach. Thallus crustose, moderately thick, to 2 mm, areolae verrucose to bullate convex, the areolae rounded to angular with mutual pressure to the center of the thallus, peripherally more discrete and all round, minutely rough, slightly pruinose, slightly shining, white ashy, gray, or usually pinkish ashy, commonly sorediate, with cephalodia on surface or more commonly between the areolae, brown ashy or reddish, surface of cephalodia botryoid, even with the surface of thallus or becoming globose; black hypothallus. Apothecia rare, formed between the areolae, adnate, level with the areolae or slightly raised, small, to 1 mm; disk black, epruinose, minutely rough, with persistent black margin; exciple brownish black; hypothecium brownish black; hymenium high, to 150 μ, upper part brownish black, covered by granular black layer; paraphyses conglutinate, slender, non or slightly thickened at tips; asci broadly clavate; spores 8, ellipsoid, thick walled, 22–27 × 9–15 μ. Pycnoconidia short, straight, acicular, 7–10 × 1 μ. Cortex C+ red, medulla C-, I-, K-. Cortex contains gyrophoric acid (GAQ tests).

A species of acid rocks. Circumpolar, arctic-alpine, in North America known from Ellesmere to Alaska, south to Adirondack Mts, NY, in the east, Minnesota and Washington.

Localities: 8, 11.

This species is difficult to distinguish from *Lecanora pelobotrya* when apothecia are not present. They have similar thalli, cephalodia, reactions, and chemistry. *L. panaeola* is usually sorediate and has usually higher, more bullated areolae, *Lecanora pelobotrya* is not sorediate and the non-fruiting areolae are only slightly convex.

5. Lecidea elegantior Magn. (= *L. elegans* (Th. Fr.) Vain., *Huilia elegantior* (Magn.) Hertel) Thallus over a black hypothallus, crustose, 1–2 mm thick, ashy or waxen, bullate areolate, the areolae to 1.5 mm broad, often fissured edges of the areolae toward the thallus margins rounded, those more central are more angular, soredia not common, with granular red to waxy brown large cephalodia containing *Gloeocapsa* algae; cortex C+ red, K-, medulla C-, K-, I-. Apothecia common, between the areolae, 1–2 mm broad; the margin thick, persistent, brownish black; epithecium brownish black; hypothecium brownish black; hymenium to 180 μ high; paraphyses coherent, slender, the tips little thickened; asci broadly clavate; spores 8, hyaline, ellipsoid, 20–30 × 10–16 μ.

A species of acid rocks, known from Scandinavia, Greenland, and Alaska, probably circumpolar but not often collected or recognized as differing from *L. panaeola*.

Locality: 8.

6. Lecidea shushanii Thoms. Thallus of large, to 1.8 mm broad, spherical globules attached by a narrow central base, the surface scabrid roughened, pale yellow, with rough red brown cephalodia between the glebae, K-, C-, P-, but containing gyrophoric acid on GE test; hypothallus black. Apothecia shining, black, adnate between the areolae or on the areolae and partly immersed, large, to 2 mm, flexuous or angular, appearing immarginate; exciple thin, black, shining; epithecium blue green, inspersed; exciple radiate, blue green laterally on the apothecium, brown black or red brown below, grading into the hypothecium; subhymenium 50 μ, red brown below to pale above; hymenium hyaline below, very blue above; paraphyses conglutinate, slightly branched, septate, not thickened above; asci clavate; spores 8, biseriate, simple, hyaline, ellipsoid, 9–12.5 × 5–7 μ.

Known only from the type: Locality 11.

In appearance and structure this species is very similar to *Lecidea panaeola* or *L. elegantior* and like them has red brown cephalodia. The spores, however, are far smaller than in those species which have spores over 20 μ long and 9 μ wide.

7. Lecidea hypocrita Mass. (*Lecidea lithospersa* Zahlbr.) Thallus lacking or forming a fine chinky to powdery chalky white crust; medulla weakly I+ blue, C-, K-, P-; hypothallus sometimes present. Apothecia scattered, with constricted base, to 2.5 mm broad, dull to shiny black, epruinose, young apothecia forming a continuous disk, characteristically becoming cracked and then breaking up into sections which sometimes regenerate to form new apothecia; epithecium very blue green (not brownish); exciple red brown, paler to the outside, KOH+ red violet; subhymenium 20 μ, pale red brown; hymenium 150–350 μ, upper part blue green, lower brown red; paraphyses strongly gelatinous, anastomosing branched, septate, 2–2.5 μ, tips enlarged to 5 μ; asci clavate, 55–65 × 12–17 μ; spores 8, simple, hyaline, ellipsoid, the tips often pointed, with gelatinous halo when young, 11–21 × 5–9 μ.

Growing on sun-exposed calcareous rocks. Known from Europe (northern and Alps), Greenland, Novaya Zemlya. North American specimens are in WIS from Alaska, Utah, and Illinois.

Localities: 1, 8, 9, 11, 14.

This species bears a close resemblance to *Lecidea auriculata* but is found on limestone, the hypothecium is red violet with KOH, and the spores are larger.

8. Lecidea auriculata Th. Fr. Thallus verrucose, dispersed to disappearing, white or ashy, K-, I+ (especially under apothecia), esorediate. Apothecia adnate, base well constricted, often aggregated, to 2.5 (rarely 4 mm) broad, becoming irregularly lobate, the disk flat or depressed convex, black, bare; margin thin, persistent, bare, black; exciple of radiating hyphae, conglutinate, a white inner zone continuing under the hypothecium, the outer part dark brown, K+ purple; hypothecium dark brown; hymenium 50–60 μ, upper part and epithecium blue brown; paraphyses septate, conglutinate, non-gelatinous, 1.5 μ, apices thickened; spores simple, hyaline, ellipsoid, 6–12 × 2.5–3.5 μ.

A species of usually acid rocks, usually in full sun. Circumpolar, arctic-alpine, in North America known from Ellesmere Island to Alaska, south to Quebec, Wisconsin, South Dakota, and Colorado.

Localities: 3, 8, 9, 11.

9. Lecidea lapicida (Ach.) Ach. Thallus moderately thick or thin, areolate, the areolae flat, rarely convex, smooth, esorediate, ashy or occasionally orange, I+ blue, K-. Apothecia round or angular, to 1.2 mm broad, adnate over the thallus, sometimes partly innate when young, the base not constricted; disk flat or becoming depressed-convex, black, bare, dull; epithecium bluish brown; the exciple thin, black, dull, slightly raised, usually persistent, in the exterior part with radiating hyphae, the interior irregular in context, brownish in the lower part of the apothecium and lacking in the center; hypothecium pale above or the lower part brownish; paraphyses conglutinate, unbranched, septate, the tips thickened to 4 μ; asci clavate; spores 8, biseriate, ellipsoid or oblong, hyaline, simple, the ends rounded, 8–13 × 4–6 μ.

On acid or calcareous rocks, usually in full sunlight, abundant in the arctic. A circumpolar, arctic-alpine species in North America growing from Ellesmere Island to Alaska, and south to Maine, Wisconsin, Minnesota, New Mexico, and California.

Localities: 1, 2, 8, 9, 11, 14. f. *ochracea* (Nyl.) Vain. with thallus orange colored and hypothecium pale was at 3.

10. **Lecidea pantherina** (Hoffm.) Th. Fr. Thallus thin or moderately thick, chinky or areolate, the areolae contiguous or rarely dispersed, flat, rarely slightly convex, smooth, ashy, white or rarely darker or orange, esorediate, I+ blue, K+ yellow turning red (norstictic acid), black hypothallus sometimes visible. Apothecia to 1.5 mm, innate in the thallus or becoming adnate and higher, the base not constricted or slightly so; the disk flat, black, bare or pruinose, dull; the exciple thin, black rarely pruinose, persistent; hypothecium of erect hyphae, hyaline or with a thin darker brownish layer in the center; exciple black or bluish black, inner part pale; epithecium blue black paraphyses conglutinate, unbranched, tips slightly thickened; spores 8, ellipsoid-oblong, hyaline, 9–15 × 4–7 μ.

A species of acid rocks. Circumpolar and arctic-alpine. In North America from Baffin island to Alaska and south to Quebec, Colorado, and Washington.

Localities: 3, 8, 11.

11. **Lecidea vorticosa** (Flörke) Körb. Thallus usually lacking, when present of very thin whitish or yellowish layer over the rock; C-, K-, I- or weakly +. Apothecia in small closely pressed together groups, to 0.9 mm broad, the base often strongly constricted; disk flat or slightly convex, black, epruinose, shining; epihymenium blue green and grading in color well into upper hymenium; exciple narrow, well defined, black, or brown black, paraplectenchymatous, the outer part somewhat columnar; hypothecium 60–120 μ, dark brown to brown black but lighter in color that the exciple; paraphyses conglutinate, sometimes branched and anastomosing, 1.7–2.5 μ thick, the tips scarcely thickened; asci cyclindrico-clavate, 40–50 × 10–14 μ; spores 8, hyaline, simple, elongate ovate, not halonate, 8.5–15 × 3.5–5.5 μ.

On acid and very seldom somewhat calcareous rocks. Circumpolar, arctic-alpine, in North America known from Ellesmere Island to Alaska, south to Quebec, New Mexico, Arizona, Washington.

Localities: 1, 2, 3, 8, 11.

The small apothecia, small spores, very dark hypothecium and the emerald epithecium and upper part of the hymenium are distinctive of this species.

12. **Lecidea crustulata** (Ach.) Spreng. (*Huilia crustulata* (Ach.) Hertel) Thallus thin, continuous to areolate, ashy glaucescent, rarely with orange tinge, sometimes but a stain on the substratum in appearance, lacking soredia, a black hypothallus sometimes apparent, I-, K-. Apothecia small, to 1 mm broad, the base slightly constricted, disk flat, black,

bare; the exciple thin, black, bare; epithecium olivaceous or brown to black; exciple brown black, K-; hypothecium pale under hymenium, the rest brown black; hymenium 70 μ, I+ blue, becoming wine red; paraphyses simple or slightly branched and anastomosing, tips slightly thickened; spores 8, biseriate, ellipsoid to fusiform-ellipsoid, 12–20 × 5–9 μ.

A species of acid rocks, granites, sandstones, gneisses. Circumpolar, arctic and temperate, in North America from Ellesmere Island to Alaska, south to Virginia, Colorado, Washington.

Localities: 1, 2, 3, 4, 9, 11.

13. **Lecidea jurana** Schaer. (*Lecidea albosuffusa* Th. Fr.) Plants of alpine and arctic, generally lacking epilithic thallus, in protected spots forming a thin whitish finely areolate crust, sometimes bluish color on calcareous substrates, K-, I+ weakly blue. Apothecia dispersed or aggregated, to 2.2 mm broad, the base much constricted; disk black, flat or slightly convex shining or covered with a white pruina, inner side of margin also strongly pruinose, both bare and pruinose apothecia on the same thallus; sometimes with an umbo in the center; epithecium brown or green (sometimes in parts of same section); exciple well developed, entirely carbonaceous; hypothecium green- or brown-black, the upper part often intense green, in thick sections appearing the same as the exciple, in thin sections appearing lighter; hymenium 70–130 μ, hyaline or pale green, occasionally brownish; paraphyses conglutinate, unbranched or very rarely anastomosing, 1.7–2.2 μ, the tips slightly thickened; spores 8, ellipsoid, simple, hyaline, halonate when young, 13–33.5 × 7–20 μ.

On calcareous rocks, limestones or sandstones. Probably circumpolar, arctic-alpine. Known from Scandinavia, the Alps, Greenland, Novaya Zemlya, Bear Island, Jan Mayen Island, Northwest Territories, and Alaska.

Localities: 1, 2, 3, 8.

14. **Lecidea macrocarpa** (DC.) Steud. (*Huilia macrocarpa* (DC.) Hertel) Thallus lacking or forming an ashy thin layer over the substratum and then I-, K-. Apothecia 1.5–3 mm broad, black, bare, flat or becoming convex; exciple thick, black, persistent or becoming excluded, irregularly radiating, dark brown in sections, K-; hypothecium brown black; epithecium brownish to brown green; hymenium 100–120 μ, brown above, hyaline below; paraphyses conglutinate, slightly branched and anastomosing, slender, 0.15 μ, tips slightly thickened to 0.2 μ, septate; asci clavate; spores 8, biseriate, hyaline, simple, ellipsoid, 15–30 × 7–11 μ.

A species of broad ecological tolerance, on sandstones, granitic, or calcareous rocks, in shade or strong sun. Circumpolar and temperate as well as arctic-alpine. In North America ranging trans-Canada to Alaska, south to North Carolina, Wisconsin, California.

Localities: 1, 2, 3, 8, 11.

Material from 1 and 2 has been determined as *Huilia macrocarpa* var. *trullisata* (Arn.) Hertel by Dr Hertel.

15. **Lecidea atrata** (Ach.) Wahlenb. (*Lecidea dicksonii* auct.) Thallus very thin, areolate-chinky or areolate, smooth, red or rusty red, dull, K-, I-; black hypothallus present. Apothecia common, small, 0.2–0.5 mm, immersed in the thallus or becoming adnate; concave or flattening, black, bare, dull, the exciple very thin, entire, usually quite prominent, black or rarely orange, persistent; epithecium olive brown or bluish brown, particolored; exciple brown, K+ purple effusion; hypothecium dark brown or brown black, K+ purple; hymenium 70–80 μ, I+ blue; paraphyses septate, branching and anastomosing, tips slightly thickened; asci broadly clavate; spores 8, biseriate, hyaline, ellipsoid, rarely subglobose also, 9–16 × 5–9 μ.

On acid rocks, particularly with some iron content, in open sun. Circumpolar, arctic alpine, in North America across Canada and Alaska, south to Adirondack Mts, New York, Wisconsin, Colorado, New Mexico, and Washington.

Localities: 8, 9, 11.

16. **Lecidea flavocaerulescens** Hornem (*Huilia flavocaerulescens* (Hornem) Hertel) Thallus moderately thick, orange or pale orange, chinky areolate, the areolae to 1 mm, thinning at the borders, a black hypothallus often present, usually sorediate with the gray green soredia in small concave soralia on the upper surface, the soralia more abundant when apothecia are absent, K-, I-. Apothecia adnate on the upper surface, large, to 1.6 mm; disk black, flat to very slightly convex, usually very bluish pruinose; the exciple thick, black, flexuous or entire, bare; epithecium brown black; exciple brown black, carbonaceous, the cells irregular; hypothecium brown black, continuous with the exciple, reddish below the hymenium; hymenium pale brownish, 100–140 μ; paraphyses conglutinate, unbranched, septate, tips much gelatinized and deep red brown; asci clavate; spores 8, biseriate, hyaline, ellipsoid, 10–33 × 7.5–12 μ.

This species is characteristically most abundant in trains of frost-riven boulders, especially those in which there may be seen water below the boulders seeping up from the permafrost. A slight shade of the sides of the boulders also seems to be preferred. It is probably circumpolar, arctic-alpine although I have seen definite records only from Europe, Spitzbergen, Bear Island, Greenland, and North America. In the latter it

occurs across Canada and in Alaska in the arctic, from northern Quebec across to the Aleutian Islands. More southerly reports seem to be based upon ochraceous plants of *Lecidea lapicida* or *L. macrocarpa*.
Localities: 8, 11.

17. Lecidea paupercula Th. Fr. Thallus thin to fairly thick, areolate, the areolae dispersed or contiguous, concave to rarely convex, red brown, rarely ashy brown, the margins of areolae ashy, lacking soredia, medulla I+ blue, K+, yellowish, hypothallus black and usually present. Apothecia adnate, the base not constricted, to 1.3 mm broad; disk flat, rarely convex, black, bare, dull; the exciple thin, black or brownish black, bare, outer edge bluish brown; epithecium brownish; hypothecium brown black, radiate; hymenium, 65–100 μ, pale, I+ blue; paraphyses conglutinate, unbranched, 1–1.5 μ, the tips scarcely thickened; asci clavate; spores 8, biseriate, hyaline, elongate ellipsoid, 7–16 × 5–6 μ.

Usually on acid rocks, granitic, sandstones, etc. Circumpolar, arctic-alpine, in North America ranging south to Labrador, Utah, Washington.
Locality: 11.

18. Lecidea atrobrunnea (Ram.) Schaer. Thallus thick or fairly thick, areolate, the areolae distributed or contiguous, convex or flat, reddish brown to ashy brown or dirty yellow, lacking soredia, medulla I+ blue, K- hypothallus black, prominent. Apothecia adnate, to 1.5 mm broad, the base becoming slightly constricted; disk flat or becoming convex, black, bare, dull; exciple thin, black, bare, dull, persistent, outer part bluish brown, inner part subhyaline, the hyphae radiating; epithecium bluish brown; hypothecium pale or partly intense brown or hyaline; hymenium bluish in upper part, 40–70 μ; paraphyses unbranched, 1.5 μ, septate, the tips slightly thickened to 3 μ; asci clavate; spores 8, biseriate, simple, hyaline, ellipsoid to subglobose, 7–12 × 3–6 μ.

A species of acid rocks. Circumpolar, arctic-alpine. In North America ranging south to Nevada, New Mexico, and California in the west. No localities in eastern Canada reported south of Belcher Islands in Hudson Bay.
Localities: 1,8.

19. Lecidea carbonoidea Thoms. Thallus large, to 13 cm across, of tiny, flat areolae 0.1–1.0 mm across, very dark sooty brown bordered by a black hypothallus and a pale gray margin about 3 areolae, 1 mm, broad; the areolae very angular, mutually pressed together, the chinks running deeply between them, the thallus from thin at the edges to 1 mm thick centrally; medulla I+ blue, K-, C-. Apothecia sunken in the areolae, usually 1 per areole, sometimes with multiple disks; disk black, concave, dull, epruinose, a thin exciple sometimes visible; epithecium brown; exciple dark brown, paraplectenchymatous, full of crystals, joining the hypothecium which is also dark brown; hymenium pale below, brown above, 65–75 μ; paraphyses conglutinate, unbranched, non-moniliform, septate, 3 μ with the tips thickened to 6 μ and brown; asci clavate, 68 × 12 μ; spores 8, biseriate, hyaline, frequently with gelatinous halo, ellipsoid, 12–12.5 × 6.5–7.5 μ.

Growing on sandstone. Known only from the type: Pitmegea R., 15 miles upstream from Cape Sabine, *Thomson* Lich. Arct. 94.

This species is close to *Lecidea athroocarpa* which has innate apothecia but the smoky thallus and smaller spores differentiate it. It also is close to *Lecidea aleutica* Degel. but the paraphyses are conglutinate, not lax, the spores biseriate, not uniseriate, and slightly larger 12–12.5 × 6.5–7.5 instead of 8.5–10.5 × 6.5, and the exciple is colored instead of hyaline. There is a resemblance to *Lecidea paupercula* but the color is a dark sooty brown instead of red brown, the apothecia are immersed instead of adnate, and the hymenium is brown above instead of blue.

20. **Lecidea picea** Lynge Thallus black, shiny, small, quite thin, angular areolate, the areolae to 0.7 mm broad, flat, the marginal areolae slightly concave; medulla K–, I–. Apothecia to 2 mm broad, adnate, at first flat with a thin exciple, later becoming convex and immarginate; the disk shining, black, epruinose; the exciple violet to violet black; the hypothecium also violet as is the lower part of the hymenium, KOH turns the violet parts of the apothecium a beautiful blue; the upper part of the hymenium blue to bluish brown; hymenium 130–160 μ, I+ blue becoming wine red; paraphyses coherent, septate, sometimes branched, the tips slightly thicker; spores ellipsoid, 5–8 × 3.5–4 μ.

Growing on acid rocks. This probably arctic circumpolar species is known from Novaya Zemlya, Spitzbergen, Siberia, and Greenland. A report from Ellesmere Island is a misidentification. It is new to Alaska.

Locality: 8.

This species is very close to *Lecidea aenea* Duf. but has flat rather than convex areolae, dark violet hymenium and hypothecium as well as exciple instead of an olive or olive black upper hymenium, colorless hypothecium and smaller spores (9–15 × 5–6 in *L. aenea*).

21. **Lecidea cinereorufa** Schaer. Thallus of small verruculose squamules, confluent into a crust which is areolately divided; brown to ashy brown, dull, hypothallus indistinct, I–, K–. Apothecia small to 1.2 mm, the base well constricted; disk flat, black, bare, dull; exciple promi-

nent, black, dark brown, or dark ashy, of irregular cells; hypothecium dark brown; epithecium bluish or bluish brown; hymenium 160–200 μ, hyaline below, bluish above; paraphyses lax, tips slightly branched and upper part of blue moniliform cells; asci cylindrical; spores 8, uniseriate, hyaline, globose, 7–12 μ.

On rocks and over mosses on rocks. An arctic-alpine species fairly common in Scandinavia, rare in North America. Reported by Anderson in Colorado and by Lowe from New York and New Hampshire. A specimen from Fort Hall Lake, Manitoba, Scotter 2930 is in WIS.

Locality: 3.

This specimen is of f. *acarosporoides* Vain. with the thallus partly squamulose, partly of small brown areolae.

22. **Lecidea instrata** Nyl. (*Lecidella instrata* (Nyl.) Choisy) Thallus crustose, brownish black, chinky subareolate, thin, I-. Apothecia 0.4–0.6 mm broad, immersed in the thallus; disk flat; exciple disappearing, pale; hypothecium hyaline; epithecium brown; hymenium 65 μ, brownish above entirely I+ blue; paraphyses coherent; spores hyaline, 10–12 × 5–7 μ.

On acid rocks. This species is known from Europe in the Alps and in Scandinavia. New to Alaska.

Locality: 8.

23. **Lecidea confluens** (Web.) Ach. Thallus powdery crumbling, thin to quite thick, chinky-areolate to becoming areolate at the margin, the areolae more or less confluent on a black hypothallus, to 1 mm broad, angular or rounded, usually chinky, depressed convex, bluish or smoky ashy to white ashy, dull, smooth, lacking soredia or isidia, K-, I+ blue. Apothecia dispersed and round or grouped and angular, to 2 mm broad, sessile, in age becoming elevated, convex, and immarginate; exciple black, smooth, becoming flexuous, brown black in section; hypothecium dark brown; epithecium olive black; hymenium 100–125 μ, hyaline below, upper part olive black, I+ blue; paraphyses conglutinate, the tips only slightly thickened; asci narrowly clavate; spores 8, biseriate, hyaline, simple, broadly ellipsoid, 6–10 × 4–7 μ.

Growing on acid rocks in full sun. Circumpolar and arctic alpine. In North America it is known across Northwest Territories and in Alberta and Wyoming.

Localities: 5, 8, 9.

24. **Lecidea speirea** (Ach.) Ach. Thallus continuous to strongly chinky, chalky white, sooty white, or bluish white, soft chalky in texture, C-, K-,

or K+ yellowish, P-, medulla I+ varying from weak to deep blue, hypothallus greenish gray. Apothecia immersed or adnate, sometimes erosion making them appear stalked, to 2.5 mm broad; exciple with a black brown outer layer, the inner part pale; disk dull black, sometimes slightly pruinose, flat; epihymenium dirty olive to brown; hypothecium brown black; hymenium 70–105 μ, hyaline; paraphyses anastomosing, the tips slightly thickened; asci clavate; spores thin walled, ellipsoid to ovate, 12–18 × 5.5–9 μ.

Growing on calcareous rocks. A circumpolar arctic to boreal species known from Labrador, Northwest Territories, and Newfoundland. New to Alaska.

Locality: 8.

25. **Lecidea umbonata** (Hepp) Mudd Thallus chalky white, frequently zonate with black hypothallus lines, continuous or weakly chinky, C-, K-, P-, the medulla I- to strongly I+ blue. Apothecia immersed to adnate, to 2.4 mm broad, the disk flat, black, frequently with a central umbo of sterile tissue; exciple with a green or black green outer layer, the interior pale; epithecium green to brown; hypothecium hyaline to brownish ocher; hymenium 50–80 μ, hyaline to roseate, rarely pale greenish, the subhymenium coppery green; paraphyses very gelatinous, simple, occasionally branching or anastomosing, 1.8–2.2 μ, the tips capitate to 4.5 μ; asci clavate; spores blunt ellipsoid, without halo, pseudo-2-celled, 8–15 × 5–8 μ.

Growing on calcareous rocks. Arctic alpine and circumpolar. Known in North America and Greenland, from Ellesmere Island, Yukon, Alberta, Alaska, south to Utah and Colorado (Hertel 1973).

Locality: 11, det. H. Hertel.

26. **Lecidea tessellata** (Ach.) Flörke Thallus moderately to quite thick, areolate, the areolae to 1.2 mm broad, continuous, angular, flat or depressed-convex, whitish to ashy or darker gray, I+ blue, K-, esorediate, a black hypothallus sometimes visible. Apothecia with the disk coming up level with the thallus (innate), between the areolae, the disk flat, black, usually pruinose, dull, the black exciple thin, persistent, angular or flexuous or circular; hypothecium hyaline or pale, the upper part of vertical hyphae, the lower part with inspersion of air; exciple with the inner part hyaline, the outer thin part brownish, the lower marginal part brown and grading into the thallus, lacking under the apothecium; epithecium black or olive black; paraphyses conglutinate, 1.5 μ thick, tips slightly thickened; spores ellipsoid or oblong to subglobose, 8–10 × 5–6 μ.

On rocks, mainly acid. Circumpolar and temperate as well as arctic-alpine. In North America ranging from Baffin Island and Devon Island to Alaska and south to Virginia, Minnesota, Arizona, and Washington.

Localities: 1, 8, 9.

The specimen distributed as *Lecidea lapicida* in Lich. Arct. 74 was determined by H. Hertel as *L. tessellata* var. *caesia* (Anzi) Arn. The variety is distinguished by a blue gray instead of yellowish to dark gray thallus, thicker, to 3 mm thick in the center, with pale instead of black green hypothallus, and finely chinky instead of slightly bullate areolae. This was new to North America in Hertel's report (1971a).

27. **Lecidea pallida** Th. Fr. Thallus of verrucules, at first dispersed, becoming conglomerate, yellowish to pale white, I-, K+ yellowish. Apothecia adnate, to 1.4 mm, semiglobose, thick, black, convex, immarginate, sometimes conglomerate, thinly pruinose when young; exciple thin, black, disappearing, outer part bluish brown, inner part blue; hypothecium thick, brown, sometimes with a blue tinge, radiate; hymenium 80–100 μ, pale blue, I+ blue; paraphyses conglutinate, unbranched, septate, tips not thickened; asci clavate; spores 8, biseriate, hyaline, elongate ellipsoid, 16–22 × 6–9 μ.

A species of soil and weathered rocks. Known from northern Europe, Jan Mayen Island, Iceland, Greenland, and in North America from Ellesmere Island and Alaska.

Localities: 3, 6.

28. **Lecidea lithophila** (Ach.) Ach. Thallus thin or moderately thick, continuous, chinky, or areolate-diffract, smooth or uneven, ashy white, rarely yellowish or orange, dull, lacking soredia, I-, K-; black hypothallus occasionally present. Apothecia immersed in the thallus to emergent, base not constricted, to 1.3 mm broad; disk flat or depressed-convex, thinly pruinose or becoming bare; exciple thin, sometimes becoming obvious, black, or rarely slightly orange, exterior black, KOH+ brownish, interior pale, grading off under the hypothecium to hyaline; epithecium olive brown, K+ brown; hypothecium hyaline, I+ blue; hymenium 55 μ, hyaline, I+ blue; paraphyses conglutinate, unbranched, septate, tips slightly thicker and darkened; asci clavate; spores 8, biseriate, hyaline ellipsoid to elongate-ellipsoid, 9–14 × 4–6 μ.

Growing on rocks in the open. A circumpolar arctic-alpine species in North America south to New York, Wisconsin, Alberta, Washington, and California.

Locality: 11.

29. **Lecidea brachyspora** (Th. Fr.) Nyl. Thallus lacking or represented only underneath the apothecia and then I+ blue. Apothecia to 2 mm broad, the margin well developed, sinuous, I+ blue, the inner portion pale; epithecium green black; hymenium 35–45 μ, hyaline; hypothecium dark brown; spores subglobose, 4–5 μ diameter or ellipsoid oblong, 6–7 × 4–5 μ.

A species of acid rocks. It is known from Scandinavia, the Arctic, and Venezuela.

Locality: 9.

This species is exceedingly close to *L. auriculata* and is distinguished mainly by the short, broad, nearly globose spores rather than the narrowly elongate spores of *L. auriculata*. The specimen was determined by Dr Hertel (1977).

30. **Lecidea diducens** Nyl. Thallus lacking or of traces beneath the apothecia and K–, I+ blue. Apothecia to 1 mm broad, the margin with black green outer layer, pale inside, I–, C+ red, K+ faintly red violet; epithecium green brown to black green; hypothecium blackish brown; hymenium 45–50 μ, upper part green brown, I+ blue; paraphyses coherent, slender, 2 μ with the tips 3–4 μ; asci broadly clavate; spores narrowly ellipsoid, 6.5–10 × 2.4–3.6 μ.

Growing on acid rocks. Apparently circumpolar, in North America known from Quebec, the Northwest Territories, and Alaska.

Locality: 8.

Differences between this and the closely related *L. auriculata* lie mainly in the C+ red reaction of *L. diducens* and the I+ blue reaction of the outer layer of the margin of *L. auriculata*. The margin is more persistent in *L. diducens* and the apothecia tend to be smaller.

31. **Lecidea melaphanoides** Nyl. Thallus thin or thick, of uneven verruculose areolae or verrucules, brownish or olivaceous, K–, with indistinct hypothallus. Apothecia dispersed or aggregated, small, to 0.4 mm; disk at first flat and with thin margin, becoming convex and immarginate; exciple radiate, of thick walled cells, exterior brown or rosy red, interior paler, K–; hypothecium dirty brown to reddish brown, sometimes particolored; epithecium bluish or bluish brown; hymenium 50 μ, hyaline, upper part bluish; paraphyses conglutinate but in KOH lax, 1.5 μ, unbranched, septate, tips thickened, gelatinous; asci clavate; spores 8, biseriate, simple or sometimes partly 1-septate, hyaline, oblong-ellipsoid, 9–14 × 5–6 μ.

On rocks. A rare species known from Scandinavia, Greenland, Northwest Territories, and Alaska.

Locality: 9.

32. **Lecidea aglaea** Sommerf. Thallus whitish, yellowish, or straw yellow, chinky areolate, the areolae convex, 1–1.5 mm broad, sometimes cracked, the hypothallus usually indistinct; the cortex 40–60 μ thick; the medulla K+ yellow, P+ yellow. Apothecia between the areolae, to 2 mm broad, black, smooth to shining; the margin soon excluded, brownish or greenish in the outer part with more or less radiating hyphae; epithecium greenish brown; hypothecium hyaline or partly pale brownish; hymenium 60–80 μ, hyaline to bluish, I+ blue; paraphyses 2–4 μ, with the tips thickened to 5–7 μ; asci clavate; spores broadly ellipsoid to ovate, 8–15 × 5–8 μ.

A species of noncalcareous rocks. It is circumpolar, arctic and alpine, in North America known from Greenland, Ellesmere Island, Alaska, the Northwest Territories, and south in the Rocky Mountains to Colorado, and in Arizona (San Francisco Peaks at ca. 4,000 m).

Locality: 8.

33. **Lecidea atromarginata** Magn. Thallus determinate, with black hypothallus border and hypothallus extending between areolae but with slight covering of hyphae giving a paler appearance there, areolate, yellowish white or white, marginally more or less lobate, dull, I-, K-, C-. Apothecia at first innate but soon adnate, sometimes raised on thallus, to 1 mm broad; disk flat or slightly convex, black, not pruinose; margin slightly paler black, thick, dull; exciple dark brown, paleing to base, outer part bluish green; epithecium olive green; hypothecium hyaline, both hymenium and hypothecium I+ blue; hymenium 50–60 μ, lower part hyaline, upper part olive green; paraphyses unbranched, 1.7–2 μ, tips thickened to 3–4 μ and darkened; asci clavate; spores 8, biseriate, hyaline, ellipsoid, 8–12 × 5–6 μ.

A species of calcareous rocks; circumpolar, arctic. Known from Siberia, Novaya Zemlya, Greenland, Ellesmere Island, Cornwallis Island, Bathurst Island, Northwest Territories, and Alaska.

Localities: 1, 9, 11.

This species is very similar in appearance to L. marginata but has a black hypothallus, a brown exciple, slightly smaller spores, and a K- thallus reaction.

34. **Lecidea leucophaea** (Flörke) Nyl. Thallus moderately thick or very thin, smooth to verruculose uneven, the areolae convex, ashy to brownish ashy, K-; a black hypothallus sometimes visible. Apothecia to 1.2 mm broad, adnate, the base broad, the margin thin, shining, brownish black or pale brown; epithecium bluish brown or rarely brownish red; hypothecium hyaline; hymenium 60–80 μ, I+ blue; paraphyses coherent,

slender with thicker tips, to 4.5 μ; asci clavate; spores hyaline, ellipsoid, 8–14 × 4–7 μ.

On acid rocks. Apparently a circumpolar arctic-alpine species reported from Greenland and in North America from Alaska (the Aleutian Islands) and Ellesmere Island.

Locality: 8.

35. Lecidea armeniaca (DC.) Fr. Thallus to 1 mm thick, areolate, the areolae to 4 mm broad, usually less, contiguous, rarely dispersed, depressed-convex or flat, reddish brown or apricot yellow to yellowish, rarely blackish brown, K+ red, P+ yellow (psoromic acid), dull or shining, smooth, lacking soredia; black hypothallus thick and areolate. Apothecia starting between the areolae and level with the thallus, becoming convex and adnate above the thallus; to 4.5 mm; disk black, shining or dull, immarginate; exciple poorly formed, 25–30 μ, blue green; hypothecium thin, brownish, poorly formed; epithecium bluish black or olive black; hymenium 110–250 μ, entirely dark, lower part blue black; paraphyses conglutinate, 1.5–2 μ, tips thickened and blue green; asci clavate; spores 8, biseriate, hyaline, ellipsoid and also subglobose, 10–12 × 4–5 μ.

A species of acid rocks, especially granitic, growing in full sunlight. Circumpolar and arctic-alpine, in North America, Baffin Island, Devon Island, north Labrador, westward to Alaska, south in the Rocky Mountains to Colorado and New Mexico, and in Washington.

Localities: 8, 9, 11.

36. Lecidea marginata Schaer. (*Lecidea elata* Schaer.) Thallus varying from isolated areolae to thick chinky-areolate crusts up to 3 mm thick, from clear yellow to chalky white, with white margins, K+ yellow (atranorin), I-, C-; gray green hypothallus seldom present. Apothecia somewhat sunken in thallus when young, soon appressed and adnate, occasionally wind and ice erosion making the apothecia almost appear stalked, sometimes surrounding the swollen young apothecia with a ring of fine thallus particles; margin dull, black; exciple when young with a black green outer narrow layer which soon disappears, leaving the exciple hyaline but becoming unclear and also invaded by algae from the nearby medulla forming a lecanorine-like margin which can cause mistakes in placing this into *Lecanora*; epithecium yellow green to black green; hypothecium hyaline; hymenium 55–75 μ, hyaline; paraphyses mainly unbranched, sometimes branched and anastomosing, 1.7–2.3 μ, tips swollen to 4 μ; asci clavate; spores 8, biseriate, hyaline, broad-ellipsoid, 8–14 × 4.5–7.5 μ.

A species of calcareous rocks. Circumpolar, arctic-alpine, in North America from Ellesmere Island to Alaska, south to Baffin Island, Alberta, Utah, Washington. Probably further south in the east but its range not well known.

Localities: 1, 14.

37. **Lecidea plebeja** Nyl. Thallus disappearing, indicated by whitish coloration of substratum or with very few whitish granules. Apothecia to 0.5 mm, adnate; disk flat to slightly convex, black, bare, dull; margin thin and becoming excluded; exciple brown black, outer part paler, continuous with the hypothecium; epithecium olive brown, pale; hypothecium olive brown to reddish brown, KOH brighter brown but not purple; hymenium 45–50 μ, olive- or yellowish-brown above, hyaline below and grading into reddish toward hypothecium; paraphyses conglutinate, 1.5–2 μ, unbranched, tips thickened to 3–5 μ and darkened; asci clavate; spores 8, biseriate, hyaline, simple, ellipsoid to subfusiform, 6–8 × 2.5–3 μ.

On rotting wood. A little known species, Europe, New York State (Adirondack Mts), and Great Slave Lake are previous reports. A specimen from California is in WIS, Jordan 696.

Locality: 14.

38. **Lecidea tornoënsis** Nyl. Thallus absent to thin, membranaceous to granulose, olivaceous brown or olivaceous green; K-, I-. Apothecia to 0.6 mm, adnate or constricted below; disk becoming strongly convex, reddish brown to brownish black, roughened, shining, immarginate; exciple scarcely formed, outer part reddish brown, inner paler, radiate; epithecium reddish brown; hypothecium reddish, brownish or hyaline; hymenium 130–180 μ, upper part brown, lower part grading into hypothecium, I+ blue; paraphyses branching anastomosing, conglutinate, tips dark brown and thickened to 6 μ; asci broadly clavate; spores 8, biseriate, hyaline, thick walled, the wall to 2.5 μ thick, broadly ellipsoid, 15–23 × 8–14 μ.

Growing on a variety of tree and shrub barks, on mosses and humus. Probably circumpolar, the range not well defined in North America as seldom collected, apparently more often collected in west but specimens from Chantry Inlet, NWT, Artillery Lake, NWT, Alaska, British Columbia, and Idaho in WIS as well as reports from New Hampshire, New York (Adirondack Mts), Washington, and California suggest a very wide range.

Localities: 1, 2, 5, 8.

39. Lecidea symmicta (Ach.) Ach. Thallus thin to moderately thick, smooth to verruculose uneven, esorediate or with scarce soredia; yellowish, I-, K+ brownish, KC+ orange brown. Apothecia commonly dense, small, to 1.5 mm, base not constricted, adnate; margin thin, pale, becoming excluded; disk convex, pale yellowish, reddish, or bluish, bare or slightly, pruinose; exciple of thin walled, radiating hyphae, yellowish; epithecium pale yellow; hypothecium hyaline; hymenium 70 μ, hyaline, I+ blue; paraphyses conglutinate, unbranched, the tips slightly thickened; asci clavate; spores 8, biseriate, hyaline, simple, oblong ellipsoid, 7-16 × 4-5 μ.

On bark and old wood. Circumpolar, boreal to temperate.
Localities: 2, 8, 14.

40. Lecidea decipiens (Hedw.) Ach. Thallus squamulose, the squamules rounded, to 8 mm broad, margins sometimes crenate, narrowly revolute, upper side flat or concave, pale red, bright red becoming darkened or white when dying, shining, smooth or chinky-areolate cracked, bare or pruinose, edges commonly white; underside white. Apothecia marginal, one or few on a squamule, to 1.8 mm broad, adnate and slightly constricted at base, black or brownish black; the margin thin and disappearing; exciple of irregular hyphae, pale; epithecium reddish brown, K+ purple; hypothecium thick, upper part pale, lower reddish; hymenium 70 μ, upper part pale reddish, I+ blue turning wine red; paraphyses conglutinate, unbranched, tips slightly thickened; asci clavate; spores 8, biseriate, hyaline, simple, ellipsoid or subglobose, 9-15 × 7-9 μ.

On calcareous soil, especially with high clay content, on old frost boils and solifluction lobe edges, etc. Circumpolar, boreal, and temperate. In North America from Baffin Island and other arctic islands to Alaska, south to New York, Oklahoma, New Mexico, Arizona, California.
Localities: 1, 11, 14.

41. Lecidea rubiformis (Wahlenb. ex Ach.) Wahlenb. Thallus of ascending squamules, the squamules rounded, concave, smooth or becoming chinky above, brownish to greenish or olive, dull or shining, margins white and commonly slightly pruinose, lower side white. Apothecia to 2 mm, solitary or conglomerate, convex, immarginate, black, slightly shining, epruinose; exciple radiate, pale brownish; epithecium reddish, K+ redder with red effusion; hypothecium red, lower part paler; hymenium 60 μ, reddish to red, I+ blue; paraphyses conglutinate, unbranched, septate, tips slightly thickened; asci clavate; spores 8, biseriate, hyaline, simple, ellipsoid to subfusiform, 10-17 × 4-7 μ.

On soil edges over rocks and on thin soil in fissures in rocks, occasionally on soil surfaces. Circumpolar, arctic-alpine. In North America: Elles-mere Island to Alaska, south to Texas, New Mexico, and California. In the east I have not seen specimens from south of Baffin Island.

Localities: 3, 8, 14.

42. Lecidea demissa (Rutstr.) Ach. Thallus squamulose, the squamules closely adnate to the substrate, contiguous and confluent, margins irregularly crenate and lobate toward the edges of the thallus, ashy red to brownish ashy, dull, the under side brownish black, over a fibrous black hypothallus. Apothecia immersed between the squamules or adnate to edges of thallus, often grouped in masses, to 4 mm broad, disk flat to slightly convex, black, sometimes shining, bare, the margin thin, black, disappearing; exciple of irregular hyphae, brownish black; epithecium brownish red; hypothecium pale, not continuous with the exciple; hymenium 70–80 μ, pale, I+ blue; paraphyses slightly branched at the base, conglutinate, tips thickened to 4–5 μ, reddish; asci clavate or partly cylindrical; spores 8, biseriate or uniseriate, hyaline, simple, oblong ellipsoid to subfusiform, 11–15 × 5–6 μ.

On sandy soil, humus, over mosses. A circumpolar arctic-alpine spe-cies, in North America from Baffin Island to Alaska south to New York, Colorado, and Washington.

Localities: 4, 11.

43. Lecidea globifera Ach. Thallus squamulose, the squamules ascend-ing, moderately thick, rounded lobate, commonly concave, smooth or becoming chinky roughened, upper side chestnut colored, olive, or pale olive, shining, underside dark. Apothecia sparse, single on lobes, to 2 mm, disk convex, black, dull, bare, the margin paler and becoming immarginate; exciple radiate, pale or reddish; epithecium brown or reddish, K+ reddish with red effusion; hypothecium brown or red, lower part paler; hymenium 50–80 μ, I+ blue becoming wine red; paraphyses conglutinate, unbranched, tips thickened; asci clavate; spores 8, bise-riate, hyaline, simple ellipsoid or globose, 7–15 × 3.5–6 μ.

On soil and on soil in crevices in rocks. Circumpolar, arctic-alpine and temperate, in North America from New England to Alaska south to Arizona and Washington in the west.

Locality: 11.

44. Lecidea uliginosa (Schrad.) Ach. Thallus crustose, of olive to red-dish brown or blackish brown granules or verrucules, compacted and the crust sometimes falsely areolate on drying. Apothecia small, to 0.6 mm broad, adnate to partly constricted below; disk flat to slightly convex, red brown to black brown with thin disappearing margin of the same color; exciple brown black or reddish black, hyphae radiately arranged; epithecium brown black; hypothecium olive brown to red brown or brown black, hymenium 80–100 μ, upper part dark olive, red brown, or brown black, pale reddish brown in lower part; paraphyses conglutinate but separable in KOH, branched, tips thickened to 4 μ and darkened; asci clavate; spores 8, biseriate, hyaline, simple, oblong-ellipsoid to ellipsoid, 10–14 × 5–9 μ.

Growing on peaty soil or old wood. A species of the entire northern hemisphere but commonest in the boreal forest.

Localities: 1, 4, 5, 7, 11.

45. Lecidea granulosa (Ehrh.) Ach. Thallus crustose, effuse, of compacted verrucules, the verrucules to 0.4 mm broad, with soralia scattered over the thallus and forming coarse granular soredia, C+ red, K+ yellow (gyrophoric acid), grayish white, grayish green, or dark olivaceous brown. Apothecia adnate, single or conglomerate; disk flat to becoming strongly convex, the margin thinning and disappearing, olive, yellowish brown, brown black, black, or even pinkish, bare; exciple pale olivaceous, the hyphae irregularly arranged; epithecium pale brownish; hypothecium hyaline to pale brownish; hymenium 60–80 μ, brownish above, pale brown below; paraphyses conglutinate in water but separating in KOH, branching and anastomosing, tips not thickened; asci cylindrico-clavate; spores 8, biseriate, hyaline, simple, ellipsoid, 8–15 × 4–7 μ.

On sandy soil, peat, old wood, occasionally old bark, the color darkening in stronger sunlight, paler in shade. A circumpolar species, also reported in the southern hemisphere, boreal and temperate, Quebec to mouth of the Mackenzie River and Alaska, south to Tennessee in the eastern mountains, Wisconsin, Colorado, Washington.

Locality: 4.

46. Lecidea berengeriana (Mass.) Nyl. Thallus of scattered to compacted verrucules, esorediate, white, grayish or greenish gray, I-, K-. Apothecia to 2 mm broad, usually less, adnate or slightly constricted below; margin thin, blackish brown, becoming immarginate; exciple reddish brown throughout or dark externally, continuous with the hypothecium, appearing paraplectenchymatous but with the cells in radiate rows; epithecium brown, rarely pale; hypothecium brown, or red brown, lower part paler, sometimes with black granules; hymenium 55–70 μ, hyaline to pale reddish brown, sometimes with black granules incorporated in it; paraphyses conglutinate, unbranched, tips thickened to 3–8 μ; asci clavate; spores 8, biseriate, hyaline, simple, oblong-ellipsoid to ellipsoid or fusiform, 9–16 × 3.5–6 μ.

On mosses, humus, and soil. A circumpolar arctic-alpine species in North America ranging from Baffin Island to Alaska, south to New York, Michigan, Wisconsin, Minnesota, Colorado, British Columbia, California.

Localities: 1, 3, 4, 6, 7, 8.

47. Lecidea fusca (Schaer.) Th. Fr. Thallus absent or thin, membranaceous or granulose with minute granules to 0.15 mm diameter, whitish to grayish or grayish green, esorediate, I-, K-. Apothecia broadly adnate, the base becoming slightly constricted; margin brown, ashy brown or black, shining, persistent or disappearing; disk pale brown, reddish brown, brown black, or black, flat becoming strongly convex, dull or slightly shining, bare; exciple radiate, outer part reddish brown, inner part paler; epithecium reddish brown; hypothecium reddish brown to brown black above, lower part paler, upper part often carbonized and with black granules; hymenium 60–80 μ, hyaline to pale reddish brown above, sometimes containing black granules, lower part rarely bluish; paraphyses conglutinate, slender, tips slightly thickened; asci clavate; spores 8, biseriate, hyaline, simple or partly 1-septate, ellipsoid, 9–17 × 3–7 μ.

Growing over mosses and humus. Range uncertain because of confusion with *L. berengeriana*. Probably circumpolar and arctic-alpine. Previously reported from New York. Specimens from Alberta, Minnesota, Isle Royale, Mich., and from Nome, Alaska, Palmer 843 (sub *Lecidea aurea*) are in WIS.

Localities: 1, 8, 11, 14.

48. Lecidea subcandida Magn. Thallus of snow white granules 0.2–0.5 mm broad, irregularly convex, dispersed or in small groups, dull, I-, K-, C-, P-. Apothecia small, to 0.6 mm broad, hemispherical, immarginate from the first; disk black, slightly pruinose, the pruina with a somewhat arachnoid distribution; exciple lacking; epithecium pale brownish; hypothecium blackish brown, the upper edge very uneven; hymenium 50–60 μ, hyaline, I+ blue; paraphyses unbranched, thin, to 1.5 μ, tips slightly thickened, very gelatinized; asci cylindrico-clavate; spores 8, uniseriate, hyaline, simple, ellipsoid, 8–12 × 3.5 μ.

On calcareous sandy soil. Previously known only from Swedish Lapland. New to North America.

Locality: 2.

49. **Lecidea assimilata** Nyl. Thallus thin to fairly thick, of verrucules either contiguous or massing to form an areolate-diffract crust, esorediate, white, ashy, or ashy brown, dull, K-, I-, hypothallus indistinct. Apothecia single or conglomerate and tuberculate, broadly adnate, the base not constricted, to 0.8 mm broad; disk black, soon convex, immarginate, dull, bare; exciple pale brownish, radiate; epithecium bluish, bluish black, or olivaceous brown; hypothecium reddish brown or brown; hymenium 50–60 μ, hyaline or brownish above, I+ blue; paraphyses slender, unbranched, tips not thickened; asci clavate; spores 8, biseriate, hyaline, simple or rarely 1-septate in part, oblong, oblong-ellipsoid or fusiform, 8–19 × 3–6 μ.

Growing on mosses and soil, particularly in late snowbank areas. Circumpolar, arctic-alpine, in North America growing from Ellesmere Island to Alaska, south to northern Quebec, Southampton Island, northern Saskatchewan, Colorado, and Washington.

Localities: 3, 4, 5, 8, 9. One specimen from 9 has a purplish hypothecium and can be placed as f. *irrubata* Th. Fr.

This species can easily be confused with *L. crassipes* (Th. Fr.) Nyl. which has a strongly constricted apothecium forming a short stipe between the granules of the thallus.

50. **Lecidea vernalis** (L.) Ach. Thallus thin and membranaceous to moderately thick, verruculose, the verrucules dispersed or compacted, convex, olivaceous whitish, olivaceous, or grayish green, esorediate, K-, I-, hypothallus indistinct. Apothecia to 1.2 mm broad, adnate, flat and marginate or convex and immarginate; disk whitish, yellowish, or reddish brown, bare; exciple radiate, pale yellow or hyaline to yellowish brown, gelatinous, upper part I+ blue; epithecium pale yellowish; hypothecium hyaline to pale yellow or yellowish brown; hymenium 60–85 μ, hyaline, paraphyses unbranched, conglutinate, tips thickened to 3–5 μ; asci clavate; spores 8, biseriate, hyaline simple or becoming 1-septate, oblong, oblong-ellipsoid, 10–20 × 3–7 μ.

Over mosses and on barks. Circumpolar, arctic-alpine and boreal, especially common in the boreal forest. In North America from Ellesmere Island to Alaska, south to Virginia, Michigan, Minnesota, Colorado, and Washington.

Localities: 1, 2, 3, 8, 9, 11, 14.

51. **Lecidea cuprea** Somm. Thallus thin to moderately thick, of verrucules which may be depressed convex to lobulate and conglomerated, whitish, K-, C-, P+ yellow (baeomycesic acid), whitish hypothallus indistinct. Apothecia to 1 mm broad, adnate, not constricted; disk convex,

cinnamon brown to reddish, margin thin, of same color as disk, disappearing; exciple radiate, outer part reddish, inner pale; epithecium reddish or yellowish; hypothecium 60–70 μ, pale yellowish or reddish brown; paraphyses conglutinate, unbranched, tips slightly thickened; asci clavate; spores 8, biseriate, hyaline, simple, oblong, 10–22 × 3–6 μ.

On sandy soils and over mosses. Circumpolar, arctic-alpine. In North America from Baffin Island and Greenland to Alaska, only in far north.

Locality: 14.

52. Lecidea ramulosa Th. Fr. Thallus quite thick, of high verrucules and torulose, isidoid structures, a very blue gray in the field, K-, C-, hypothallus indistinct, whitish. Apothecia dispersed or grouped, the base constricted; disk flat to becoming convex, black, margin thin, black, becoming immarginate; exciple radiate, outer part bluish brown to olivaceous, inner part pale; epithecium brown; hypothecium brownish to reddish brown, lower part paler; hymenium 50–60 μ, brownish to reddish brown, I+ blue; paraphyses conglutinate, unbranched, tips not thickened; asci clavate; spores 8, biseriate, hyaline, simple, oblong-elongate, 9–18 × 3–6 μ.

This species grows particularly in cold moist seepages below snowbanks, in drainage rills, late snowbank areas, etc., on soil. It is high arctic and circumpolar, apparently common in the Arctic archipelago and occurring across the North Slope of Alaska, and the north of the Northwest Territories.

Localities: 4, 9, 13.

53. Lecidea limosa Ach. Thallus thin to moderately thick, cartilaginous-membranaceous to verruculose, the verrucules minute, to 0.15 mm broad, white to olivaceous, I-, K-, hypothallus arachnoid, white. Apothecia to 1 mm, adnate; disk flat to convex, brown black to black, translucent-olivaceous, margin soon disappearing; exciple not radiate, hyphae irregularly arranged; epithecium bluish black; hypothecium pale olive to brown above, lower part hyaline; hymenium 60–80 μ, lower part with scattered granules and bluish, upper part bluish- or greenish-black; paraphyses conglutinate, unbranched, 4–6 μ thick; asci clavate; spores 8, biseriate, hyaline, simple, subfusiform to ellipsoid, 10–16 × 3–5 μ.

On soil or over mosses. Range of this species is uncertain because of problems in determinations. Lich. Arct. 11 from the Kaolak River was distributed under this name but consists of misdetermined specimens of *Mycoblastus alpinus*. Lowe reported this species from the Adirondack Mountains, NY. It has been reported from Ellesmere Island by Darbishire and from Port Clarence near Teller, Alaska, by Nylander; a number of reports from Greenland; no specimens examined from Alaska.

3. MYCOBLASTUS Norm.

Thallus crustose, smooth to wrinkled or warty, attached by hapters to substrate; cortex not well developed, algae distributed in upper layers of thallus. Apothecia middle sized to large, often conglomerate, adnate to sessile; disk becoming very convex, black, shining; exciple soon disappearing; hypothecium hyaline to red brown or brown; hymenium bluish above, hyaline below; epithecium blue black; paraphyses unbranched, capitate; asci clavate; spores 1 or 2 per ascus, large, simple, thick walled. Algae: *Trebouxia*.

1 Apothecium with red color in lower part of hypothecium   1. *M. sanguinarius*
  Apothecium not red inside, hypothecium hyaline   2. *M. alpinus*

1. **Mycoblastus sanguinarius** (L.) Norm. Thallus of flat granules or a more or less continuous wrinkled, lumpy or chinky crust, shining, white to yellowish white, or gray, I-, K+ yellow, C-, P- (containing atranorin and caperatic acid). Apothecia to 3 mm, single or confluent, adnate or sessile; disk soon high convex, black, shining, rough; exciple not well differentiated, dark; epithecium blue black; hypothecium brownish in upper part, lower part blood red K+ purple; hymenium 90–100 μ, lower part hyaline, upper part brilliant greenish blue; hymenium I+ greenish; asci I+ blue; paraphyses conglutinate, unbranched, 2 μ, the tips capitate, 3–5 μ, blue above; asci clavate; spores 1 per ascus, thick walled, simple, hyaline but contents sometimes granular and brownish, 46–100 × 23–46 μ, walls to 8 μ thick, a gelatinous halo sometimes visible.

On tree twigs, trunks, old wood, and in the tundra over mosses in the open. Circumpolar, arctic and boreal, in North America it ranges from Ward Hunt Island north of Ellsemere Island to Alaska, and south to North Carolina, Michigan, Idaho, and Oregon.

Localities: 3, 8, 9, 11, 14.

2. **Mycoblastus alpinus** (Fr.) Kernst. Thallus of flat granules or a more or less continuous, chinky, or wrinkled crust, white to yellowish white, or gray, K+ yellow, I-, P-. Apothecia to 3 mm, single or confluent; disk soon high convex, immarginate, black, shining; epithecium blue black; hypothecium hyaline both above and below; hymenium 90–100 μ, lower part hyaline, upper part brilliant greenish blue; paraphyses conglutinate, unbranched, the tips capitate, upper part blue, asci clavate, spores 1 or 2 per ascus, thick walled, simple, hyaline, 46–100 × 24–40 μ.

On tree twigs, bark of trunks, old wood, over mosses and soil. Circumpolar, arctic and boreal, in North America from Devon Island to Alaska and south to Washington.

Localities: 2, 14.

The differences are so minor it is possible this species may be only a variation of the preceding.

4. CATILLARIA (Ach.) Th. Fr.

Crustose lichens with granulose, chinky, or areolate thalli without well-developed cortex, algae in upper layers, lacking rhizinae. Apothecia small, immersed, adnate, or sessile; disk flat to convex, yellowish, pinkish, brown, or black; exciple of same color as disk, often disappearing; hypothecium hyaline to brown; hymenium hyaline to brownish; paraphyses unbranched, capitate; asci clavate; spores 8, hyaline, rarely simple, usually 1-septate, ellipsoid, oblong-ellipsoid, or fusiform. Algae: *Myrmecia, Gongrosira, Cystococcus, Trentepolia.*

1 On rocks; spores 7–10 × 2.5–3.5 μ 1. *C. chalybeia*
  On moss and humus; spores 12–17 × 5–6 μ 2. *C. muscicola*

1. **Catillaria chalybeia** (Borr.) Mass. Thallus crustose, thin, continuous or minutely areolate, ashy black, brownish black, or sooty white, to disappearing, sometimes with black hypothallus. Apothecia small, to 0.3 mm, broadly adnate; disk black, bare, dull, slightly scabrid, flat or becoming convex; margin usually persistent; exciple dark reddish brown; epithecium dark; hypothecium red-brown; hymenium 65–70 μ, upper part very dark, lower part blue, center hyaline; I+ blue, becoming wine red; paraphyses lax, capitate; asci clavate; spores 8, biseriate, hyaline, 1-septate, ovoid, 7–10 × 2.5–3.5 μ.

Growing on acid rocks, especially on marine and lake shores. Known from Europe, Africa, Greenland, North and South America. Fink reported it from Maine, Ohio, and California. Specimens in WIS are from Indiana, Wisconsin, Saskatchewan, and Alaska.

Locality: 14.

2. **Catillaria muscicola** Lynge Thallus crustose, varnish-like, clay colored, submembranaceous to minutely granulose-verruculose. Apothecia numerous, large, to 1.5 mm, flat and thick marginate, becoming very convex and immarginate, in age recurved; disk minutely roughened, black, concolorous with the margin; exciple dark brownish red, radiate, outer part palisade-plectenchymatous and thick; epithecium pale; hypothecium indistinct, dark brownish red HCl+ pale violet; hymenium narrow, 50 μ, pale bluish, I+ reddish or blackish; paraphyses conglutinate but easily separating in KOH or HCl, slender, unbranched or rarely branched, tips not much thickened; asci saccate; spores 8,

irregularly arranged, hyaline, 1-septate, the center scarcely constricted, fusiform, 12–17 × 5–6 μ.

Over mosses and humus. This species is known only from Greenland, the Northwest Territories, and Alaska.

Locality: 1.

5. TONINIA Mass. em. Th. Fr.

Thallus crustose to squamulose or subfruticose, with pseudoparenchymatous upper layer, distinct algal layer and medulla, attached across the underside, hypothallus not prominent. Apothecia small to large, adnate or sessile; disk flat to convex, black; exciple of same color as disk, disappearing; hypothecium hyaline to dark brown, hymenium hyaline to brownish; paraphyses conglutinate or lax, unbranched, tips thickened to clavate; asci clavate or cylindrico-clavate, spores 8, hyaline, 1-septate to 7-septate, ellipsoid to oblong or fusiform.

1   Squamules flattened, thallus white to gray, K-, containing zeorin, spores 1–3-septate   1. *T. lobulata*
    Squamules bullate, high convex to becoming subfruticose on stipes composed of the rhizoid-like hypothallus, K+ or -, lacking zeorin; spores simple or 1-septate, rarely 3-septate   2

2 (1) Thallus rusty brown, exciple and epithecium red brown   2. *T. tristis*
    Thallus silvery pruinose, greenish gray to olive brown under pruina, exciple pale to dark brown; epithecium blue black or purplish black, K+ violet   3. *T. caeruleonigricans*

1. **Toninia lobulata** (Somm.) Lunge Thallus squamulose, the squamules to 1.5 mm broad, crenulate and laciniate, becoming confluent, flattened, greenish gray, pale yellowish or whitish; hypothallus indistinct. Apothecia adnate, the base slightly constricted, partly conglomerate; disk flat to very convex, becoming immarginate, black; exciple red brown inside, exterior paler; radiate; epithecium brown to blackish; hypothecium reddish brown, K+ purple; hymenium 70–80 μ, upper part brownish or olive to bluish, I+ blue turning reddish; paraphyses conglutinate, unbranched, upper part colored and thickened; asci clavate; spores 8, biseriate, hyaline, 1–3-septate, fusiform or oblong-elongate, 13–24 × 4–6 μ.

On calcareous soil and over mosses and humus. Circumpolar, arctic-alpine, in North America from Ellesmere, Baffin and Devon Islands, to Alaska, south to northern Quebec, Churchill in Manitoba, Wyoming, Colorado, and Washington.

Localities: 1, 7, 8, 9, 10, 13.

From the arctic literature it is difficult to determine whether one has *T. lobulata* or *T. cumulata* as both may have simple, 1-septate, or 3-septate spores. However the chemotaxonomy simplifies the problem considerably. *T. cumulata* produces atranorin and is K+ yellow, *T. lobulata* produces zeorin and is K-. The test used is the GAo-T solution which yields yellow, curling and branching slender crystals in the presence of atranorin, and clear, double-ended, hexagonal, pointed crystals, sometimes with a broader band around the middle, in the presence of zeorin. The presence of these substances in these species has not previously been reported in the literature.

2. **Toninia tristis** (Th. Fr.) Th. Fr. (*Toninia tabacina* (Mass.) Flag.) Thallus of swollen, bullate and convolute squamules to 3 mm broad, solid or fistulose, olive brown, reddish brown, 'tobacco brown', epruinose, a fimbriate and rhizoid-like hypothallus may extend into the soil below. Apothecia adnate, the base well constricted, the black margin disappearing; disk flat to convex, black, bare; exciple dark brown; hypothecium dark reddish brown; epithecium dark red brown; hymenium 40–60 μ, brown, I+ blue turning wine red; paraphyses conglutinate, unbranched, tips thickened and darkened; asci clavate; spores 8, irregular in ascus, hyaline, 1-septate, oblong to fusiform, 11–18 × 4–6 μ.

On calcareous soil, on rocks, and over mosses. This species is known from Europe, Greenland, and western North America. In the latter from Alaska to Artillery Lake, NWT, Saskatchewan, and south to New Mexico and Arizona.

Localities: 3, 8, 11.

3. **Toninia caeruleonigricans** (Lightf.) Th. Fr. Thallus of bullate verrucules, to subfruticose squamulose, silvery pruinose, greenish gray to olive brown under the pruina; attached to the substratum by rhizoidal hyphae which continue the stalk-like base into the ground. Apothecia adnate, the base well constricted, black marginate; disk flat or depressed-convex, black, bare or bluish pruinose; exciple pale to dark brown or bluish black, radiate; epithecium blue black or purplish black, K+ violet; hypothecium red brown above paler below; hymenium 75 μ, upper part brown to blackish, I+ blue becoming wine red; paraphyses conglutinate, unbranched, the tips thickened; asci clavate; spores 8, irregular in ascus, hyaline, 1-septate, fusiform, 14–25 × 2–4 μ.

On calcareous soil and thin soil of cracks in rocks, very common. Circumpolar arctic and boreal; in North America from Ellesmere Island to Alaska, south to Quebec, New England, southern Ontario, Wisconsin, New Mexico.

Localities: 1, 3, 8, 11, 14.

6. BACIDIA (De Not.) Zahlbr.

Thallus crustose, uniform, continuous, varnish-like to verruculose, sometimes dispersed and disappearing. Apothecia sessile or adnate, with only proper exciple which is usually radiate and sometimes palisade-plectenchymatous. Spores 8, hyaline, 3–many-septate, acicular, bacilli-form-acicular, or pointed-ellipsoid. Hymenium usually I+ blue. Algae: *Protococcus*.

1   Spores acicular, bacillar, or fusiform-acicular   2
      Spores oblong or ellipsoid with pointed ends   5

2 (1) Hymenium blue with I; thallus not lemon yellow   3
      Hymenium I-; thallus lemon yellow with definite edge   4. *B. alpina*

3 (2) Hypothecium dark, red brown; epithecium blue; spores 10–15-celled 4
      Hypothecium pale brownish or hyaline; thallus thin, verruculose; on humus and bark   3. *B. beckhausii*

4 (3) Growing on mosses; spores with ends pointed; hypothecium red brown to violet brown   1. *B. bagliettoana*
      Growing on bark; spores with rounded ends; hypothecium purple   2. *B. subincompta*

5 (1) On moss, earth, bark, or bones; spores symmetrically ended   6
      On rocks   10

6 (5) Hypothecium dark, red brown to violet brown; apothecia black, convex, immarginate   5. *B. melaena*
      Hypothecium hyaline or pale brownish   7

7 (6) Apothecia pale, pink or darkening to yellow brown, sometimes clustered; spores 1–5-septate, 12–24 × 4–6 μ 6. *B. sphaeroides*
      Apothecia dark brown to black   8

8 (7) Hymenium brown above   9
      Hymenium and epithecium blue   9. *B. trisepta*

9 (8) Spores 6–8 μ wide; thallus gray, verrucose   7. *B. obscurata*
      Spores 5–6 μ wide; thallus obsolete   8. *B. microcarpa*

10 (5) Spores tailed, one end narrower than the other, 6–8-celled, 35–50 × 5.5–7 μ; thallus gray brown, verruculose areolate; hypothecium brown black   10. *B. lugubris*
      Spores not tailed, 3-celled, 14–22 × 3–4.5 μ; thallus gray or lacking; hypothecium black; hymenium pale rose violet, upper part blue   11. *B. coprodes*

1. **Bacidia bagliettoana** (Mass. & De Not. in Mass.) Jatta (*Bacidia muscorum* (Sw.) Mudd) Thallus crustose, verruculose to verruculose-granular, white, ashy, olive, or glaucescent K-, C-, P-. Apothecia 0.3–1 mm, sometimes confluent, flat but soon becoming convex, narrowly attached; exciple soon disappearing; disk black, dull or slightly shining, minutely roughened, rarely pruinose; proper exciple reddish, to 60 μ thick, radiate, the outer part forming a palisadeplectenchyma; epithecium bluish or bluish black; hypothecium 60 μ, red brown or violet brown, paler toward the hymenium; hymenium 40–65 μ, hyaline below, bluish above; paraphyses slender, coherent, gelatinous, 2 μ, the tips thickened to 5 μ; asci 50–55 × 14 μ; spores 8, long-acicular, 10–15-septate, 30–42 × 2–3 μ.

On humus and over mosses, sometimes on soil. A circumpolar arctic and temperate species, reported as common in the arctic from Scandinavia, Novaya Zemlya, Bear Island, Greenland, Baffin Island, the Northwest Territories, and the Bering Straits region.

Localities: 1, 2, 7, 8, 9.

2. **Bacidia subincompta** (Nyl.) Arn. Thallus verrucose to subisidiate, ashy, ashy bluish or dark, rarely lacking. Apothecia to 0.9 mm, the base constricted, soon convex; the margin thin or disappearing, the disk black; the exciple violet or with the outer part pale, the hyphae radiating; upper part of the hypothecium purple or reddish brown, the lower part paler; upper part of the hymenium blue with black granules in the epithecium; the hymenium 50–55 μ, I+ blue turning wine red, spores bacillar with the ends usually rounded, 5–7-septate, 24–38 × 2.5–3.5 μ.

Growing on bark. A probably circumpolar arctic species known from northern Europe, Siberia, and the Reindeer Preserve, NWT.

Locality: 8.

3. **Bacidia beckhausii** Körb. Thallus very thin, verrucose or uneven-verrucose, sometimes lacking, whitish or ashy green. Apothecia very small, 0.15–0.5 mm, convex; the margin disappearing early, black, pruinose or not; exciple black or livid, or pale, soon disappearing, cells radiate, outer part violet red; epithecium brownish or reddish; hypothecium pale or yellowish; hymenium lower part hyaline, upper part blackish; purplish, or brownish; epithecium, upper part of hymenium, and outer part of exciple KOH+ violet; paraphyses coherent, unbranched, slender, 1 μ, gelatinous; asci clavate, 36–60 × 8 μ; spores 8, bacillar with blunted ends, straight or slightly curved, 3–7-septate, 17–26 × 1.5–3 μ.

Normally found on bark and rotting wood but found on weathered caribou bone near Point Barrow. Ranging widely in northern Europe

and North America. Darbishire reported it from Ellesmere Island.
Localities: 4, 5.

4. **Bacidia alpina** (Schaer.) Vain. (*Arthroraphis citrinella* (Ach.) Poelt) Thallus crustose, margin subeffigurate, the crust abruptly limited, sometimes verruculose, lemon yellow, continuous, non-sorediate or the center becoming farinose-sorediate. Apothecia small, 0.3–1.5 mm, rare, black, sparse and dispersed or conglutinate, broadly attached; disk black, dull, rough, with persistent black margin; exciple outer part almost carbonaceous, dark reddish black, inner part bluish green; epithecium olive green or bluish green; hypothecium pale, lower part brownish; hymenium 100–150 μ; paraphyses slender, 1.7 μ, lax, not gelatinous; asci cylindrical or narrowly clavate, 85 × 10–13 μ; spores 8, hyaline, acicular, straight, tips acute, base slightly attenuate, 6–10-septate, 35–54 × 3–4 μ.

On soil and over humus. Known from Scandinavia, Siberia, Novaya Zemlya, Jan Mayen Island, Ellesmere Island, and Port Burwell.
Localities: 1, 4, 8, 11.

Arctic reports of *Bacidia flavovirescens* are generally this species according to Lynge.

5. **Bacidia melaena** (Nyl.) Zahlbr. Thallus crustose, of very small verrucules, appearing coarsely granular sorediate, dispersed or disappearing, ashy glaucescent or olive green. Apothecia small, 0.2–0.4 mm, sometimes a few confluent, at first marginate but soon high convex and immarginate; disk rough but shining black; exciple reddish inside and inner part KOH+ violet reddish, outer part hyaline to greenish and KOH–; epithecium brownish or olivaceous; hypothecium reddish brown or violet brown, KOH+ violet reddish; hymenium 36–70 μ, hyaline below, olivaceous to pale brownish above; paraphyses gelatinous, strongly coherent, tips thickened to 4 μ; asci 40–50 × 18 μ; spores 8, usually biseriate, hyaline, ellipsoid-pointed, tips somewhat rounded, 3-septate, 14–28 × 4–6 μ.

On rotten wood, humus, and over mosses. A circumpolar species reported from Greenland, Ellesmere Island, Labrador, and Newfoundland. Undoubtedly to be found across northern Canada.
Localities: 1, 2, 8, 14.

6. **Bacidia sphaeroides** (Dicks.) Zahlbr. Thallus crustose, ashy gray or glaucescent, verruculose, rarely granular-sorediate. Apothecia abundant, to 1.2 mm broad, base constricted, at first with pale white margin, this disappearing and the apothecia becoming high convex and usually

more or less coarsely tuberculate; disk pink when fresh, turning to yellowish in the herbarium, sometimes brownish but usually pale colored; exciple radiate, pale yellowish, outer part forming a palisade-plectenchyma; epithecium pale; hypothecium hyaline, pale yellowish, or brownish; hymenium 60–70 μ, hyaline; paraphyses 1–1.5 μ, tips slightly clavate, conglutinate to slightly lax, easily separable in KOH, little gelatinized; asci clavate, 60–70 × 10–14 μ; spores biseriate, hyaline, pointed-ellipsoid, 3-septate, 12–24 × 4–6 μ.

On mosses, humus, rotten wood, occasionally on mossy rocks, and on bases of trees. Circumpolar, boreal and temperate, in the arctic reported from Novaya Zemlya, Scandinavia, Russia, Siberia, Greenland, Yukon, and Alaska.

Localities: 2, 3, 8, 9, 14.

7. **Bacidia obscurata** (Somm.) Zahlbr. (*B. fusca* auct.) Thallus crustose, verruculose, ashy glaucescent, white, or brownish glaucescent, more or less continuous. Apothecia 0.7–2.0 mm broad, margin soon disappearing; disk high convex, brown black to black, shining, epruinose, minutely rugose; exciple radiate, lower side of apothecium with palisadeplectenchyma, pale or red brown; epithecium pale or brown; hypothecium hyaline to light brown (not dark red brown as in *B. melaena*); hymenium 90–120 μ, hyaline to brownish, upper part brownish; paraphyses unbranched, very coherent, 1–2 μ, tips thickened, gelatinous; asci clavate, 60–90 × 12–16 μ; spores 8, biseriate, hyaline, ellipsoid with acute tips, 3-septate, 16–34 × 6–8 μ.

On mosses, soil, old wood, and occasionally bark. Widely distributed in the northern hemisphere and northward to Kamchatka, Greenland, Labrador, Ellesmere Island, Newfoundland, and Northwest Territories.

Localities: 2, 8. Also Cheena Ridge near Fairbanks.

8. **Bacidia microcarpa** (Th. Fr.) Lettau Thallus crustose, white to ashy white, verrucose to disappearing. Apothecia small, 0.2–0.7 mm broad, the base constricted, black, bare, dull; exciple red brown, columnar, extending under the apothecium; epithecium brown; hypothecium pale to brownish; hymenium brown, 60–70 μ, I+ blue; paraphyses coherent, gelatinous walled, septate; spores 8, 1–3-septate, fusiform with rounded tips, hyaline, 16–30 × 5–6 μ.

Usually growing over mosses, reported from calcareous slate and in this area found on old caribou bones. A probably circumpolar boreal species reported from Greenland, Labrador, Novaya Zemlya, Bear Island, Ohio, and Europe.

Localities: 5, 8.

9. **Bacidia trisepta** (Naeg.) Zahlbr. Thallus crustose, verrucose, the small verrucules tending to be scattered, ashy- or olive-green. Apothecia 0.2–0.9 mm, soon high convex, almost spherical; disk black, rarely livid, dull; margin soon disappearing; exciple radiate, inside violet brown, outer part pale, or exciple entirely pale, base vertically plectenchymatous; epithecium black or blue; hypothecium pale brownish to hyaline; hymenium 40–70 μ, entirely or with upper part blue; paraphyses very coherent, very gelatinous, unbranched, tips only slightly thickened; asci clavate; spores 8, biseriate, hyaline, oblong to fusiform, 3-septate, slightly curved, 12–30 × 5–7 μ.

On old wood and bark of trees. Circumpolar, temperate, reported north to Labrador and Newfoundland in North America where its northern limits are not fully known.

Locality: 5.

10. **Bacidia lugubris** (Somm.) Zahlbr. Thallus crustose, verruculose areolate, areoles convex, ashy brown, whitish brown, or dove brown, esorediate. Apothecia adnate or sessile, margin persistent, often flexuous, black; disk flat, black; exciple dark brown, paraplectenchymatous, not strongly radiate, inner part paler, reddish brown; epithecium red brown; hypothecium upper part paler reddish brown, lower part darker; hymenium pale, upper part reddish brown; paraphyses very lax, unbranched, 1.5 μ, tips 3–4 μ, not gelatinous; asci clavate, 55–60 × 15–20 μ; spores 8, multiseriate, clavate, one end 'tailed,' 5–7-septate, 35–50 × 5–7 μ.

Growing on acid rocks. Arctic-alpine, reported from Europe, North American, and Africa. Reported from Ellesmere Island by Darbishire; and I have seen specimens from Ungava Peninsula.

Localities: 8, 9, 11.

11. **Bacidia coprodes** (Körb.) Lettau Thallus crustose, verrucose uneven to subareolate or lacking, esorediate or partly sorediate, glaucescent or ashy, whitish glaucescent or blackish. Apothecia to 0.8 mm, at first flat and marginate, soon convex and immarginate; exciple dark outside, violet within, radiate; epithecium and upper part of hymenium greenish or bluish; hypothecium lower part dark, upper part red violet to pale; hymenium 50–55 μ, I+ blue then wine red, upper part bluish or greenish, lower hyaline or red violet; paraphyses coherent, septate, 1.5 μ, the tip capitate 4–5 μ; asci cylindrico-clavate 50–52 μ; spores uniseriate or biseriate, fusiform, hyaline, 3-septate, 11–19 × 3–5 μ.

On calcareous rocks in Europe, Novaya Zemlya, and Bear Island.

Locality: 8.

7. LOPADIUM Körb.

Thallus crustose, granulose, warty, or coralloid, undifferentiated into layers. Apothecia sessile; disk concave to flat or slightly convex, light colored or dark; proper exciple thin to thick, of same color as disk, of coherent, more or less radiate hyphae; hypothecium hyaline or brown to black; hymenium I+ blue; paraphyses unbranched or slightly branched, coherent; asci clavate; spores 1-8, hyaline or brownish, muriform with many cells.

1	Spore 1 per ascus	2
	Spores 8 per ascus	3. *L. fecundum*
2 (1)	Thallus coralloid-isidioid	1. *L. coralloideum*
	Thallus granular crustose	2. *L. pezizoideum*

1. **Lopadium coralloideum** (Nyl.) Lynge Thallus crustose, verruculose and coralloid-isidioid; the isidia short-cylindrical, brownish olive to ashy olive. Apothecia top-shaped to cup-shaped, short stalked, the base narrowed; margin prominent, persistent, black or dark olive brown; disk black, concave, shining, epruinose; epithecium dark brown; hypothecium brown or slightly paler; exciple brown, interior paler; hymenium 180 μ, I+ blue, upper part brown; paraphyses scarcely branched, slender, tips thickened to 2-3 μ, coherent; asci clavate; spore 1, oblong, tips rounded, muriform, with numerous cells, hyaline to brownish, 82-115 × 20-40 μ.

On mosses, humus, plant remains, and rocks, rarely on barks. Circumpolar, arctic, known from Greenland, Baffin Island, Ungava Peninsula, Dubawnt Lake, Artillery Lake, Alaska.

Localities: 4, 6.

2. **Lopadium pezizoideum** (Ach.) Körb. Thallus verruculose crustose, sometimes subsquamulose, dark olive brown, olivaceous, or greenish olive. Apothecia to 2 mm, top-shaped or peltate; margin black, smooth, entire; disk black, dull or shining, concave to flat, often minutely tuberculate; epithecium brown; exciple radiate, dark brown on exterior, paler within; hypothecium brown or brownish above, paler below; hymenium 120-170 μ, the asci I+ blue, becoming wine red, the rest I+ yellow, upper part brownish; paraphyses scarcely branched, coherent, slender, the tips thickened to 2-3 μ and darkened; asci clavate; spore 1, oblong, muriform with numerous cells, hyaline to brownish, 48-135 × 20-46 μ.

On mosses, humus, and bark of willow and alder in arctic, on tree barks along the Pacific coast. Circumpolar arctic-alpine but also on tree barks in oceanic forests. In North America south to New Brunswick, Ontario, Manitoba, Idaho, and Washington. Fink reported it from Massachusetts and Florida; these are probably incorrect.

Localities: 1, 2, 3, 6, 7, 8, 9, 11.

3. **Lopadium fecundum** Th. Fr. Thallus crustose, verruculose-granulose, the granules sometimes conglomerating, brownish to ashy green, unchanged when wet, K-. Apothecia sessile; disk becoming convex, black; margin entire, black, becoming excluded; exciple brown; hypothecium brown or reddish brown; hymenium I+ blue; paraphyses slender, becoming gelatinous-conglutinate, the tips becoming greenish black capitate; asci saccate; spores 8, muriform with many cells, hyaline, oblong, 22–40 × 10–18 μ.

Growing over mosses. A species reported from Scandinavia, Greenland, Siberia, and Colorado (Anderson 1967). New to Alaska.

Locality: 10.

8. RHIZOCARPON (Ram.) Th. Fr.

Crustose, usually with black, continuous or radiating black hypothallus; thallus areolate, verrucose, or continuous, corticate with a pseudocortex of compacted gelatinized cells, medulla loose, lacking lower cortex, lower side grading into hypothallus. Apothecia black, circular or angular, upon the hypothallus and between the areoles; margin lecideine, black, a subhymenium dark and in many species extending down to the hypothallus; hymenium hyaline, greenish, or reddish, upper part greenish, reddish, reddish black, or greenish black; paraphyses anastomosing branched, coherent, clavate or capitate, upper part often darkened; asci clavate or cylindrical; spores (1–)2–8, uniseriate or biseriate, halonate with a gelatinous epispore, 2-celled to more or less muriform, hyaline or dark brown.

Species with 2-celled brown spores can easily be mistaken for species of *Buellia* but members of the genus lack the gelatinous epispore and have paraphyses which are unbranched or branched only at the tips instead of anastomosing branched.

1	Thallus yellow, containing rhizocarpic acid (UV+ yellow, hexagonal and rectangular very yellow plates in GE and GAo-T tests)	2
	Thallus brown, ashy, or white, UV-, lacking rhizocarpic acid	9
2	(1) Spores 1-septate	3
	Spores more than 1-septate	7

3 (2) Spores small, 9–18 μ; hymenium low, 50–100 μ; a black epihymenium; upper part of hymenium dark reddish or rarely dark greenish in KOH; tips of paraphyses rounded; medulla I-   4

Spores larger, 18–32 μ; hymenium 100–150 μ; epihymenium indistinct; upper part of hymenium KOH+ brownish green to green; paraphyses clavate   6

4 (3) Free-living; disk dull sooty; white medullary tissue below subhymenium

     5

Parasitic on *Sporastatia*; disk shining; dark subhymenium connected with prothallus   3. *R. pusillum*

5 (4) Medulla K+ yellow, containing stictic acid   1. *R. superficiale*

Medulla K+ red, containing norstictic acid   2. *R. crystalligenum*

6 (3) Medulla I+ blue   4. *R. eupetraeoides*

Medulla I- or faintly blue   5. *R. inarense*

7 (2) Ripe spores few-septate, many only transversely   6. *R. intermediellum*

Ripe spores muriform   8

8 (7) Hymenium greenish to hyaline, upper part green to brown, rarely red

     7. *R. riparium*

Hymenium hyaline, upper part red   8. *R. geographicum*

9 (1) Thallus composed of umbilicate red brown areoles with blackish gray margins and resembling apothecia; spores 8 per ascus, 1-septate

     9. *R. rittokense*

Thallus non umbilicate, smooth, chinky, farinose, verrucose, flat-areolate, or glebulose-areolate   10

10 (9) Spores 1-septate   11

Spores muriform   18

11 (10) Spores remaining hyaline or only finally darkening   12

Spores soon darkening   14

12 (11) Thallus white subfarinose; apothecia with pruinose false-thalloid margin; disk pruinose   10. *R. chioneum*

Thallus ashy gray, gray brown or brown   13

13 (12) Medulla I+ blue, areolae gray to gray brown; epithecium brown, K+ very red; spores 18–30 × 8–14 μ 11. *R. polycarpum*

Medulla I-, areolae gray brown to red brown or ochraceous and thin to nearly disappearing; spores 11–29 × 9–14 μ 12. *R. hochstetteri*

14 (11) Thallus white   15

Thallus gray, gray brown, brown, or orange   16

15 (14) Thallus farinose, medulla K-, cortex UV+, apothecial margin pruinose; spores 12–16 × 6–10 μ, epithecium K+ purple   10. *R. chioneum*

Thallus areolate, medulla K+ red (containing norstictic acid) or K- (with psoromic acid), cortex UV+, margin of apothecium epruinose; spores 18–32 × 10–15 μ, epithecium K+ brown green to green

     4. *R. eupetraeoides*

16 (14) Medulla K+ red (norstictic acid), thallus gray, verrucose or smooth, apothecia convex, epithecium green black, K-, exciple K+ red; spores 21–30 × 9–12 μ   13. *R. copelandii*

Medulla K-   17

17 (16) Thallus gray to brown, areolate to bullate-areolate, epithecium red brown, K+ violet, exciple K+ slightly yellow; spores 22–38 × 12–16 μ   14. *R. badioatrum*

Thallus orange, areolate, epithecium blue black, exciple K+ violet; spores 12–15 × 7–7.5 μ   15. *R. alaxensis*

18 (10) Spores hyaline   19

Spores very soon dark   20

19 (18) Spores 12–14 × 5.5–7.7 μ; apothecia convex in botryose clusters, margin pruinose or not, thallus chalky white   16. *R. cumulatum*

Spores 21–30 × 12–15 μ; apothecia flat, not botryose-clustered, margin and disk pruinose, thallus chalky white   17. *R. umbilicatum*

20 (18) Spores 1–2 per ascus, 24–69 × 16–33 μ; thallus gray to brown gray, high verrucose, sometimes pruinose, medulla I-, P- or P+ red, K- or K+ red, apothecia becoming convex, epithecium purplish brown, K+ violet   18. *R. disporum*

Spores 4–8 per ascus   21

21 (20) Medulla I+ blue, thallus gray, gray brown, rarely whitish   22

Medulla I-, thallus chalky white, apothecia soon convex and heaped in compound groups, epithecium pale olive; spores 12–14 × 5.5–7.5 μ   16. *R. cumulatum*

22   Medulla K+ red (norstictic acid), C-; thallus gray to whitish, verrucose; epithecium reddish brown to olive brown, K-; spores 20–24 × 9–15 μ   19. *R. eupetraeum*

Medulla K-, C+ red (gyrophoric acid); thallus gray brown to pale gray, bullate verrucose; epithecium reddish brown to olive brown, K+ violet; spores 24–46 × 11–18 μ   20. *R. grande*

1. **Rhizocarpon superficiale** (Schaer.) Vain. Thallus on black hypothallus, dispersed or close, marginal areoles sometimes subradiate, areoles angular and concave to flat or convex, yellowish white, yellow, or greenish yellow, dull or glossy to scabrous; medulla K+ yellow, P+ red (stictic acid), I-. Apothecia round or angular, higher than areoles, flat to slightly convex; margin persistent, black; disk black, minutely papillate 'sooty'; exciple dark red brown, K+ violet; epithecium with dark grains; hymenium 70–100 μ, hyaline to brownish, upper part reddish or greenish; paraphyses coherent, little branched, tips capitate; asci clavate; spores 8, 1-septate, brown, with thin halo, 11–18 × 6–8 μ.

On exposed rocks, usually acidic. Circumpolar and arctic-alpine, in North America from Greenland and Ellesmere Island to Alaska, south to

New Hampshire in the east and Colorado in the west.
Localities: 2, 9, 11, 14.

2. **Rhizocarpon crystalligenum** Lynge Thallus on black hypothallus, areolae 3-6 mm broad, dispersed or contiguous, flat or slightly convex; margins slightly crenate, yellow to greenish yellow, medulla K+ red, P+ orange red, I- (containing rhizocarpic and norstictic acids). Apothecia to 1 mm, rounded, higher than areolae; thin margins may disappear; disk black; epithecium reddish; margin violet brown, radiate; hypothecium reddish to violet brown, hymenium 80-95 μ, I+ blue; paraphyses coherent; spores 1-septate dark with thin halo, 13-18 × 7.5-8.5 μ.

Growing on acid rocks. Known from the American arctic but the range uncertain due to lack of chemically tested material.
Locality: 8.

3. **Rhizocarpon pusillum** Runem. Thallus parasitic on *Sporastatia*, very small, to 1 cm broad; prothallus indistinct; areolae 0.2-0.6 mm broad, flat to convex, angular or rounded-angular, whitish yellow to yellow, dull, smooth to slightly farinose, medulla white; K-, P+ yellow, I-; containing rhizocarpic and psoromic acids. Apothecia to 0.7 mm broad, angular, flat to convex, without distinct margin; exciple reddish brown, K+ violet; subhymenium brown, K-; hymenium 70-100 μ, hyaline to brownish; epihymenium dark, containing dark grains; paraphyses rounded tipped; asci clavate; spores 8, dark brown, 1-septate, the center constricted, small, 9-14 × 4-6 μ.

Growing over the thallus of *Sporastatia testudinea*, on rocks. Reported from Europe in the Alps and Pyrenees. New to North America. I am indebted to Dr H. Hertel for calling this species to my attention.

Locality: 11. Another specimen was collected at Divide Camp nunataks, St Elias Mts, Yukon, David Murray 2382.

The comments of R.A. Anderson (1965) comparing this species with *R. effiguratum* (Anzi) Th. Fr. which he reported from California, Colorado, Montana, and Wyoming should also be taken into account. The only difference lies apparently in the I+ blue reaction of the medulla in *R. effiguratum*, a character found variable by Anderson.

4. **Rhizocarpon eupetraeoides** (Nyl.) Blomb. Thallus on black hypothallus, of scattered or grouped areoles, the areoles angular or rounded, slightly or very convex, whitish yellow, yellow, or orange yellow, surface farinose to smooth; medulla K+ red or K-, P+ yellow or P-, I+ blue. Apothecia rounded, flat to slightly convex; margin black; disk smooth to slightly rough, black; exciple dark reddish black, K+ red; epithecium dark; hymenium 100-140 μ, upper part green to greenish blue, lower

part hyaline; paraphyses branched, coherent, tips clavate; asci clavate; spores 8, 1-septate, dark, halonate, 18–32 × 10–15 μ.

On acid rocks. Circumpolar and arctic-alpine, in North America from Maine, Northwest Territories, and Alaska.

Locality: 11.

5. **Rhizocarpon inarense** (Vain.) Vain. Over black hypothallus, areolae to 1 mm, dispersed, angular to rounded, slightly convex, whitish yellow to grayish yellow, rarely bluish green, dull; medulla K+ red, P+ yellowish, margin K+ reddish; hymenium K- or + green. Apothecia to 1.1 mm, rounded very convex, black; margin persistent; epithecium indistinct, greenish; hypothecium dark brown; hymenium 110–150 μ, hyaline to greenish, upper part dirty green to blue green; spores dark, with halo, 21–30 × 10–12 μ.

Mainly a species of basic rocks. It is circumpolar arctic.

Locality: 8.

This species contains rhizocarpic and norstictic acids ± psoromic acid ± gyrophoric acid.

6. **Rhizocarpon intermediellum** Räs. Black hypothallus present or not, areolae contiguous to dispersed, to 0.7 mm, irregularly angular, more or less convex, yellow to greenish yellow, dull or glossy, medulla K-, P-, I-. Apothecia to 0.5 mm, rounded, concave or flat, margin distinct, epihymenium indistinct, brownish; margin brownish; hypothecium brownish; hymenium 50–90 μ, hyaline or greenish, upper part brownish red; spores with few septae, 1–4 transverse and sometimes 1 longitudinal, dark, with halo, 12–21 × 6–10 μ.

Growing on calcareous rocks. A circumpolar arctic-alpine species.

Locality: 8.

7. **Rhizocarpon riparium** Räs. Black hypothallus at margins, sometimes between areoles; areoles nearly continuous, angular or irregular, flat, sometimes divided, grayish green, greenish yellow to yellow, smooth, dull; medulla K-, P+ yellow or P-, I+ blue, sometimes C+ red (gyrophoric acid). Apothecia angular or round, flat or slightly concave; margin thin or thick, black; disk black, smooth; exciple brown, K+ brownish red; epithecium indistinct, brownish to dirty green; hymenium 120–200 μ, hyaline or faint green, upper part brownish to dirty green; paraphyses hyaline above, clavate; asci clavate; spores 8, muriform with many cells, halonate, dark, 24–40 × 11–15 μ.

Growing on acid rocks. Circumpolar, arctic-alpine and in South America. In North America south to New York, New Hampshire, Minnesota, South Dakota, and Arizona.

Localities: 8, 9.

8. Rhizocarpon geographicum (L.) DC. Black hypothallus present and sometimes conspicuous; areoles contiguous or dispersed, more or less angular, flat to convex, bright yellow, greenish yellow, rarely grayish yellow, smooth, dull or glossy, medulla K-, C-, or C+ red (gyrophoric acid), P+ yellow (psoromic acid), I+ blue. Apothecia between the areoles, angular to rounded, flat; margin thin, distinct or not; disk black, smooth; exciple reddish brown, dark, K+ reddish violet; epithecium indistinct, reddish; hymenium 150-180 μ, hyaline below, upper part reddish brown to reddish violet; paraphyses clavate, tips pale; asci clavate; spores 8, muriform with few to many septa dark brown, halonate, 20-40 × 10-22 μ.

Growing on acid or calcareous rocks. A circumpolar arctic-alpine species, in North America south to New Hampshire, New York, Minnesota, Colorado, and California.

Localities: 1, 2, 3, 8, 9, 11, 14.

9. Rhizocarpon rittokense (Hellb.) Th. Fr. Black hypothallus present; areoles umbilicate, concave, flattish, or convex, round, brown, shining, often pruinose toward margin, margin thick, dark, subpulverulent, lower side dark; medulla K-, C-, P-, I-. Apothecia round, convex, margin thin, subpersistent; disk black, becoming rugulose and convex, epruinose; exciple black outside, paler to interior, K- or K+ violet; epithecium brown; hymenium 100-140 μ, hyaline; paraphyses coherent, branched, tips clavate; asci clavate to saccate; spores 8, 1-septate, halonate, becoming brown to black, 20-24 × 10-15 μ.

Growing on acid rocks. Circumpolar, arctic, the range including Greenland, Baffin Island, Devon Island, Ungava Peninsula, and Alaska.

Localities: 3, 8, 11.

10. Rhizocarpon chioneum (Norm.) Th. Fr. Blue black hypothallus often present at margins of thallus; thallus thick, farinose, chinky-areolate, white; medulla K-, C-, P-, I-. Apothecia with false thalloid margin around the proper margin; sessile to raised, sometimes on small pedicels by erosion of thallus; margin thin, black, pruinose, disappearing; disk flat to slightly convex, black, sometimes pruinose, smooth; exciple black or purplish black; epithecium purplish black; hymenium 100-125

μ, hyaline below, violet red above; paraphyses coherent, thin, capitate; asci clavate; spore 8, 2-celled, constricted, hyaline but darkening, halonate, 12–23 × 8–12 μ, often aborted.

Growing on calcareous rocks. Probably circumpolar, arctic, in North America from Ellesmere Island, Cornwallis Island, Alaska, and Alberta. Localities: 1, 2, 3, 8, 11, 14.

11. **Rhizocarpon polycarpum** (Hepp) Th. Fr. Thallus over black hypothallus, areolae to 0.8 mm broad, usually contiguous and angular, sometimes crescent shaped around the apothecia, pale brownish to dark brownish or grayish brown, flat to slightly convex, the medulla K-, C-, KC-, P-, I+ blue. Apothecia brown black to black, to 0.8 mm broad, rounded angular to very angular, slightly constricted at base, sessile, flat or concave dull or shining; the margin usually moderately thick; epithecium brown or violet brown, K+ red to red violet; hypothecium brown to black; hymenium 110–130 μ, I+ blue, hyaline or faint greenish blue above; paraphyses coherent, branched, the tips thickened; spores hyaline to slowly darkening, 1-(rarely 3-)septate, with halo, 17–30 × 7–14 μ.

Growing on acid rocks. A circumpolar arctic-alpine species. Locality: 8.

12. **Rhizocarpon hochstetteri** (Körb.) Vain. Black hypothallus sometimes present, very thin thallus uniform, subcontinuous to chinkyareolate, rarely verrucose to dispersed or lacking, ashy, bluish ashy, reddish ashy or ochraceous, sometimes slightly pruinose, esorediate, medulla K-, C-, KC-, P-, I-. Apothecia to 1.2 mm, black, slightly higher than the thallus, the disk black, bare, flat, smooth; the margin thin, persistent; epithecium bluish black or olivaceous brown; exciple reddish brown, radiate, hypothecium reddish brown; hymenium 100–110 μ, upper part bluish or olivaceous, paraphyses slightly capitate, coherent, spores remaining hyaline or slightly darkening, 11–29 × 9–14 μ.

Growing on acid rocks. Circumpolar and arctic, probably also boreal, but its range is not adequately known. Locality: 8.

13. **Rhizocarpon copelandii** (Körb.) Th. Fr. Black hypothallus present; thallus areolate, areoles rounded or angular, convex or flat; usually slightly shining, ashy gray, medulla K+ yellow turning red (norstictic acid), C-, P+ yellowish, I-. Apothecia between areoles, flat to convex; margin thin to thick, persistent; disk black, dull or shining, slightly roughened; exciple red brown, K+ red with needle-like crystals produced; epithecium green black; hymenium 140 μ, pale to brownish,

greenish above; paraphyses coherent, tips dark, capitate; asci clavate; spores 8, 1-septate, soon dark, bluish black, halonate, 20–30 × 9–12 μ.

Growing on acid rocks. Circumpolar, arctic, in North America from Ellesmere Island to Alaska, south in Labrador, Manitoba, and Saskatchewan.

Localities: 9, 11.

14. Rhizocarpon badioatrum (Flörke) Th. Fr. Black hypothallus prominent; thallus areolate to verrucose, brown to dark brown, red brown, olivaceous, or ashy, little pruinose, esorediate; areoles flat to convex, contiguous to dispersed, round or angular, smooth; medulla K-, C-, I-. Apothecia dispersed or aggregated, round or rounded-angular; margin thin and persistent or disappearing; disk black, flat, papillate; exciple dark red brown, K+ yellow mist; epithecium reddish; hymenium 130 μ, pale below, reddish above; paraphyses coherent, capitate; asci saccate; spores 8, 1-septate, brown, constricted, halonate, 22–38 × 12–16 μ.

On acid rocks. Circumpolar arctic, boreal and temperate. In North America from Baffin Island to Alaska, to New Hampshire, Minnesota, and Colorado.

Locality: 3.

15. Rhizocarpon alaxensis Thoms. Thallus rusty ochraceous, thin areolate, the areoles contiguous but not forming a continuous thallus, nonradiate, areoles broad, angular, flat to slightly convex; medulla K-, C-, P-, I-. Apothecia immersed and level to slightly above the thallus, single and round or grouped and angular, concave; margin thick, black, shining, persistent; disk concave, black, shining; exciple dark red brown; epithecium blue black; hymenium 37–50 μ, upper part blue, lower hyaline; paraphyses clavate, tips dark; asci broadly clavate to saccate; spores 8, biseriate, 2-celled, brown, halonate, 12–15 × 7–7.5 μ.

On calcareous rocks. Known only from the type material.

Locality: 9.

16. Rhizocarpon cumulatum Thoms. Hypothallus lacking or thin, white; thallus chalky white, dull, farinose or not, becoming chinky-areolate, thick, contiguous, well defined; medulla K-, C-, P-, I-. Apothecia at first flat, soon convex and in compound heaps, botryose; margin black, pruinose or not, sometimes with false thalloid margin; disk black, dull, pruinose; exciple pale above and at sides, red brown under the apothecium; epithecium pale olive green; hymenium 90–100 μ, hyaline; paraphyses branched, intricate, non-capitate; asci clavate; spores 8,

3-septate transversely, the center cells 1-septate longitudinally, very soon brown, halonate, 12–14 × 5.5–7.5 μ.

On calcareous rock. Known only from the type material.

Locality: 1.

17. **Rhizocarpon umbilicatum** (Ram.) Jatta Hypothallus pale lead gray; thallus thin or thick, continuous to chinky or areolate, white or bluish gray when thin, chalky, areolae angular, small, flat to slightly convex, K-, C-, I-. Apothecia more or less immersed in thick thalli, adnate on thinner thalli; margin thick, whitish pruinose; disk black, smooth, epruinose; exciple dark, radiate; epithecium olivaceous; hymenium 110–180 μ, hyaline, upper part olivaceous; paraphyses coherent, branched, tips thickened; asci narrowly saccate; spores 8, biseriate, 1–2-septate longitudinally, 3–4-septate transversely, hyaline becoming dark, with thick halo, 18–30 × 11–15 μ.

On calcareous rocks. Circumpolar, arctic-alpine, in North America from Ellesmere Island to Alaska.

Locality: 2.

18. **Rhizocarpon disporum** (Naeg. ex Hepp) Müll. Arg. Black hypothallus usually present; thallus verruculose, verrucules flat or usually convex, ashy, ashy brown, pruinose, surface smooth or slightly chinky, medulla K- or K+ red, C-, P- or P+ red, I-. Apothecia adnate or between the areoles; margin thin, black, disappearing; disk flat or becoming convex, black, smooth or slightly roughened; exciple non-radiate, dark, purplish brown, inside paler; epithecium purple brown or brown; hymenium 115 μ, pale to brown; paraphyses coherent, scarcely branched; asci clavate; spores 1 or 2 per ascus, soon brown, muriform with numerous cells, 26–70 × 14–30 μ.

Growing on acid or calcareous rocks in strong light. Circumpolar, arctic-alpine. In North America south to Adirondack Mountains, Minnesota, South Dakota, Colorado, and Arizona.

Localities: 1, 3, 8, 9, 11.

19. **Rhizocarpon eupetraeum** (Nyl.) Arn. Black hypothallus present; thallus areolate, areolae verruculose, the verrucules rounded, constricted at base, ashy, smooth, not pruinose; medulla K+ yellow, turning red (norstictic acid), C-, P+ yellow, I+ blue. Apothecia between the verrucules, round, flat; margin persistent or disappearing; disk flat, black, smooth; exciple radiate, reddish brown; epithecium reddish brown; hymenium 120–160 μ, brownish, upper part reddish or hyaline; paraphyses coherent, clavate; asci clavate; spores 8, dark, halonate,

muriform, 3–4-septate transversely, 1–2-septate longitudinally, 20–34 × 9–15 µ.

On acid rocks. Known from Europe and North America. In North America it is arctic-alpine, and ranges south to Vermont, New York, Wisconsin, Minnesota, and Idaho.

Localities: 8, 11.

20. **Rhizocarpon grande** (Flörke) Arn. Black hypothallus prominent; areolae verruculose, the base constricted and the areolae becoming glebulose, smooth, ashy, ashy brown, sometimes white and sometimes violet tinged; epruinose, esorediate; medulla K-, C+ red (gyrophoric acid), P-, I+ pale blue. Apothecia between the areolae, angular or rounded; margin thin or disappearing; disk flat, sometimes convex, black, dull; exciple brown; epithecium reddish brown or olivaceous brown, usually K+ violet; hymenium hyaline or brownish below, upper part brown; paraphyses coherent, capitate; asci clavate; spores 8, muriform, usually many-celled, dark, halonate, often aborted, 24–46 × 11–18 µ.

On acid rocks and boulders. Circumpolar, arctic, boreal and temperate, in North America south to Tennessee, Michigan, South Dakota, and Colorado.

Localities: 2, 3, 8.

STEREOCAULACEAE

Primary thallus crustose, granulose, to squamulose, usually with cephalodia containing *Nostoc* or *Stigonema* algae. Pseudopodetia fruticose, arising from the primary thallus, with a central more or less solid central medullary layer, surrounded by a more or less cartilaginous outer layer, the symbiotic algae in phyllocladia or granules strewn over the surface and usually with scattered cephalodia interspersed. Apothecia terminal or lateral, pale brown or blackening, convex, immarginate or soon so; spores eight, acicular or fusiform, transversely septate.

1 Hypothecium dark; pseudopodetia with granular surface, no phyllocladia; apothecia black, lacking any exciple.   1. *Pilophorus*
  Hypothecium hyaline; pseudopodetia with phyllocladia; apothecia pale or brown, with marginal exciple which soon disappears.   2. *Stereocaulon*

1. PILOPHORUS (Tuck.) Th. Fr.

Primary thallus granulose, crustose, and with cephalodia containing

Nostoc. Erect thallus a pseudopodetium unbranched or umbellate-branched, bearing apothecia at the tip; medulla loose, K+ yellow. Apothecia terminal, black, with black hypothecium; no exciple is developed; the margin consisting only of dense sterile paraphyses; a columella of vegetative tissue extends upward within the apothecium in *P. robustus* and one other species; asci cylindrical; spores 8, hyaline, simple, ellipsoid or elongate-ellipsoid. Pycnoconidia at the tip of the pseudopodetia, flask-shaped, the conidia formed on terminal cells of the conidiophores, sickle-shaped, 5 × 1 μ.

One species reaches the North Slope.

1. **Pilophorus robustus** Th. Fr. Primary thallus gray green; granulose, evanescent. Pseudopodetia 1–2.5 cm tall, irregularly or umbellately branched, covered with loose granules, the interior hyphae of the stalk strongly gelatinized, mainly longitudinally oriented; the contents atranorin and zeorin. Apothecia on the tips of the pseudopodetia to 2.5 mm broad, 1.5 mm high, margin of hymenium continuing below its attachment to the stalk and reflexed; a large columella of vegetative tissue within the apothecium; hymenium 200 μ; spores rounded when young, spindle-shaped when mature, 18–24 × 4–6.5 μ.

Growing on rocks. This is an arctic and boreal circumpolar species known from Scandinavia, Spitzbergen, Siberia, and Alaska.

Locality: 8.

2. STEREOCAULON Schreb.

Primary thallus crustose, areolate or warty, granulose, sorediate, or squamulose, with pseudopodetia arising from the primary thallus, usually branched, center solid with a central medullary layer surrounded by a more or less cartilaginous outer layer, often with a tomentum covering it; algae are in the phyllocladia which vary from granulose to squamulose, digitate, subcoralloid, or peltate. With cephalodia on the primary thallus or the pseudopodetia containing *Nostoc* or *Stigonema*. Apothecia pale or black brown, at first with a light margin, usually becoming convex and immarginate, outer side with a palisade-like plectenchyma, a central loose medullary layer, hypothecium hyaline or yellowish, hymenium mostly I+ blue, paraphyses simple or slightly branched, usually capitate, asci with thickened tips, spores 8, acicular or fusiform, 4 to many celled with only transverse walls. Pycnidia mostly imbedded at the tips of phyllocladia, fulcra exobasidial, pycnoconidia bacillar to filiform, straight to bent, usually 8–12 × 0.5 μ.

1   Pseudopodetia on rocks, tightly attached by holdfasts, little or not tomentose   **2**

Pseudopodetia on earth, sand, or clay, between gravel, only occasionally on rocks and then on soil in cracks; pseudopodetia loosely attached, easily lifted, usually with distinct, often thick, tomentum   **5**

2 (1) Pseudopodetia tipped by capitate soralia, P+ yellow, containing atranorin and lobaric acids; phyllocladia becoming peltate   1. *S. symphycheilum*

Pseudopodetia not capitate-sorediate; phyllocladia granular to coralloid but not becoming peltate   **3**

3 (2) Phyllocladia not coralloid by with sparse, granular, whitish phyllocladia, tips much branched and cauliflower-like, base naked
2. *S. botryosum*

Phyllocladia coralloid, elongate cylindrical, simple to pinnately or irregularly branched or coalescing into squamules   **4**

4 (3) Small, the pseudopodetia 1.5–2.5 cm tall; phyllocladia minute, 0.2–0.4 mm long, congestedly branched with coralloid divisions 0.1–0.15 mm diameter   3. *S. subcoralloides*

Taller; phyllocladia coarser; pseudopodetia 2–8 cm tall, 1–3 mm thick; phyllocladia dactyliform, to 1 mm long, 0.15–0.3 mm thick
4. *S. dactylophyllum*

5 (1) Phyllocladia at least in part rounded-foliose or rounded squamulose, margins more or less crenulate or incised, in part becoming granulose; pseudopodetia thickly arachnoid-tomentose, distinctly dorsiventral even when appearing partly erect; cephalodia small; apothecia very small, numerous, sunken; containing atranorin plus stictic or lobaric acid   5. *S. tomentosum*

Phyllocladia neither foliose nor squamulose, mainly dactyliform, attached together at the base; apothecia granulose, warty or small squamulose but in the latter case not rounded; apothecia reaching over 1 mm in diameter   **6**

6 (5) Pseudopodetia either very fragile or clearly dorsiventral   **7**

Pseudopodetia not exceedingly fragile, mainly erect or plainly ascending, at least the fruiting pseudopodetia not decumbent; phyllocladia more or less equally distributed around the pseudopodetia, at least on the erect fruiting stalks   **9**

7 (6) Pseudopodetia forming confused entangled cushions, occasional pseudopodetia lying irregularly here and there, very fragile, with very thin or no tomentum, not clearly dorsiventral, white with a rosy tinge; phyllocladia coarse in comparison with the pseudopodetia, running together, whitish; apothecia flat, large, frequent   6. *S. rivulorum*

Pseudopodetia prostrate, forming low, flat cushions, underside only exceptionally with phyllocladia (the prostrate forms of *S. paschale* with small granular or long finger-shaped phyllocladia)   **8**

8 (7) Pseudopodetia always closely pressed to the substratum, partly inter-
grown, with blackish tomentum; phyllocladia squamulose, forming an
almost continuous blue or slate gray crust; apothecia small, high
convex, blackish brown; cephalodia unknown   7. *S. saxatile*
Pseudopodetia somewhat ascending with distinct main axis; with
whitish or pale rosy tomentum; phyllocladia likewise whitish, at most
whitish gray, not blue gray nor dark gray, thick warty, always clearly
swollen; apothecia finally quite flat, clear red brown; cephalodia white,
spherical   8. *S. alpinum*

9 (6) Pseudopodetia with black tomentum, not rigid, frequently irregularly
bent, at the base distinctly thinner, dying, the mats easily lifted; phyllocladia botryose clustered, mostly very small granular, little more than
0.1 mm broad, seldom somewhat elongate and short irregularly
finger-shaped; cephalodia abundant and distinct, dark olive brown,
flat, granular; containing atranorin and lobaric acid   9. *S. paschale*
Pseudopodetia clearly or even strongly white tomentose, strong rigid,
phyllocladia larger, not botryose clustered; cephalodia indistinct or
almost spherical   10

10 (9) With abundant phyllocladia projecting beyond the tomentum; the
phyllocladia elongate, cylindrical, or tips clavate and irregularly
massed; apothecia flat and with distinct margin   10. *S. glareosum*
Phyllocladia smaller, granulose, in part almost hidden by the dense
tomentum; apothecia terminal, soon swollen and split

11. *S. incrustatum*

1. **Stereocaulon symphycheilum** Lamb. Primary thallus disappearing, of
minute peltate squamules with pale margins; pseudopodetia firmly attached to rock, prostrate and strongly dorsiventral, short, 3–15 mm tall,
simple or sparingly branched, the dorsal surface more or less continu-
ously corticate, the ventral ecorticate, white to pale ochraceous, smooth
and glabrous to finely subpubescent but not tomentose, mainly tipped
by conspicuous globose capitate soralia which are pulverulent or granulose and greenish white to gray; phyllocladia beginning as nodules, more
or less crenate and becoming peltate as in *S. vesuvianum*; K+ yellow, P+
sulphur yellow, containing atranorin and lobaric acid; cephalodia more
or less abundant, verruculose-tuberculate or scabrid, dark brown to
brown black, without cortex, containing *Stigonema* algae. Apothecia rare,
usually terminal, to 2.5 mm broad, flat to becoming pileate-convex,
margin disappearing; disk brown blackish; exciple 50–90 μ, colorless to
pale brown, of conglutinate, radiating hyphae, occasionally with granules in striae between the hyphae; hymenium 50–60 μ, pale to dark
brown in the upper part; paraphyses discrete, 1–2 μ thick, simple or
branched, the tips capitate, 3–4 μ, brown; asci clavate, I+ blue; spores
(4–)6–8 in ascus, elongate-fusiform, straight, 3-septate, 20–35 × 3–4.5 μ.

Growing on rocks. It was reported by Lamb (1961) from Finland, Sweden, and numerous localities in Alaska, including Lake Peters in the Brooks Range. The similar species, *S. vesuvianum*, is P+ orange and contains stictic acid instead of lobaric acid (stictic acid is lacking in an acid deficient strain).

2. **Stereocaulon botryosum** Ach. em. Frey Primary thallus farinose, disappearing. Pseudopodetia firmly attached to rocks in dense tufts, mainly radiating from a center, the base 0.5–1.5 mm thick, orangish, branching from the base upwards, the tips becoming fastigiate and forming cauliflower-like masses; phyllocladia glaucous to grayish white, very dense in upper parts, granular or verrucose, 0.2–0.3 mm broad, rarely flattened, usually bare below and tomentose above; cephalodia rare, inconspicuous, tuberculose, containing *Nostoc*; pseudopodetia K+ yellow, containing atranorin and or lobaric acid. Apothecia terminal on elongated pseudopodetia or level with the surface of the tuft, sometimes corymbose, dark brown, 1.5–3 mm, flat to becoming convex and immarginate, lower side tomentose; hypothecium dense; hymenium grayish white, upper part yellow brown; paraphyses more or less free, 1–1.5 μ, tips clavate 3–4.5 μ; asci cylindrico-clavate; spores 4–6, fusiform, one end narrower, 3-septate, 20–31 × 4.5 μ.

Growing on rocks. An arctic-montane species in the northern hemisphere and into Antarctica in the southern hemisphere. South to Labrador, Alberta, and Washington in North America.

Locality: 11.

3. **Stereocaulon subcoralloides** (Nyl.) Nyl. Pseudopodetia loosely attached to stone, caespitose, erect, 6–15 mm tall, 0.5–1 mm thick, base dark, upper part pale, lacking tomentum, few branches on lower part of axis, upper part with numerous small branches; phyllocladia few in lower part, very abundant in upper, sometimes almost confluent, grainlike to coralloid, rarely incised squamiform; cephalodia numerous, among the phyllocladia on the upper branches, olive brown, tuberculate, containing *Stigonema*; pseudopodetia K+ yellow, inner part K-, containing atranorin and lobaric acid. Apothecia few, terminal, 1.5–2 mm broad, disk flat, dark brown with pale margin, becoming convex or divided and with margins reflexed; hypothecium hyaline; hymenium 45–50 μ, I+ pale blue, upper part dark brown; paraphyses coherent, 1.5–1.8 μ, tips brown, capitate, 3–3.5 μ; asci broadly clavate; spores 8, narrowly ellipsoid with blunt ends, 3–6-septate, mainly 4-septate, 27–37 × 3–4.5 μ.

Growing directly on rocks and outcrops. Known from Scandinavia, northern Quebec, and Alaska.

Locality: 11.

4. **Stereocaulon dactylophyllum** Flörke Primary thallus disappearing; pseudopodetia erect, 2–8 cm tall, firmly attached to rocks, simple at base, branching upwards with spreading long and short branches, the base 1–3 mm thick, anisotomic with the main branches distinct, naked to white, roseate, or gray, partly blackish, tomentose especially at the base, several pseudopodetia forming tufts or cushions; phyllocladia originating as short side branches which then branch repeatedly, mostly more or less dactyliform or as hands with many fingers, coralline, cylindrical, to 1 mm long, 0.15–0.3 mm thick, usually tapered toward the apices, occasionally fusing into crust-like masses; cephalodia rare, immersed in the dark tomentum, warty granular, gray or brownish gray, containing *Stigonema*; pseudopodetia K+ yellow, P+ orange, containing atranorin and stictic acid. Apothecia common, terminal, 1–2 mm broad, flat with pale margin, becoming convex with revolute margin; disk light to dark brown, older apothecia segmenting into smaller swollen units; hypothecium indistinct; hymenium 50–75 μ, I+ or -, or only asci blue; paraphyses weakly gelatinous, 1.4–1.8 μ, tips brown, capitate to 4 μ; asci clavate; spores 8, to 5-septate, 40–72 × 8–14 μ.

Growing on acid rocks. It is a disjunct amphi-Atlantic boreal species occurring in Greenland, and Labrador to North Carolina, and inland in Ungava and Ontario.

Locality: 11.

5. **Stereocaulon tomentosum** Fr. Primary thallus disappearing; pseudopodetia loosely attached to the substratum, caespitose, erect or decumbent, clearly dorsiventral, 3–8 cm tall, 1–2 mm thick, little branched at base, upper part with many recurved short branches, more or less anisotomic, underside thickly tomentose, upper side with abundant phyllocladia which are crenate squamulose, ashy gray, overlapping and thickly covering the pseudopodetia; cephalodia inconspicuous, concealed in the tomentum of the underside, more or less spherical, containing *Nostoc*; pseudopodetia K+ yellow, P+ yellow or orange red, containing either stictic acid (typical strain) or P-, with lobaric acid (Sasakii strain). Apothecia usually abundant toward the ends of the branches but borne laterally, to 1.2 mm broad; margin pale and disappearing; disk dark brown, slightly convex to convex; hypothecium grayish; hymenium 45–50 μ, I+ dark blue; paraphyses coherent 1.5–1.8 μ, tips brown, capitate, 3–4 μ; asci cylindrico-clavate; spores 8, cylindrico-fusiform, somewhat spirally twisted in ascus, usually 3-septate, rarely to 7-septate, 20–35 × 2.5–3 μ.

This species grows on earth in open places, occasionally on gravels. It is circumpolar boreal montane in the northern hemisphere, extending in

South America to the tip of the continent. The lobaric acid strain is amphi-Beringian, extending into Alaska.

Localities: 8, 9, 14.

Stereocaulon tomentosum and *S. paschale* give much trouble in identifications. In *S. tomentosum* the apothecia are numerous, lateral, small and convex, the phyllocladia more squamuliform, appressed to the pseudopodetia, the pseudopodetia dorsiventral with the phyllocladia on the upper side, the cephalodia pale, inconspicuous, and containing *Nostoc*; in *S. paschale* the apothecia are few, terminal, larger and flatter, the phyllocladia spreading, digitately divided, the pseudopodetia radiate with the phyllocladia to all sides, the cephalodia dark, obvious, and containing *Stigonema*. Except in northwest America where the lobaric acid strain of *S. tomentosum* may occur as well as the typical strain, they may be distinguished by the lobaric acid in *S. paschale* and stictic acid in the typical strain of *S. tomentosum*.

6. **Stereocaulon rivulorum** Magn. Primary thallus disappearing; pseudopodetia loosely or firmly adhering to the ground, erect or decumbent, forming low lax tufts, whitish with a bluish tinge, 0.5–1.2 mm thick, bare or with a whitish to roseate thin tomentum, very fragile, sparingly branched with no distinct main axis; phyllocladia glaucous white, granular or somewhat elongated, broader toward the apices, dispersed in groups along the pseudopodetia leaving spaces showing on the main axis, conglomerate and angular crustose on some decumbent pseudopodetia; cephalodia infrequent, on the underside of dorsiventral pseudopodetia, inconspicuous, tuberculiform with granular surface, of the same color as the tomentum or brownish violet, containing *Nostoc*; pseudopodetia K+ yellow, containing atranorin and lobaric acid. Apothecia usually numerous, terminal or sublateral, 1–3 mm broad, disk dark brown with long persistent pale margin, lower side flat, white and tomentose; hypothecium hyaline; hymenium 60–70 μ; paraphyses coherent, 1.5 μ, tips brown, capitate, to 4.5 μ; asci cylindric-clavate; spores 8, cylindrico-fusiform with one end narrower, 3-septate, 25–37 × 3–3.5 μ.

Growing in low places below permanent snowbanks and along streams where it is occasionally inundated, circumpolar, arctic-alpine, in North America: in northern Canada and Alaska.

Localities: 10, 11.

7. **Stereocaulon saxatile** Magn. Primary thallus disappearing; pseudopodetia forming tufts about 1 cm thick, 10–15 cm broad, decumbent, densely packed, richly branched toward the outer ends, 3–4 cm long,

1 mm thick, dorsiventral, the lower side grayish tomentose; phyllocladia distinctly dorsiventral, gray or bluish white, underside darker, squamulose or verrucose-squamulose to subcoralloid; cephalodia unknown; pseudopodetia K+ yellow, containing atranorin and lobaric acid. Apothecia rare, in the ends of the marginal pseudopodetia, cap-like or divided into clusters of convex sections; disk blackish brown; hypothecium pale; hymenium 50 μ, I+ blue; paraphyses 1-2 μ, tips capitate, to 4 μ, brown; asci cylindrico-clavate; spores 8, fusiform, 3-septate, 25-33 × 2-3 μ.

Growing on rocks, stones, and outcrops in the open. It is known from northern Europe and North America.

Locality: 7.

8. Stereocaulon alpinum Laur. in Funck Primary thallus disappearing; pseudopodetia dorsiventral or erect, caespitose, firmly attached to the ground, 1-4 cm tall, 0.5-1 (-2) mm thick at the base, the base pale brownish or blackish, sparingly branched, lower part without phyllocladia but with a grayish, whitish, or roseate tomentum; phyllocladia crowded toward the apices, whitish gray with a bluish tinge or whitish, turning pale dirty yellowish in the herbarium, granuliform or united and somewhat flattened, lobate, 0.5 mm broad, indistinctly stalked, appressed to the pseudopodetium; cephalodia on the lower side, hemispherical, grayish white, partly concealed within the tomentum, containing *Nostoc*; central axis K-, phyllocladia K+ yellow, P+ yellow, containing atranorin and lobaric acid. Apothecia terminal on the pseudopodetia or grouped on the upper branches, immarginate, disk dark brown 1-1.5 mm broad, convex or irregularly swollen, lower side with the center tomentose, the margin bare; hypothecium hyaline; hymenium 40-50 μ, yellowish, I+ blue; paraphyses slightly coherent, 1.7-2.0 μ thick, tips brownish, capitate 3-4 μ; asci elongate-clavate; spores 8, cylindrical with one end narrower, slightly bent or straight, 3-septate, 27-38 × 2.5-3 μ.

Growing on ground, especially under a late snow cover. Circumpolar, in Canada and Alaska in North America.

Localities: 1, 2, 3, 4, 5, 8, 11.

9. Stereocaulon paschale (L.) Hoffm. Primary thallus disappearing; pseudopodetia very loosely attached to the substratum or becoming free as the base dies off, crowded or dispersed among mosses, 2-6 cm tall, 0.6-2 mm thick at the base, naked below with a white to roseate tomentum above; phyllocladia few toward base, crowded toward the apices, whitish or grayish white, sometimes roseate tinged, crowded in small stalked clusters, verruciform-coralloid or rarely verruciform-squamulose; cephalodia tuberculose, dark olive brown, containing *Stigonema* or rarely *Nostoc*; pseudopodetia K+ yellow, P+ yellow, containing atranorin and lobaric acid. Apothecia somewhat rare, terminal on the main or on short side branches, 1–3 mm broad; disk flat or slightly convex, dark brown; hypothecium hyaline; hymenium 45–50 μ, I+ dark blue; paraphyses loosely coherent, upper part occasionally branched, 1.5–2 μ, tips brown, capitate, 3–4 μ; asci clavate; spores 6–8, cylindrical, 3-septate, often slightly bent, 25–30 × 2.5–3 μ.

Growing on earth, mainly among mosses, when on rocks on an accumulation of humus or soil. Circumpolar, arctic-montane in the northern hemisphere, extending southward in the western hemisphere to Antarctica.

Localities: 1, 3, 11.

10. **Stereocaulon glareosum** (Savicz) Magn. Primary thallus composed of dense clusters of phyllocladia over the ground; pseudopodetia firmly attached to the ground, slightly tufted or more or less dispersed, pale, 1–2.5 cm tall, base to 1 mm thick, tapering to 0.5 mm at apices, entirely white or grayish white tomentose, sparingly branched; phyllocladia abundant at the base as well as on the substratum, more sparse above, rarely granular, usually cylindrical or papilliform, 0.3–1 mm long 0.2–0.3 mm thick, cortex continuous for long stretches on the pseudopodetia; cephalodia abundant among the phyllocladia on the branches, verruciform or globose, pale brownish violet or rose whitish, containing *Nostoc*; pseudopodetia K+ yellow, containing atranorin. Apothecia rare, at the apices or lateral on upper branches, with thin white margin; disc flat to convex, sometimes dividing, dark brown; hypothecium hyaline; hymenium 45–50 μ, grayish, only the asci I+ blue; paraphyses somewhat free, 1.5 μ, brownish, tips brown, clavate, 4.5 μ; asci clavate; spores 8, cylindrical-fusiform, slightly bent, 3-septate, 22–42 × 2.5–3.5 μ.

Growing on bare soil, frost boils, or among mosses, on acid soils. Circumpolar, in North America known from Alaska and the Canadian Rocky Mountains.

Localities: 2, 3, 4, 7, 8.

11. **Stereocaulon incrustatum** Flörke Primary thallus disappearing, of small granules; pseudopodetia in thick or loose tufts, ascending at the circumference, erect centrally, simple or with few branches, 1.5–3 cm tall, 0.8–1.2 mm thick, with thick whitish gray, ashy, or roseate tomentum, for long stretches without phyllocladia; phyllocladia irregularly distributed along the pseudopodetia, a little thicker toward the apices, whitish gray or bluish gray, verruciform, very small, 0.08–0.15 mm

thick, often concealed by the tomentum; cephalodia mainly where the phyllocladia are sparse, covered by the tomentum or larger, to 2 mm broad, dark brown, containing *Nostoc*; pseudopodetia K+ yellow, P-, containing only atranorin. Apothecia common, terminal, single or confluent, immarginate; disk dark brown; hypothecium pale; hymenium 60–65 μ, I+ blue; paraphyses somewhat coherent, 1–1.5 μ, tips capitate, brown, 2–3.5 μ; asci cylindrico-clavate; spores 4–6, cylindrical-fusiform, straight, 3-septate, 35–44 × 2.5–3 μ.

Growing on sandy soils. Circumpolar, arctic-alpine. In North America across extreme northern Canada and in the Arctic archipelago, and in Alaska.

Localities: 1, 2, 9, 11, 14.

BAEOMYCETACEAE

Thallus crustose, granulose, squamulose or marginally foliose, attached by medullary hyphae or rhizines. Apothecia on distinct stipes of dense or loose hyphal strands, stipes often containing algae; hypothecium and exciple not distinct from the interior of the stipe, hyaline, lacking algae; spores 8, fusiform or ellipsoid, hyaline, 1–4-celled. Algae: *Cystococcus*.

1. BAEOMYCES Pers.

Thallus crustose, granulose, squamulose, or marginally foliose, attached by medullary hyphae or rhizines. Cortex with one or more paraplectenchymatous layers or composed of interwoven hyphae more or less parallel to the upper surface or lacking, algal layer continuous. Apothecia round, finally swollen, often multiparted, on more or less distinct stipes of dense or loose strands, often containing algae, from the base partly or entirely overgrown by an algal layer with a cortex similar to that of the thallus; hypothecium and exciple not distinct from the interior of the stipe, hyaline and lacking algae; asci cylindrical; paraphyses simple or the upper part sparingly branched, slender; spores 8, fusiform or ellipsoid, hyaline, 1–4-celled. Pycnidia immersed in warts of the thallus, fulcra endobasidial; conidia short, bacilliform.

1   Apothecia pink, soon swollen and emarginate, thallus crustose, whitish or
     with pale rosy tinge, verrucules of the sterile thallus to 1 mm thick,
     farinose sorediate, containing baeomycesic acid   1. *B. roseus*
     Apothecia brownish, concave to flat, marginate, later swollen and with
     reflexed margins   2

2 (1) Thallus partly sorediate, partly of small squamules, the squamules some-
　　　times raised and attached by the margins　　　　　　　　　　　3
　　　Thallus with the margins distinctly foliose-squamulose, the center
　　　squamulose, containing stictic acid, K+ yellow　　　4. *B. placophyllus*

3 (2) Thallus containing stictic acid, K+ yellow, squamules small　　2. *B. rufus*
　　　Thallus containing norstictic acid, K+ yellow turning red, squamules
　　　larger, to 2 mm long　　　　　　　　　　　　　　3. *B. carneus*

1. **Baeomyces roseus** Pers. Thallus gray, white, or roseate, crustose,
smooth or powdery, covered with spherical or flattened verruculose
warts with a narrowed base, in fertile plants the warts to 0.1–0.3 mm
broad, in sterile ones to 1 mm; podetia 2–6 mm tall, 1 mm thick, smooth
or fissured, whitish, K+ yellow, P+ yellow, containing baeomycesic acid.
Apothecia pink, 1–4 mm broad, spherical emarginate or with a very
narrow margin; hypothecium not sharply defined from the interior of
the podetium; hymenium 100 μ, hyaline, I+ blue; paraphyses 2–2.5 μ,
simple, little thickened above; asci cylindrical; spores 8, fusiform, mainly
1-celled, occasionally indistinctly 2-celled, 12–26 × 2.5–3 μ.

　　On clay soils. In the arctic usually only the sterile thallus is present
and chemotaxonomy becomes essential to identify the thallus. It is
circumpolar and temperate, in North America mainly in the eastern
United States, but in Canada ranging in rare stands from Labrador
westward to Alaska.

　　Locality: 3.

2. **Baeomyces rufus** (Huds.) Rebent. Thallus crustose, powdery sorediate, greenish or whitish gray, partly squamulose, the squamules to 1
mm broad, dorsiventral, somewhat raised, sometimes imbricate, with a
warty but smooth cortex; podetia short, rarely to 6 mm tall, the lower
part or to just below the apothecium covered with an algal zone, the
stalk flattened or cylindrical, almost always fissured; podetia and thallus
K+ yellow, P+ orange, containing stictic acid. Apothecia concave or flat,
becoming slightly convex; disk red brown, seldom reddish or pink;
hypothecium distinct, at edges gradually merging into a palisade plectenchyma; hymenium 90–105 μ, I-; paraphyses slender, 1–1.8 μ, the tips
scarcely thickened; asci cylindrical; spores 8, often indistinctly 2-celled,
hyaline, ellipsoid, 8–13 × 2.5–4.5 μ.

　　Growing on soil and over rocks in moist places, especially with seep-
age water. Circumpolar in the boreal forest, in North America from
Newfoundland to Alaska, south to New England, New Mexico, and
California.

　　Locality: 11.

3. **Baeomyces carneus** (Retz.) Flörke Thallus squamulose or partly areolate-squamulose, the squamules 0.2–2 mm long, 0.2–0.7 mm broad, flat or partly convex; ascending or adnate, ± sorediate with labriform soralia, whitish or ashy, the margin crenulate, underside whitish, occasionally with rhizinae; K+ yellow turning red, P+ orange, containing norstictic acid. Apothecia to 2 mm broad, solitary or clustered, sessile to subsessile or long stipitate, the stipes to 5 mm tall and furrowed; disk flat to convex, edges sometimes reflexed, reddish brown, sometimes slightly pruinose; pale margin disappearing; hypothecium brownish; hymenium 70–100 μ, hyaline or brownish; asci cylindrical; paraphyses lax, slender, 1–3 μ, slightly branched at the tips which are only slightly thickened; spores 8, spindle-shaped, simple or 1-septate, hyaline, 3–11 × 2–2.75 μ.

On clay soils and frost boils, sometimes on rocks. Circumpolar, arctic and boreal. In North America across Canada and in Alaska, south to Maine, New Hampshire, Minnesota. A comparatively rare species.

Locality: 3

4. **Baeomyces placophyllus** Ach. Thallus forming rosettes, the center squamulose, the margins lobate-foliose, the marginal lobes close to the substratum and raised only at the tips, the main lobes to 5 mm broad, gray green, when dry whitish or ochraceous, somewhat pruinose, the margins with pale whitish isidia or papillae 0.2 mm wide, broadening into squamules; podetia large, to 6 mm tall and 2 mm thick, deeply fissured, naked or covered with squamules; K+ yellow, P+ orange, containing stictic acid. Apothecia to 4 mm broad, one or few per podetium; the margin often reflexed, disk soon swollen, light ochraceous brown to brownish red; hypothecium pale; hymenium 90–150 μ, I-; paraphyses slender, simple or branched toward the tips; asci cylindrical; spores 8, fusiform, 1-celled, hyaline, 8–14 × 2–4 μ.

On sand, clay soil, or humus. Circumpolar, arctic, and boreal, in North America south to New England, the region of Great Slave Lake, and Alaska. A rare species.

Localities: 2, 9.

CLADONIACEAE

Plants with a crustose or squamulose to small foliose primary thallus and erect podetia which bear the apothecia. Podetia hollow and lined with a cartilaginous inner layer. Apothecia usually on the tips or laterally on the podetia, with proper exciple, becoming convex, spores 8, hyaline, simple, ellipsoid.

1. CLADONIA

Primary thallus persistent or disappearing, crustose, squamulose, or foliose, with an upper cortex of more or less dense, conglutinate hyphae. Podetia arising from the upper surface of the primary squamules, erect, subulate, cup-forming, or branched; soon becoming hollow, with an outer cortex which may persist or disintegrate, a medullary layer with algae in the outer part, dense or loose, the inner part forming a cartilaginous layer which may be continuous or forming a network. Apothecia at tips of podetia or on the primary squamules, lecideine, red, pale brown, or dark brown, usually immarginate and convex; exciple radiate; hypothecium pale to reddish; hymenium 30–70 μ thick; paraphyses slender, 2–5 μ, simple or sparingly branched, not thickened or more or less clavate; asci cylindrical to clavate; spores 8, biseriate, fusiform-ovoid to ovoid, simple, hyaline. Pycnidia at apices of podetia, margins of cups, sides of podetia, or on primary squamules; pycnospores cylindrical, filiform straight or curved, 5–14 × 0.5–1 μ. Algae: *Trebouxia*.

1   Podetia much branched and interwoven   2
          Podetia little or not branched, awl-shaped, cup-shaped, or with few branches   12

2 (1) Surface of podetia decorticate and arachnoid, squamules lacking, cups lacking   3
          Surface of podetia corticate, at least in part, and not arachnoid; squamules sometimes present; cups present or lacking   8

3 (2) Podetia P-, lacking fumarprotocetraric acid   4
          Podetia P+ red (fumarprotocetraric acid) or P+ yellow (psoromic acid)   5

4 (3) Podetia forming rounded heads, tips of branches divergent, the branches in dichotomies and equally branched polytomies around open axils; pycnidial jelly red; plant containing usnic acid and accessory perlatolic and pseudonorrangiformic acids   1. *C. stellaris*
          Podetia not forming equally branched systems, having a main branch with smaller side branches; pycnidial jelly colorless; containing usnic acid and accessory rangiformic acid   2. *C. mitis*

5 (3) Thallus forming rounded heads, branching in equal dichotomies or polytomies; containing usnic acid and psoromic acid (P+ yellow)   3. *C. aberrans*
          Thallus not forming rounded heads, branching unequally, P+ red (fumarprotocetraric acid)   6

6 (5) Podetia K+ yellow (atranorin), lacking usnic acid (KC-), gray or whitish gray; branching mainly tetrachotomous   4. *C. rangiferina*
          Podetia K-, lacking atranorin, containing usnic acid (KC+ yellow) yellowish gray to glaucescent greenish   7

7 (6) Branching mainly trichotomous or tetrachotomous   5. *C. sylvatica*
     Branching dichotomous   6. *C. tenuis*

8 (2) Plant K-, lacking atranorin, usnic acid present   9
     Plant K+ yellow, containing atranorin, lacking usnic acid; podetia branching mainly dichotomously, pale gray, older parts dying, surface with scattered areoles over a loose cortex but not arachnoid
     11. *C. thomsonii*

9 (8) P+ red (fumarprotocetraric acid); podetia with branches nearly vertical, surface slightly dull arachnoid, areolate in places   7. *C. alaskana*
     P-, podetia with branches divaricate, surface glossy or dull but not arachnoid   10

10 (9) Lacking hypothamnolic acid   11
     With hypothamnolic acid (S-shaped crystals in GE test)
     10. *C. pseudostellata*

11 (10) Often with distinct but small cups, the cups denticulate; cortex smooth or verruculose; containing usnic and barbatic acids
     8. *C. amaurocraea*
     Lacking cups; branches somewhat spinescent, containing usnic acid with accessory squamatic acid   9. *C. uncialis*

12 (1) Podetia forming cups   13
     Podetia not forming cups, awl-shaped with pointed tips or tipped with apothecia which form heads on the stalk-like podetia   33

13 (12) Apothecia and pycnidia red   14
     Apothecia and pycnidia brown   19

14 (13) Podetia sorediate   15
     Podetia lacking soredia   17

15 (14) Podetia slender, tall, broadening at tips into shallow cups, farinose sorediate, containing usnic acid and bellidiflorin   16
     Podetia short, cups flaring from near the base, granulose-sorediate, containing usnic acid and zeorin   14. *C. pleurota*

16 (15) Containing zeorin, UV-, podetia regular, little split   12. *C. deformis*
     Containing squamatic acid, medulla UV+, podetia with split and torn sides   13. *C. gonecha*

17 (14) Cups short, regular, little squamulose, dying bases not strong yellow
     18
     Cups taller, very squamulose, dying bases very yellow; containing usnic and squamatic acids and bellidiflora   17. *C. bellidiflora*

18 (17) Podetia smooth to with broad, low verruculae; containing usnic and barbatic acids   15. *C. coccifera*
     Podetia very rough with smaller, more protuberant verruculae; containing usnic and squamatic acids and bellidiflorin
     16. *C. metacorallifera*

19 (13) Podetia sorediate   20
         Podetia lacking soredia   26

20 (19) Cups with the edges inrolled, opening to the interior of the podetium, farinose sorediate; containing only squamatic acid   18. *C. cenotea*
         Cup edges not inrolled, interior not opening into the podetium   21

21 (20) Podetia with the soredia in rounded patches toward the upper part, a large part of the base corticate, the cups small on top of very long cylindrical podetia   19. *C. cornuta*
         Podetia mostly covered with soredia, only a small part of the base corticate   22

22 (21) Podetia very slender and tall, the cups small, poorly developed; podetia brownish white to whitish, P+ red (fumarprotocetraric acid); soredia very finely farinose   20. *C. subulata*
         Podetia stouter, shorter, the cups well developed, yellowish, greenish, greenish gray, P+ or P-, soredia coarser, granulose or farinose   23

23 (22) Cups trombone-shaped, short, abruptly flaring on top of slender podetia; soredia farinose   21. *C. fimbriata*
         Cups goblet-shaped, gradually flaring from the base; soredia granular   24

24 (23) Podetia P+ red, KC-, containing fumarprotocetraric acid, inner part of podetia opaque white when soredia are shed; apothecia dark brown   25
         Podetia P-, KC+ yellow, lacking fumarprotocetraric acid, containing usnic acid, inner part of podetia translucent when soredia are shed; apothecia pale brown   24. *C. carneola*

25 (24) Cups tall and stout, 20–40 mm tall, stalk to 3 mm thick   22. *C. major*
         Cups short and stout, to 10 mm tall, stalk to 1 mm thick   23. *C. chlorophaea*

26 (19) Podetia proliferating from the centers of the cups   27
         Podetia proliferating from the margins of the cups   28

27 (26) Podetia K-, P+ red; containing fumarprotocetraric acid   25. *C. verticillata*
         Podetia K+ yellow, P+ red; containing atranorin and fumarprotocetraric acid   26. *C. lepidota*

28 (26) Podetia K+ yellow, containing atranorin   27. *C. ecmoycna*
         Podetia K-; lacking atranorin   29

29 (28) Cups with inside closed to the interior   30
         Cups with the interior open to the inside of the podetium   32. *C. crispata*

30 (29) Cups much split and irregular   28. *C. phyllophora*
        Cups regular, not split   31

31 (30) Podetia tall, with cups flaring rapidly toward tips of podetia, podetia
        olive green   29. *C. gracilis* var. *dilatata*
        Podetia short, cups flaring gradually from the base; podetia bluish
        green or gray green   32

32 (31) Primary squamules thin, not rosette-forming   30. *C. pyxidata*
        Primary squamules thick, leathery, forming rosettes around the
        podetia   31. *C. pocillum*

33 (12) Podetia with red apothecia, K-, P-, lacking thamnolic acid, surface of
        podetium very squamulose   17. *C. bellidiflora*
        Podetia with brown apothecia or apothecia lacking   34

34 (33) Podetia K+ yellow or yellow turning red, containing atranorin   35
        Podetia K-, lacking atranorin   38

35 (34) Sides of podetia torn and fissured, K+ yellow, podetia tipped with large
        dark brown apothecia   33. *C. cariosa*
        Sides of podetia not torn and fissured, podetia with or without
        apothecia   36

36 (35) Esorediate, cortex continuous, P+ red (fumarprotocetraric acid), K+
        yellow   27. *C. ecmocyna*
        Cortex sorediate-granulose   37

37 (36) P+ yellow, K+ yellow turning red (norstictic acid)   34. *C. acuminata*
        P+ yellow, K+ yellow, containing psoromic acid   35. *C. norrlini*

38 (34) Podetia P+ yellow, containing psoromic acid; surface with peltate
        squamules   36. *C. macrophylla*
        Podetia P- or P+ red, lacking psoromic acid, if squamules present they
        are not peltate   39

39 (38) Podetia sorediate   40
        Podetia lacking soredia   45

40 (39) Podetia P+ red, containing furmarprotocetraric acid; medulla UV–
        41
        Podetia P-, lacking fumarprotocetraric acid; medulla UV+   42

41 (40) Podetia whitish or brownish white, mostly decorticate, not squamu-
        lose, branches rising vertically   20. *C. subulata*
        Podetia olive green, corticate and squamulose, only tips sorediate,
        branches widely divergent   37. *C. scabriuscula*

42 (40) Podetia gray in color, lacking usnic acid, KC-   43
        Podetia yellow in color, containing usnic and barbatic acids   44

43 (42) Containing perlatolic acid (tufts of long slender needles in GE)
        38. *C. decorticata*
        Containing squamatic acid (short asymmetric hexagonal crystals in
        GE, brown dendritic crystals in K_2CO_3)   39. *C. glauca*

44 (42) Podetia tall and slender, 20–80 mm   40. *C. cyanipes*
         Podetia short, less than 20 mm   41. *C. bacilliformis*

45 (39) Podetia P+ red, containing fumarprotocetraric acid   46
         Podetia P-, lacking fumarprotocetraric acid   47

46 (45) Podetia 30–80 mm tall, to 1 mm thick   29. *C. gracilis* var. *gracilis*
         Podetia 50–100 mm tall, to 2.5 mm thick   29. *C. gracilis* var. *elongata*

47 (45) Podetia yellowish, KC+ yellow, containing usnic and barbatic acids
         (latter rhombic crystals in GE, no crystals in K_2CO_3)
                                       8. *C. amaurocraea*
         Podetia olive-green, KC-, lacking usnic acid, containing squamatic acid
         (thick, elongate hexagons in GE, brown dendritic crystals in K_2CO_3)
                                       42. *C. subfurcata*

1. **Cladonia stellaris** (Opiz) Pouz. & Vezda (*Cladonia alpestris* (L.) Rabenh.)
Primary thallus crustose, soon disappearing and rarely seen, of dispersed or glomerate granules or verrucules, 0.16–0.28 mm in diameter, pale yellow. Podetia 5–10(–18) cm tall, 1–1.5(–2.5) mm in diameter; forming compact rounded, subglobose heads, these growing singly or in groups or in mats; the branching very dense, isotomic, rarely dichotomous, mainly polytomous, (3–)4–6 branches in a whorl, the branches mainly equal in size but sometimes forming sympodia, spreading widely and showing little tendency to curve in one direction, the axils almost always perforate, the ultimate branchlets in small whorls of (3–)4–6 members around a perforate axil; pale yellowish to whitish, darker gray in deeper woods, the base blackening; the surface dull, arachnoid to subtomentose with the dispersed algal areolae vaguely defined, the older parts becoming translucent as the outer layers disperse; P-, K-, KC+ yellow; containing usnic acid and accessory perlatolic and pseudo-norrangiformic acids. Apothecia small, on the tips of subcorymbosely grouped branchlets at the tips of the podetia; brown or dark brown. Pycnidia at the tips of branchlets and containing a red jelly.

     This species grows on soil and humus, occasionally on old wood and stumps and thin soil over rocks. It is common in lichen woodlands. It is circumpolar, in North America ranging across Canada and Alaska and south to West Virginia, the Great Lakes region, Wyoming, and California.

     Localities: 8, 11.

2. **Cladonia mitis** Sandst. Primary thallus unknown. Podetia in large mats or in tufts; intricately, loosely branched 30–70(–100) mm tall and 1–1.5(–4) mm in diameter; the anisotomic branching typically in whorls of 3 or 4, sometimes dichotomous, forming a sympodium which has

more remote nodes than in *C. arbuscula*, each node with 1–2 short branches ramified in turn in the same unequal fashion, toward the tips the branching pattern becoming equal, the sterile branchlets ending with 2–5 points all recurved to one side, or in the case of unequal branches with some recurved and others spreading; the axils mainly open; the fertile branchlets mainly in short corymbs, the entire appearance less tufted and less well combed than in *C. arbuscula*, clear yellow or with a bluish cast varying toward glaucescent, rarely to sulfur yellow, the extreme tips scarcely browned; the surface dull, scarcely arachnoid-tomentose, smooth or verruculose with age; mainly impellucid, rarely semipellucid; normally lacking soredia; K-, P-, KC+ yellow; containing usnic acid plus accessory rangiformic acid. Apothecia small, borne singly or more or less in corymbs at the tips of the branches; brown. Pycnidia at the tips of the branches; the jelly colorless.

This species grows on sandy soils, humus, in bogs and on sand plains. It is especially common in lichen woodlands. In the northern hemisphere it is circumpolar, ranging across Canada and Alaska south to Virginia, the Great Lakes states, Colorado, and Oregon.

Localities: 3, 8, 11, 14.

3. **Cladonia aberrans** (Abb.) Stuck. Primary thallus unknown. Podetia to 10 cm tall, to 1.5 mm in diameter; forming compact, rounded, subglobose heads which grow singly, in clumps or in mats; branching dense, isotomic, rarely dichotomous, mainly polytomous, 3–6 branches in a whorl, mainly equal in size but sometimes forming sympodia, spreading widely and showing little tendency to curve in one direction, axils perforate, ultimate branchlets in small whorls of 3–6 branches around a perforate axil; pale yellowish to whitish, the base blackening; surface dull, arachnoid to subtomentose, the dispersed algal areoles vaguely defined, older parts becoming translucent; K-, KC+ yellow, P+ yellow, containing usnic and psoromic acids. Apothecia small, on the tips of ultimate branchlets, brown. Pycnidia at tips of branchlets and containing red jelly.

Growing on soil and humus. A circumpolar species known from Europe, North America, and Asia. It is often considered a P+ yellow, psoromic acid containing, strain of *C. stellaris*.

Locality: 14.

4. **Cladonia rangiferina** (L.) Web. Primary thallus soon disappearing and seldom seen; crustose, thin, of subglobose verrucules 0.22–0.4 mm in diameter; ashy white. Podetia in tufts or in extensive mats; usually robust, 5–10(–15) cm tall and 1–1.6(–3) mm in diameter; the branch-

ing anisotomic dichotomous, trichotomous or tetrachotomous, the branches rebranched, the tips of sterile branchlets with whorls of 2–3 branchlets around an open axil and recurved to one side, the tips of fertile branches in a corymbose arrangement; the axils open; ashy gray, more gray than the color of the other Cladinae, becoming whitish or pale greenish in shaded places, occasionally blackened in the sun, the base blackening on dying; the surface markedly arachnoid-tomentose with the greenish areolae scattered and more or less verruculose, especially toward the base, the young parts more or less smooth; usually lacking soredia; usually not pellucid; K+ yellow, KC-, P+ red, containing atranorin and fumarprotocetraric acid. Apothecia small; solitary or in corymbs at the tips of the branchlets; dark brown or black. Pycnidia at the tips of the branchlets; the jelly colorless.

Growing on sandy soils, soils rich in humus, over rock outcrops and abundant in lichen woodlands. A circumpolar arctic and temperate species, in North America trans-Canadian and in Alaska, south to Georgia, Alabama, Arkansas, Montana, and Oregon.

Localities: 1, 2, 3, 6, 8, 11.

5. Cladonia arbuscula (Wallr.) Rabenh. (*C. sylvatica* (L.) Hoffm.) Primary thallus usually disappearing and seldom seen, crustose, of subglobular verrucules 0.12–0.48 mm in diameter, pale yellow. Podetia forming tufts or in larger mats; robust, 50–100(–150) mm tall and 1–2(–4) mm in diameter; the branching typically anisotomic, tetrachotomous or trichotomous, rarely dichotomous, the whorls of unequal branching so that the podetium forms a sympodium with unilateral growths, especially in the lower part, the upper part showing more equal branching, the uppermost branchlets terminated by 2–3(–5) short, stout points which are recurved all toward one side, the fertile branchlets in a corymbose arrangement; the axils mainly open, sometimes widely so; pale yellow, becoming greenish glaucescent in more moist and shady habitats; the surface dull, little arachnoid-tomentose, smooth or rarely verruculose, lacking soredia (except in a rare form), impellucid; K-, KC+ yellow, P+ red, containing usnic and fumarprotocetraric acids plus accessory ursolic acid. Apothecia small, solitary or grouped at the tips of corymbose branchlets; dark brown. Pycnidia at the tips of the branchlets; the jelly colorless.

A species of sandy soil, or humus soil in bogs, lichen woodlands and open tundras. Circumpolar, arctic and temperate, in North America trans-Canada and in Alaska, and south to Maryland, Wisconsin, and Washington.

Localities: 3, 8, 11.

6. **Cladonia tenuis** (Flörke) Harm. No primary thallus. Podetia in mats, 5–8 cm tall, to 1.5 mm diameter, branching anisotomic dichotomous to form sympodia, tending to curve to one side, axils perforate or not, pale yellow or greenish yellow, P+ red, KC+ yellow, K-, containing usnic and fumarprotocetraric acids, the surface dull, little arachnoid, more or less smooth, the greenish areolae very slightly visible. Apothecia small, at the tips of branchlets, brown. Pycnidia at branchlet tips containing red jelly.

A species of soil and over rock outcrops, known from Europe, the Himalayas, eastern North America, and western North America. In the latter it is mainly near the northwest coast and in the Aleutian Islands.
Locality: 8.

7. **Cladonia alaskana** Evans Primary thallus unknown. Podetia forming compact tufts; dying at the base and growing from the apices; branching irregularly by dichotomies or in whorls of 3 or 4, the branches ascending closely to each other; axils open or closed; podetial wall often split; 40–80 mm tall and up to 2 mm in diameter; the surface not at all glossy, somewhat arachnoid under a lens; pale gray tinged with yellowish or brownish tints, darker in older parts; the cartilaginous layer forming a continuous hollow cylinder around the central canal and forming a series of irregular bulges projecting into this canal; this layer sharply delimited from the outer medullary layer in which the algal clusters are located near the periphery in a single discontinuous layer covered by the cortex; few small squamules on the podetia, up to 1 mm long and undivided or little lobed, rounded at apex; branches bearing the apothecia larger than the sterile branches and irregularly divided and split in the upper part, thus perforate at the apices but without actual cups; K+ yellowish turning to gray, KC+ yellow, P+ red; containing usnic and fumarprotocetraric acids plus accessory ursolic acid. Apothecia 1–2 mm in diameter; clustered or confluent around the apical perforations; pale to dark brown. Pycnidia at the apices of the podetia.

This species grows on soil. It is North American and known from northern Alaska eastward to Aberdeen Lake on the Thelon River in the Northwest Territories. Quite probably it will eventually be collected in Siberia and then would be known as an amphi-Beringian species.
Locality: 11.

8. **Cladonia amaurocraea** (Flörke) Schaer. Primary thallus rarely seen, usually disappearing, of small squamules up to 1.7 mm broad and long; crenate or digitate incised; ascending; flat; sparse or in tufts; upper side glaucescent; underside white; esorediate. Podetia from the older dead

parts or from fragments of old podetia lying on the ground, rarely from the primary squamules; in groups or in mats, the marginal podetia often more or less decumbent and the tips then curved upwards; 15–120 mm tall and up to 1.5(–3.5) mm in diameter; dichotomously or sympodially branched, occasionally branching irregularly or from the margins of the cups in fascicles; often cupless and cylindrical; sometimes with narrow cups which flare rapidly; the interior membrane closed, or open into the interior by a single or several small holes; the axils are closed or slightly perforated; esorediate; the cortex continuous or more commonly dispersed areolate, the areolae little raised to rarely verruculose; shining or dull; impellucid; yellowish or yellowish glaucescent to glaucescent or olive green; the margins of the cups with tapering spine-like proliferations and the tips of the sterile branches also tapering to spinescent tips; inner cartilaginous part of the medulla continuous, not sharply delimited from the outer part of the medullary layer; K-, KC+ yellowish, P-; containing usnic and barbatic acids. Apothecia solitary or clustered at the tips of the podetia; middle sized, up to 2(–3.5) mm in diameter; brown or yellowish brown or reddish brown. Pycnidia on the margins of the cups or the tips of the podetia.

Growing on peaty, sandy, or bog soils, sometimes among boulders or on hummocks. A circumpolar species in North America across Canada and Alaska south to New York, Michigan, Wisconsin, and Minnesota.

Localities: 1, 2, 4, 6, 7, 8, 9, 11, 13, 14.

9. Cladonia uncialis (L.) Web. in Wigg. Primary squamules rarely seen, usually disappearing; small, up to 1 mm long; broad crenate or incised crenate or rarely incised; ascending; flat; sparse or tufted; the upper side yellowish glaucescent; the underside white. Podetia appearing to arise from above the remains of dead podetia, or from fragments of other podetia; the base dying and the growth continuing from the apices; 20–80(–110) mm tall and 1–1.5(–4) mm in diameter; subcylindrical; cupless; dilated at the axils; dichotomously or sympodially or verticillately branched, repeatedly branched, the whole forming dense mats or tufts; most of the axils usually perforate; the ultimate branchlets spinelike and in small whorls around open perforations; the inner cartilaginous layer continuous; the inner surface smooth; the outer surface smooth and shiny, sometimes with the greenish areolae projecting slightly; light green or deep or pale yellowish; the podetia very brittle; K-, KC+ yellowish, P-; containing usnic acid plus accessory squamatic acid. Apothecia small; up to 0.8 mm in diameter; brown; at the apices of the branches in usually a radiate or subcymose arrangement of the branchlets; pycnidia at the tips of the branchlets.

A species of sandy soils and among mosses in bogs. Cosmopolitan, growing trans-Canada and in Alaska, south to Georgia and Arkansas, and in Washington.

Localities: 1, 2, 3, 4, 8, 11, 13, 14.

10. **Cladonia pseudostellata** Asah. Primary thallus lacking. Podetia in tufts or mats, 30–60 mm tall, 1–4 mm diameter, cupless, sympodially branched with the upper branches whorled, the branchlets short, dichotomous or in whorls, spinescent, the axils usually perforate, surface smooth, straw colored to yellowish, containing usnic and hypothamnolic acids, K-, KC+ yellowish, P-; inner cartilaginous layer continuous, 40–50 μ thick, the outer cortex 30–40 μ thick.

This species grows on soils rich in humus. It is known from Japan and Alaska, in the latter being known from the Aleutian Islands, the Brooks Range, the Alaska Range, and Wickersham Dome northwest of Fairbanks.

Locality: 8.

11. **Cladonia thomsonii** Ahti Primary thallus unknown. Podetia dying at the base and growing from the apices; in dense tufts; about 50–70 mm tall and 2–4(–5) mm in diameter; the upper branches quite slender, up to 1.2 mm in diameter; cupless; subcylindrical dichotomous or subsympodial in branching; the axils perforate; tips brown, obtuse; the outer cortex smooth, subcontinuous or chinky to areolate or verruculose, rarely with dispersed areas, dull, ashy white or scarcely yellowish or brownish, the dead part becoming dark brown; the cortex of subdistinct hyphae and about 15–25 μ thick; the medullary layer with scarce glomerules of algae, the inner cartilaginous layer well developed, continuous and distinctly limited from the interior of the medulla; K+ yellow becoming orange, KC-, P-; containing atranorin plus an unknown acid. Apothecia and pycnidia unknown.

Growing on gravelly tundras. An Alaskan species distinguished by different secondary substances from *C. wainii* Savicz.

Localities: 7, 8, and Dexter Creek 6 mi N of Nome, *Pegau 84*.

12. **Cladonia deformis** (L.) Hoffm. Primary squamules usually disappearing, small to middle-sized, 2–4 mm long, incised crenate or lobate, flat to convex or involute, upper side glaucescent to yellowish, underside white to pale brown to dark, esorediate or the underside sparingly sorediate. Podetia arising from the upper side of the squamules, 25–85 mm tall and up to 5 mm thick on the stalk; the cup gradually flaring, moderately or little expanded, the margin entire or dentate to prolifer-

ate with blunt or cupbearing proliferations, the cups usually imperforate, the major portion of the podetium, except for the base, being densely covered with fine, farinose soredia, the soredia quite yellowish; the basal corticate portion continuous to chinky, yellowish glaucescent, the decorticate portion opaque; K-, KC+ yellow, P-; containing usnic acid, zeorin, and accessory bellidiflorin. Apothecia scarlet or pale, on the margins of the cups or on short marginal proliferations. Pycnidia on the margins of the cups or on the apices of the marginal proliferations.

Growing on soil rich in humus, among mosses, on rotting logs, tree bases, and over rocks with thin soil. A circumpolar species in the arctic and boreal forest, south in North America to West Virginia and Colorado in the mountains.

Localities: 1, 2, 3, 8, 11, 13, 14.

13. Cladonia gonecha (Ach.) Asah. Primary squamules larger than in *C. deformis*, 5-10 mm long and 2.5-5 mm wide, incised or lobate, upper side yellowish glaucescent, underside white to pale brown or dark, sometimes sorediate. Podetia about the same length as in *C. deformis*, about 25-85 mm tall, but stouter and the cups more irregular, the marginal proliferations more prominent and the sides more often fissured and split; the surface covered, except at the base, with farinose yellowish soredia; the basal corticate area as in *C. deformis*; K-, KC+ yellowish, P-; containing usnic and squamatic acids and accessory bellidiflorin. Apothecia and pycnidia similar to those of *C. deformis*.

Growing on soils rich in humus, on rotting wood and tree bases and on thin soil over rocks. A circumpolar northern species growing trans-Canada and in Alaska, south to New York, Michigan, Colorado, and British Columbia.

Localities: 3, 8, 11.

14. Cladonia pleurota (Flörke) Schaer. Primary squamules persistent or disappearing; small to large, 1-7 mm long, up to 5 mm broad, irregularly crenate-incised to lobate; upper side yellowish to olivaceous or pale glaucescent, underside pale or brownish toward the base; esorediate or with scattered granules below. Podetia varying to 40 mm tall but usually much less; cup-bearing, the cups flaring quite soon and gradually from the base, goblet-shaped, regular and entire or dentate to proliferate from the margin, the proliferations bearing apothecia or rarely small cups, very rarely with small cups proliferate from the center; the inside of the cups and the upper part of the podetia with granular soredia, the base corticate and the cortex continuous or areolate to verruculose; K-, KC+ yellow, P-; containing usnic acid, zeorin and accessory bellidiflorin.

Apothecia scarlet, on the margins of the cups or on stipes from the margins of the cups, convex. Pycnidia usually on the margins of the cups, rarely on the upper side of the primary thallus.

Growing on soil and on thin coating of soil over rocks. Circumpolar and also in southern hemisphere. It ranges south to South Carolina, Texas, Colorado, and California.

Localities: 2, 8, 11.

15. Cladonia coccifera (L.) Willd. Primary squamules persistent or disappearing; small, or becoming as large as 12 mm × 5 mm, irregularly crenately incised or lobed; yellowish to glaucescent or olive colored, underside white or rarely yellowish, the base becoming orange to blackish brown; esorediate. Podetia becoming up to 50 mm tall but usually shorter; the cups goblet-shaped, flaring gradually, often oblique at the mouth, simple or with marginal proliferations which bear apothecia or more rarely cups in a second tier; corticate, the cortex yellowish to glaucescent, subcontinuous at the base and becoming verruculose areolate above, the verrucules contiguous or dispersed; esorediate, with or without squamules, the decorticate part opaque, white or yellowish white; K-, KC+ yellow, P-; containing usnic and barbatic acids. Apothecia large, scarlet. Pycnidia occurring especially on the margins of the cups but also on the tips of proliferations from the margins.

Growing on soil, especially upon humus in muskegs, over rocks, and in rotting wood. A circumpolar arctic and boreal forest species ranging south to Quebec, Michigan, and Colorado.

Localities: 1, 2, 3, 4, 5, 6, 7, 8, 11, 13.

16. Cladonia metacorallifera Asahina Primary squamules persistent, small, 1-2 mm long, 2-4 mm broad, crenate or incised, becoming involute, sparse or dense, appressed or ascending; esorediate; upperside yellowish to yellowish green, underside white. Podetia cup-forming, the cups flaring gradually from the base and globlet-shaped or from a long base with an abrupt flare or more trumpet-shaped; yellowish; esorediate, but most of the surface covered with crowded verruculae which are smaller and more protuberant than in *C. coccifera*, many showing a tendency to rise from the deeper tissues and give rise to minute, more or less appressed squamules; margins of the cups subentire, rarely proliferate; K-, KC+ yellow, P-; containing usnic, squamatic, didymic acids and bellidiflorin. Apothecia scarlet, on marginal stipes or short proliferations. Pycnidia on the margins of the cups.

Growing on earth containing humus. Described from Asia, South America, and in North America in Alaska, Alberta, and Saskatchewan.

Localities: 7, 8, 14.

17. **Cladonia bellidiflora** (Ach.) Schaer. Primary squamules commonly disappearing, small or middle-sized. Podetia 20–50 mm tall, the base commonly dying; squamulose, the squamules small to middle-sized, 2–5 mm long, crenate or laciniate, ascending or erect, flat or involute, sparse to abundant, upper side yellowish or pale to glaucescent; lower side white, becoming brown to black toward the base. Podetia cupless or with short and narrow cups which are quite regular or oblique, margins subentire or dentate to radiate proliferate, esorediate; cortex continuous to chinky or areolate discontinuous, smooth or verruculose, squamulose, the squamules splitting off the cortex, the cortex yellowish or yellowish glaucescent, the decorticate part white to faintly yellowish, opaque. K-, KC+ yellow, P-; containing usnic and squamatic acids and bellidiflorin. Apothecia scarlet, on proliferations or on the edges of the cups. Pycnidia on the apices of proliferations or the margins of the cups.

Growing on soil and humus. A circumpolar species also growing in the southern hemisphere. In North America it is trans-Canadian and Alaskan, south to Newfoundland in the east, California in the west.

Localities: 1, 4, 6.

18. **Cladonia cenotea** (Ach.) Schaer. Primary squamules persistent or disappearing; small, 1–3(–5) mm long and 1 mm broad; thick; irregularly or subdigitately divided; the lobes crenate; ascending; flat or involute; scattered or tufted; the upper side glaucescent or pale to olive green or brownish; the underside white; esorediate or the underside becoming sparsely sorediate. Podetia arising from the upper side of the primary squamules; the base persistent or dying; 5–100 mm tall and up to 5 mm in diameter; usually cup-bearing, the cups flaring gradually or abruptly, distinctly perforate and the margins inrolled into the perforation, the margins usually oblique, proliferate, with the proliferations acute or bearing small secondary cups, the podetia not usually branched below the cups; the cortex disintegrating over most of the upper part of the podetium, becoming farinose-sorediate; whitish or variegated or brownish or greenish with the fine soredia; the lower portion corticate and more or less squamulose; the dying parts becoming black; K-, KC-, P-; containing squamatic acid. Apothecia small; livid brown or waxy pale brown; on the margins of the cups or the tips of the irregular proliferations, subsolitary, rarely conglomerate. Pycnidia on the margins of the cups.

On soil rich in humus, old wood, stumps, and thin soil over rocks. An arctic-alpine circumpolar species also reported from Australia. In North America it is trans-Canadian and Alaskan, and south to West Virginia, Wisconsin, Colorado, and Oregon.

Locality: 3.

19. **Cladonia cornuta** (L.) Hoffm. Primary squamules usually disappearing; middle-sized, 3–8 mm long, 1–4 mm broad; irregularly lobed, sinuate or crenate or incised-crenate; ascending; upper side glaucescent or olive green; underside white; esorediate or rarely with the underside sparsely granulose. Podetia arising from the upper side of the primary squamules; tall and slender, 20–120(–160) mm tall and up to 2.5(–5) mm in diameter; usually cupless and cylindrical or tapering; cups, when present, very narrow, 2–4 mm broad, gradually or abruptly flaring, the interior usually closed, the margins proliferating or simple; the tips tapering; the cortex continuous to subareolate with smooth areoles; the upper part of the podetium becoming sorediate in definite patches with farinose soredia or the entire top becoming covered with soredia; usually without squamules but in one form with squamules similar to the primary squamules on the podetia; K-, KC-, P+ red; containing fumarprotocetraric acid. Apothecia on the margins of the cups or rarely on the apices of podetia; brown or reddish brown or pale; middle-sized, up to 4 mm broad; subentire or perforated or conglomerate. Pycnidia on the margins of the cups.

Growing on sandy soil or more commonly on soil rich in humus, on rotting logs and wood, especially common in sheltered spots as willow thickets. A northern circumpolar species also known in the southern hemisphere. In North America south to the northern border states, Pennsylvania, and in Colorado.

Localities: 1, 3, 8, 13, 14.

20. **Cladonia subulata** (L.) Wigg. Primary squamules persistent or often disappearing; small; upper side whitish glaucescent to blackening, the underside white. Podetia from the upper side of the primary squamules; tall and slender, 30–100 mm tall and up to 3.5 mm thick; cylindrical; cupless or with irregular cups which are formed by circles of long proliferations; with a slight amount of cortex at the base, or more usually totally decorticate and covered with very fine farinose soredia; white to ashy or pale glaucescent or with brownish variegation in the ashy coloring; K-, KC-, P+ red; containing fumarprotocetraric acid. Apothecia rare; sessile on the margins of the cups or else on stipes from the margins or on the tips of the podetia; dark brown, brown or reddish brown. Pycnidia on the margins of the cups or more often the tips of the proliferations or podetia.

Growing on soil rich in humus, peats, and rotting wood and logs. A circumpolar species also known in the southern hemisphere. It grows southward in North America to North Carolina, Wisconsin, Iowa, Colorado, and California.

Localities: 8, 11, 14.

21. **Cladonia fimbriata** (L.) Fr. Primary squamules persistent or disappearing; middle-sized, 2–10 mm long; digitately or irregularly lobed; crenate to sinuate or incised on the margins; flat to involute or concave above; the upper side glaucescent to olive green or pale glaucescent; the underside white; esorediate, or sparsely granulose sorediate below. Podetia from the upper side of the primary squamules; 10–20 mm tall, the stalk 1–2 mm thick; cup-bearing; the cups flaring rapidly to form trumpet shapes, entire margined or rarely proliferate, the interior closed; the surface sometimes slightly corticate at the base but mainly decorticate and farinose sorediate; K- or K+ brownish, KC-, P+ red; containing fumarprotocetraric acid. Apothecia brown; on the margins of the cups or on short stipes from the margins. Pycnidia from the margins of the cups.

Growing on soil, rotten logs, and earthen banks. A circumpolar arctic and temperate species also known from the southern hemisphere. In North America the range is from Alaska to Labrador, south to North Carolina, Iowa, Utah, and California.

Localities: 2, 8, 11, 13, 14.

22. **Cladonia major** (Hag.) Sandst. This species is similar to the preceding, C. fimbriata except that the podetia are 20–40 mm tall and the stalks are up to 3 mm thick; K- or K+ yellowish in young parts, KC-, P+ red; containing fumarprotocetraric acid.

On earth, earthen banks, and rotting wood. A species reported from Europe and North America, in the latter ranging from Alaska to Newfoundland, south to North Carolina, Wisconsin, Nebraska, Colorado, and California.

Localities: 3, 7.

23. **Cladonia chlorophaea** (Flörke) Spreng. Primary squamules persistent or disappearing; small to middle-sized, 4–7(–15) mm long, almost as broad; incised crenate or laciniately lobed, margin crenate; ascending; more or less involute-concave; esorediate or with sparse granulose soredia on the underside; upper side glaucescent or olive glaucescent or pale glaucescent; dull to rarely shining; underside white, darkening toward base. Podetia cup-bearing; the cups flaring gradually, goblet-shaped, regular or irregular and with proliferations from the edges; the base of the podetium corticate; the cortex areolate or verruculose; glaucescent or ashy to olivaceous, the upper part of the podetium and the inner surface of the closed cups decorticate and granulose sorediate, the decorticate areas becoming white, if sterile the margins of the cups usually entire; K-, KC-, P+ red; containing fumarprotocetraric acid. Apothecia sessile, or stipitate from the margins of the cups; brown.

Pycnidia on the margins of the cups or on the primary squamules.

A species growing on a wide variety of substrates, earth, rotting logs, tree bases, thin soil over rocks, and humus. A cosmopolitan species found in all states and Canada.

Localities: 1, 3, 8, 9.

24. **Cladonia carneola** Fr. Primary squamules persistent or disappearing; middle-sized to large, up to 13 mm long and 10 mm wide but usually much less, broadly lobed or narrowly laciniate, crenate, flat or convolute, ascending; upper side yellowish glaucescent or yellowish to olive green; underside white or yellowish, darkening toward the base; esorediate or with the underside sparsely granulose. Podetia arising from the primary squamules, up to 40 mm tall but usually much less; cup-bearing, the cups tapering from the base much like those of *C. deformis* or more abruptly flaring; entire margined or with proliferations from the margin, the interior of the cup not opening into the interior of the podetium; the base or varying heights of the podetium corticate, the cortex subcontinuous to areolate; the rest of the podetium yellowish farinose sorediate or rarely granulose sorediate, usually esquamulose or with a few squamules at the base; yellowish, or yellowish glaucescent; K-, KC+ yellow, P-; containing usnic acid plus accessory barbatic acid. Apothecia middle-sized to large, up to 6 mm broad, on stipes or sessile around the margins or from the centers of the cups; pale, waxy yellowish or pale brown. Pycnidia from the margins of the cups or on the short stipes around the edges of the cups.

Growing on soil rich in humus and on rotting wood. A northern circumpolar species ranging in North America from Alaska and trans-Canada south to the American border, and in Washington and Oregon.

Localities: 7, 13.

25. **Cladonia verticillata** (Hoffm.) Schaer. Primary squamules persistent or disappearing; varying from small to large, depending on the form and the substratum, up to 8 mm long and 4 mm broad; irregularly wedge-shaped or lobed, the lobes crenate or slightly incised, flat or convolute, often ascending, rarely caespitose; the upper side olive green or reddish or brownish glaucescent or slaty green; underside white, or black toward the base; esorediate. Podetia from the upper side or the margins of the primary squamules, up to 50 mm tall and 3 mm in diameter, flaring quickly at the tops into short, broad cups, up to 9 mm across, shallow and with small, pointed or cup-bearing proliferations growing from the centers of the closed cups; the margins entire or with apothecia or pycnidia; sometimes with several tiers of cups, each arising

from the center of the previous tier; the cortex continuous or chinky areolate, the areolae smooth, subcontiguous, the narrow interspaces white, dull, whitish, green to olivaceous or ashy or bluish green or even quite brown; esorediate; with or without squamules; K-, KC-, P+ red; containing fumarprotocetraric acid. Apothecia sessile or on short stipes; small to middle-sized, usually less than 3 mm broad; brown or reddish brown; rounded; broader than the stipes which support them. Pycnidia in the centers of the cups or on the margins.

Growing on soil, earthen banks, on thin soil over rocks, and rotting wood. A cosmopolitan species found in all of North America except Florida.

Localities: 3, 8, 11.

26. **Cladonia lepidota** Nyl. Primary squamules persistent or disappearing; small to large, up to 20 mm long reported but much less in American material; rounded to wedge-shaped; crenated or entire; flat, the upper side whitish glaucescent to glaucescent or olive green; underside white, blackening at the base; esorediate. Podetia arising from the upper side of the primary squamules; 60–100(–150) mm tall, up to 1.5 mm broad, rarely 2 mm; with narrow cups, flaring gradually or abruptly, at the tips 2–5 mm broad, with tiers from the centers or from the margins of the cups; inner surface of cups closed or becoming cribrose; shallow; the margins becoming split or denticulate; cortex continuous or becoming areolate-chinky, the areoles slightly raised, more or less dispersed, the narrow decorticate part more or less distinctly subtomentose, whitish glaucescent to pale ashy brown, with the base blackening but with white and black corticate spots at the base as in *C. phyllophora*; esorediate; squamulose and with grouping of squamules toward the edges of the cups; K+ yellow, KC-, P+ red; containing atranorin and fumarprotocetraric acid. Apothecia on the margins of the cups or on short stipes, brown or reddish brown. Pycnidia on the margins of the cups or denticulations.

Growing on soil rich in humus, in moist places among boulders and near late snow banks in the tundras. A far northern and alpine species, circumpolar, in North America known from northern Canada, Alaska, Quebec, Labrador, and in Colorado.

Localities: 3, 7, 8, 9, 11.

27. **Cladonia ecmocyna** (Ach.) Nyl. Primary squamules disappearing. Podetia with the base dying and growth continuing from the top, with occasional branching to form large tufts; 50–100 mm tall or more; up to 2.5–5 mm in diameter; subsimple or with submonopodial branching;

cupless or with narrow cups, with marginal or with occasional central proliferations which are long extended; the cortex continuous or sub-continuous, or areolate with smooth areolae, dull or nitid, glaucescent or whitish or pale to olive green, or brownish in strong light; K+ yellow; KC-, P+ red; containing atranorin and fumarprotocetraric acid. Apothecia brown; on the margins of the cups. Pycnidia on the margins of the cups.

This species prefers moister habitats than the closely related C. gracilis, occurring near late snow banks and in bogs with soils of high humus content. It is circumpolar and in South America. The North American range is arctic-alpine south to Maine, New Hampshire, and Colorado. Localities: 1, 3, 8, 9, 11.

28. **Cladonia phyllophora** (Ehrh.) Hoffm. Primary squamules persistent or disappearing; small to middle-sized, rarely large, 2–5 mm long, rarely up to 15 mm, the breadth up to 4 mm, irregularly broadening from the base, crenate or incised, sinuate, flat or involute, ascending; upper side glaucescent or olive glaucescent; underside white or the base becoming black; esorediate. Podetia from the upper surface of the primary squamules; 8–80 mm tall and up to 4 mm in diameter; cup-bearing, the cups irregular, rarely regular, sometimes scarcely recognizable as cups; the interior closed, the margins commonly proliferate, the sides with a continuous cortex or becoming areolate with small, hardly elevated areoles, the part between the areoles decorticate and subtomentose, in the older parts the patches of cortex appearing as whitish or pale spots in a dark background; esorediate, with or without squamules, if present, the squamules small and similar to those of the primary type; K-, KC-, P+ red; containing fumarprotocetraric acid. Apothecia small to middle-sized, up to 2 mm in diameter, on the tips of the proliferations from the margins of the cups, sometimes fastigiately clustered but usually not; dark brown, rarely pale brown or reddish brown. Pycnidia on the margins of the cups or the tips of the proliferations.

Growing on sandy soil or soil rich in humus. A northern circumpolar species, in North America trans-Canadian and Alaskan south to West Virginia, Wisconsin, Minnesota, Colorado, and Oregon.

Locality: 7.

29. **Cladonia gracilis** (L.) Willd. Primary squamules persistent, or persistent only at the base of the podetia, or disappearing; small to middle-sized, rarely large, 2–5(–10) mm long, 0.8–6 mm broad; irregularly lobed or crenate or sinuate, flat or convolute, ascending; upper side glaucescent or olive green, more greenish than in C. verticillata which tends to be

bluish; underside white. Podetia arising from the upper side of the primary squamules, very variable, the size depending partly upon the variation collected, up to 140 mm tall; cup bearing in most forms, subulate in some, the inner membrane of the cups closed, cups shallow and flaring quite rapidly, usually proliferate from the margins, the proliferations bearing apothecia, or cups, or occasionally subulate; esorediate and with a smooth cortex which is continuous or of slightly dispersed flat areolae which are separated by smooth white bands or lines, varying in color from greenish glaucescent to olive or in sun forms to brownish green or brown, the base becoming black but not spotted; squamulose or not; K-, KC-, P+ red; containing fumarprotocetraric acid. Apothecia varying from small to large; on the margins of the cups or on marginal proliferations; dark brown. Pycnidia on the margins of the cups.

Growing on soil, sandy or high in humus, in bogs, thin soil over rocks, and on rotting wood. An almost cosmopolitan species represented in the forested regions of North America by var. *dilatata* and in the tundras additionally by vars. *gracilis* and *elongata*. The species ranges south to Virginia, Kansas, and Colorado.

Localities: var. *gracilis* 1, 2, 3, 4, 5, 11, 13; var. *dilatata* (Hoffm.) Vain. 14; f. *amaura* (Flörke) Aigr. 7; var *elongata* (Jacq.) Fr. 2, 11.

Var. *gracilis* has tall slender unbranched podetia tapering to points and about 30–80 mm tall; var. *elongata* is similar but taller and thicker, 50 to 100 mm tall and 2.5 mm thick; f. *amaura* is similar to var. *gracilis* but deeply browned, probably a sun exposure ecotype; var. *dilatata*, the forest type, has cups which usually are proliferated at the margins and bear apothecia.

30. **Cladonia pyxidata** (L.) Hoffm. Primary squamules persistent, rarely disappearing; small to medium-sized, 2–7(–15) mm long and up to 4 mm broad; irregularly lobed or incised, the tips rounded, the sides crenate or sinuate; quite thick; ascending or appressed; upper side glaucescent to pale olive green or brownish; underside white, darkening at the base; esorediate. Podetia growing from the upper side of the primary squamules, 4–40(–70) mm tall; simple or with short marginal proliferations bearing apothecia, cup-bearing, the cups flaring gradually and goblet-shaped; quite deep, the interior closed and decorticate in part with small peltate squamules covering the interior as well as the outer side; slaty gray to olive green or with brownish shades; K-, KC-, P+ red; containing fumarprotocetraric acid. Apothecia uncommon, small to quite large, borne on the margins of the cups or on short stipes on the margins; brown or reddish brown, rarely pale. Pycnidia on the margins of the cups or on the upper side of the primary squamules.

On mineral soil, humus, rotting logs, and over wood, usually among mosses. A cosmopolitan species, abundant northwards in North America and ranging south to South Carolina, Kentucky, Wisconsin, Nebraska, Colorado, and California.

Localities: 1, 2, 8, 13.

31. Cladonia pocillum (Ach.) O. Rich. Primary squamules thick, contiguous, forming little appressed rosettes around the groups of podetia which are usually sterile; tips rounded, upper side glaucescent to olive green or brownish; underside white, darkening at the base, esorediate. Podetia growing from upper side of the squamules toward the center of the thallus, to 20 mm tall, simple or rarely with short marginal proliferations bearing apothecia, cup-bearing, the cups flaring gradually and goblet-shaped, the interior closed and with peltate squamules lining the cup, the exterior also with peltate squamules which are coarse, slaty gray to olive green, or brownish; K-, KC-, P+ red; containing fumarprotocetraric acid. Apothecia brown, on the margins of the cups or on short stipes on the margins. Pycnidia on the margins of the cups or the upper side of the primary squamules.

A species of calcareous earths and in the open. Probably cosmopolitan but not always differentiated from C. *pyxidata*. More common than C. *pyxidata* in the north and on the Great Plains.

Localities: 1, 2, 3, 4, 5, 6, 8, 9, 10, 11, 12, 13, 14.

32. Cladonia crispata (Ach.) Flot. Primary squamules persistent or disappearing, to 4 mm long, 0.5 mm broad, digitate lobed, crenate edged, ascending, flat or involute, scattered or in tufts; upper side greenish to olive brownish; underside white or the base brown. Podetia to 110 mm tall, 5 mm diameter, usually less, cupless or with cups, the cups perforate and open directly into the interior of the podetium, flaring abruptly, or cupless and the branches radiately or sympodially developed; the axils commonly perforate; tips blunt or subulate, or with brown apothecia, the cortex shining or dull, continuous, smooth or with slight areolae, not translucent, bluish or pale olive green to brown or reddish brown, the dying parts black, K-, C-, KC-, P-, containing only squamatic acid.

This species grows on earth, fallen logs, old stumps and over rocks. It is arctic to temperate and circumpolar.

Locality: 8.

33. Cladonia cariosa (Ach.) Spreng. Primary squamules persistent, small to middle-sized, 1-7 mm long and 1-2 mm broad, rarely larger, a little coarser and larger than those of C. *capitata*, irregularly divided,

crenate, incised or rounded, flat or concave, sometimes reflexed, confluent or sometimes caespitose; upper side white or blusih green or olive
green; underside white or darkening toward the base; esorediate. Podetia from the margins or the upper surface of the primary squamules,
7–26 mm tall, up to 2 mm in diameter, cylindrical; cupless, terminated by
apothecia, somewhat branched toward the top, the sides markedly
fissured; cortex continuous at first but soon becoming verruculose-
areolate and fissured, the decorticate parts exposing the inner medullary
layer which also becomes fissured and torn, the whole appearance
suggesting half-dead podetia, this resemblance heightened by the pale
white to bluish green or olive green color; K+ yellow, KC-, P- or rarely
P+ red; containing atranorin. Apothecia on the tips of the podetia
and branches, dark brown and quite large, larger than in C. *capitata*.
Pycnidia on the primary squamules.

Growing on soils, especially when rich in humus. A circumpolar arctic
and temperate species ranging trans-Canada, in Alaska, and south to
Connecticut, Wisconsin, Arizona, and New Mexico.

Locality: 14.

34. Cladonia acuminata (Ach.) Norrl. Primary squamules persistent; up
to 7 mm long and 3 mm broad, narrowly elongate, sinuate-edged,
becoming crenate, involute-concave; upper side glaucescent; the underside white; the margins and underside of the primary squamules sometimes becoming granulose sorediate. Podetia 20–50 mm tall and up to 2.5
mm in diameter, cylindrical, cupless, the upper part branching irregularly or subfastigiately or simple; the sterile apices blunt and the fertile
ones becoming dilated upwards; sides becoming chinky and sulcate, or
subcontinuous corticate, sparsely granulose sorediate or esorediate,
verruculose-areolate or becoming decorticate, densely or loosely
covered with squamules similar to the primary squamules; dull, not
transparent where decorticate, the decorticate part chalky white, the
granules and verrucae whitish- to ashy-glaucescent; K+ yellow to
orange or red, KC-, P+ yellow; containing atranorin and norstictic acid.
Apothecia reddish brown, large, commonly perforate or sublobate.

Growing on soil rich in humus. A rare northern species of Europe and
North America as well as Greenland, ranging trans-Canada and in
Alaska, south in Maine, Vermont, and Pennsylvania.

Locality: 1.

35. Cladonia norrlini Vain. Primary squamules persistent, small to
large, subwedge-shaped, sinuate margined, becoming involute concave,
ascending; upper side bluish green; underside white, lacking soredia or

with granulose soredia on the margins and underside. Podetia to 45 mm tall, 3 mm diameter, cylindrical, cupless, toward the tips simple or irregularly branched, the branches ascending, the sides becoming entirely decorticate or the base verruculose-areolate, granulose sorediate or farinose sorediate, dull, the decorticate areas chalky white, the granules ashy or whitish blue green, K+ yellow, P+ yellow, containing atranorin and psoromic acid.

Growing on soil rich in humus and on this soil over rocks. A circumpolar boreal and alpine-arctic species.

Locality: 8.

36. **Cladonia macrophylla** (Schaer.) Stenh. Primary squamules persistent, large or rarely middle-sized, length and breadth 3–8 mm, irregularly lobed, crenate or entire or lobate-incised, concave or involute; upper side glaucescent or rarely olive green; underside white or darkening toward the base; esorediate. Podetia arising from the primary squamules, solitary or rarely several per squamule, 10–60 mm tall, up to 5 mm in diameter but usually less than the maximum; cylindrical, cupless, commonly tipped with apothecia; those podetia with apothecia with fewer branches than those which are sterile, the branches widely spreading, the fertile ones thickening, the sterile ones blunt or subulate, the sides more or less fissured as in *C. cariosa*; esorediate to verruculose, granulose above but not sorediate; cortex areolate to verruculose with *peltate* squamules which are usually dispersed and up to 1 mm broad, the decorticate part of the podetium opaque, ashy to pale brown, the verrucae and squamules glaucescent to olive green to brownish green; K-, KC-, P+ yellow; containing psoromic acid plus an unknown acid. Apothecia brown, confluent or conglomerate or fissured-lobate, sometimes with squamules among the disks. Pycnidia on the apices and sides of the podetium.

Growing on soils with high humus content and on thin soil over rocks. A circumpolar, far northern species ranging in Greenland, Newfoundland, Quebec, Keewatin District, to Alaska, with a disjunct station in the Adirondack Mts of New York.

Localities: 3, 4, 8, 11.

37. **Cladonia scabriuscula** (Del. in Duby) Nyl. Primary squamules disappearing; middle-sized, 2–5 mm long and as broad; irregularly lobed; crenate; ascending; the upper side glaucescent; the underside white; sorediate or not. Podetia 30–110 mm tall, up to 2–5 mm in diameter; mainly dichotomously branching, to some extent sympodial, the branches cylindrical, dilating a little at the axils; the axils open or closed;

the apices usually subulate; partly or entirely isidiose-sorediate or sorediate, the soredia accompanied by minute spreading and appressed squamules, the apices granulose-sorediate or verruculose-granulose; squamulose or not; the cortex chinky to areolate becoming decorticate, the interspaces dull, white; whitish glaucescent to ashy olive or ashy bluish and variegated; K-, KC-, P+ red; containing fumarprotocetraric acid plus accessory ursolic acid. Apothecia small; dark brown; sometimes on corymbose branchlets, at the tips of the podetia. Pycnidia on the tips of the branches.

Growing on earth, old cut banks, and soil over rocks. A circumpolar species reported from all continents except Africa. In North America it grows trans-Canada, in Alaska, south to Pennsylvania, the Great Lakes vicinity, and in Washington and California.

Localities: 2, 3.

38. Cladonia decorticata (Flörke) Spreng. Primary squamules usually persistent; small, up to 4 mm by 2 mm, irregularly or palmately divided, crenate, usually becoming concave; upper side glaucescent or olive green or whitish glaucescent; underside white, darkening toward the base; esorediate or sparingly granulose. Podetia from the upper side of the primary squamules, 10-40 mm tall, up to 2.5 mm in diameter, cylindrical, cupless, simple or subsimple, dichotomous to trichotomous; the tips bearing apothecia dilated, those which are sterile blunt or subulate; the sides entire or fissured, sparingly granulose sorediate; cortex verruculose, the verrucules dispersed, partly developing into small squamules, the upper part of the podetium with reflexed, squarrose squamules similar to the primary squamules, the lower part becoming entirely squamulose, the part between the squamules decorticate; dull and not translucent, white or ashy or ashy brown, the verrucules and squamules white or ashy or olive glaucescent, K-, KC-, P-; containing perlatolic acid. Apothecia middle-sized to large or often small, confluent or lobate-fissured, dark brown or red brown. Pycnidia on the upper side of the primary squamules and the base of the podetia.

Growing on sandy soils and on soils rich in humus. A circumpolar and arctic species, trans-Canadian and Alaskan ranging south to Connecticut, New York, Wisconsin, and Colorado.

Locality: 1.

39. Cladonia glauca Flörke. Primary squamules persistent or disappearing; middle-sized, or small to large, 1-5 mm long and 1 mm broad; irregularly or subdigitately lobed or partly incised; the lobes crenate-incised; ascending; flat or involute; sparse or tufted; the upper side glaucescent or pale greenish white; the underside white and esorediate or sparsely sorediate. Podetia arising from the upper side of the primary squamules; sometimes in small tufted groups; the base persistent or dying; 25–80(–100) mm tall and up to 2.5 mm in diameter; cylindrical; cupless or rarely with minute cups up to 3 mm broad at the apices of the usually simple podetia, the rare branches suberect; the apices attenuate or blunt; the axils perforated or closed; entirely decorticate and farinose sorediate, or rarely with a small corticate area at the base; opaque; impellucid; whitish or ashy glaucescent or rarely ashy brown or variegated; lacking squamules or with a few at the base; K-, KC-, P-; containing squamatic acid. Apothecia small; sometimes aggregated; at the tips of the podetia; commonly peltate and perforate; brown or yellowish brown. Pycnidia on the podetia, at the edges of the axils and the cups, and also on the sides and apices of the podetia.

Growing on soil rich in humus on heaths and in bogs. A circumpolar boreal species which ranges south to Massachusetts and Connecticut in North America.

Locality: Chena Ridge, Fairbanks. Not collected on North Slope.

40. **Cladonia cyanipes** (Sommerf.) Nyl. Primary squamules persistent or disappearing; small or middle-sized, up to 6.5 mm long and 1 mm broad, usually narrowly laciniate, crenate, flat or involute concave, ascending; upper side esorediate; yellowish or yellowish glaucescent, the margins and the underside farinose or with granular soredia; the underside white or pale yellowish. Podetia 20–80 mm tall, slender, cupless; simple or sparingly branched, apices subulate or with rudimentary cups; distinctly yellowish or with a bluish tinge toward the base; the base slightly corticate or the podetium usually entirely decorticate and covered with farinose soredia; K-, KC+ yellowish, P-; containing usnic and barbatic acids. Apothecia at the tips of the podetia; small, pale brownish or pale yellowish. Pycnidia at the tips of the podetia.

A species which grows on soil rich in humus, on humus, and on rotting wood. A circumpolar arctic-alpine species which grows southward in the mountains to the White, Green, and Adirondack Mountains in the east and in Washington and Colorado in the west.

Locality: 9.

41. **Cladonia bacilliformis** (Nyl.) Vain. Primary squamules usually persistent; small, up to 3 mm long and 0.3 mm broad, subentire or crenate to incised-crenate, more or less concave, ascending or appressed; upper side yellowish; underside yellowish or yellowish white, farinose sorediate. Podetia short, 5–20 mm tall, up to 2 mm thick, unbranched,

truncate or with narrow or obsolete cups, the margins of the cups with slender proliferations; thickly sorediate with farinose soredia, yellowish or strongly yellow; esquamulose; K-, KC+ yellowish, P-; containing usnic and barbatic acids. Apothecia small, up to 1 mm; pale yellowish to pale brown. Pycnidia on the upper side and margins of the primary squamules and also on the tips of the podetia.

Growing on soil over rocks, usually with humus, and on rotting wood. A circumpolar far northern species in North America known from Alaska and Colorado.

Localities: 2, 8, 14.

42. **Cladonia subfurcata** (Nyl.) Arn. Primary squamules disappearing; middle-sized, 1–5 mm long; irregularly crenate-lobate or laciniate; ascending; flat; tufted; esorediate; upper side glaucescent to olive or pale glaucescent, underside white. Podetia growing from the upper side of the primary squamules; the base dying and growth continuing from the apices; 15–110 mm tall and 1–1.5(–2.5) mm in diameter; subcylindrical and somewhat thickened at the axils; cupless; branching in sympodia, dichotomously or trichotomously; the axils perforated or not; the branchlets divaricate or ascending, usually short, the apices obtuse or shortly attenuate, erect, branching widely or subradiately spinulose around a perforate axil; forming tufts or small mats; esorediate; the cortex partly subcontinuous, partly contiguous-areolate, smooth or low rugose; lacking squamules or with a few at the base; dull or shining, not pellucid; pale or olive glaucescent or reddish brown to brown, also variegated at times, the dying parts usually spotted, the decorticate portion black and the spots of cortex pale to olive green; K-, KC-, P-; containing squamatic acid. Apothecia small, up to 0.7 mm in diameter; at the apices of the branchlets. Pycnidia at the apices of the branchlets; containing reddish material which turns violet with KOH.

Growing on soil rich in humus. A circumpolar arctic and boreal species also found in southern South America.

Localities: 7, 9, 10, 11.

UMBILICARIACEAE Körb.

Foliose lichens attached in the center by an umbilicus to the substratum, monophyllus or polyphyllus; heteromerous, both upper and lower sides with cortex, lower side variously smooth or areolate-warty, sometimes with rhizinae, lamellae, or trabeculae. Apothecia usually sessile to adnate on the upper side of the thallus, occasionally immersed in the upper surface, occasionally stipitate, lecideine with proper margin which is usually carbonaceous; asci with 1 to 8 colorless or brown, 1–2-celled or muriform spores. Pycnoconidia endo- or exobasidial. Algae: *Trebouxia*.

Because these are so commonly recognized by their appearance as the 'rock tripes' a departure from the format of this report is taken and they are keyed out in one key instead of generic keys. In addition to the species collected by the author the many additional species collected on the North Slope by G.A. Llano and reported upon in his thorough monograph 'A Monograph of the Lichen Family Umbilicariaceae in the Western Hemisphere,' (Navexos P-831, obtainable from the Smithsonian Institution), are added for the benefit of the user.

KEY TO THE UMBILICARIACEAE OF THE NORTH SLOPE

1   Thallus pustulate (blistered) above, black-charred below
                                   Lasallia pensylvanica
      Thallus smooth, reticulate ridged, vermiform ridged, or areolate above but not pustulate; underside various   2

2 (1) Thallus with isidia above; underside dark, smooth to areolate-papillate
                                    Umbilicaria deusta
      Thallus lacking isidia   3

3 (2) Upper side with raised network or with vermiform ridges   4
      Upper side smooth, lacking vermiform or netted ridges   9

4 (3) Upper side with a raised network of ridges which may fade toward the margin, the center often with white pruinose crystalline granules over the umbo   5
      Upper side with vermiform, not net-like, ridges   8

5 (4) Underside lacking rhizinae (or rarely with few scattered toward margins)   6
      Underside with cylindrical to flat, branched rhizinae, pink to light brown; apothecia of omphalodisk type   *Omphalodiscus virginis*

6 (5) Apothecium a gyrodisk with concentric gyri; proper margin continuous; lower surface pruinose only toward margins; thallus 120–230 μ thick (rhizinae rare and scattered toward margins)
                                    Umbilicaria proboscidea
      Apothecium an omphalodisk (with a central sterile column passing into secondary fissures); proper margin continuous   7

7 (6) Lower surface light gray with a dark spot around the umbilicus, pruinose, smooth to chinky; thallus 300–500 μ thick; rarely with rhizinae   *Omphalodiscus krascheninnikovii*
      Lower surface dark brown to black or brown blotched, epruinose, smooth   *Omphalodiscus decussatus*

8 (4) Underside dark gray to dove gray   *Umbilicaria hyperborea*
       Underside with a black center spot around the umbilicus, peripherally
       dark gray to dove gray   *Umbilicaria arctica*

9 (3) Lower side with rhizinae, lamellae, or trabeculae   10
       Lower side smooth, lacking extra structures as rhizinae, trabeculae, or
       lamellae; apothecia of leiodisk type with smooth to cracked disk and
       proper margin   19

10 (9) Lower side with thick black felty mat of hairs, the hairs a mixture of
       cylindrical and ball-tipped types; thallus, thick, rigid, gray above
                                                            Umbilicaria vellea
       Lower side with plate-like trabeculae and/or rhizinae   11

11 (10) Lower side with rhizinae, lacking trabeculae   12
        Lower side with trabeculae, with or without rhizinae   14

12 (11) Apothecia gyrose   13
        Apothecia leiodisk (with smooth disk surrounded by proper margin);
        upper side dark brown to black; lower side black, verrucose
                                                            Agyrophora scholanderi

13 (12) Upper side pitted, gray to light brown; lower side light brown to pink;
        rhizinae light brown; apothecia stipitate   *Umbilicaria cylindrica*
        Upper side not pitted, brown to olive brown; lower side black to reddish
        black; rhizinae black; apothecia sessile (rare)   *Umbilicaria caroliniana*

14 (11) Rhizinae lacking, trabeculae covering entire underside, underside black;
        apothecia with radiating gyri and no margin   *Actinogyra mühlenbergii*
        Rhizinae present as well as trabeculae and lamellae   15

15 (14) Trabeculae over entire underside; thallus flat, margins finely perforate
        to fenestrate; underside dark brown, lamellae granulose to papillose
                                                            Umbilicaria torrefacta
        Trabeculae mainly close to the umbilicus, fading marginally   16

16 (15) Upper cortex chinky-areolate, brown to dark brown; underside with
        cylindrical to flat, tangled, elongate rhizinae; apothecia angular,
        immersed to adnate   *Umbilicaria angulata*
        Upper cortex smooth, not chinky-areolate, apothecia not angular, with
        concentric gyri and persistent margin   17

17 (16) Underside black; rhizinae ball-tipped, black   18
        Underside brown, rhizinae pale, lower cortex areolate, rhizinae fibril-
        lose; upper side ashy to dark brown; margins becoming powdery
        sorediate   *Umbilicaria hirsuta*

18 (17) Thallus monophyllous, thick, 300–500 μ, rigid, white to violet brown,
        densely rhizinose below with long brown rhizinae interspersed with
        short ball-tipped black ones   *Umbilicaria vellea*
        Thallus becoming polyphyllous, thinner, 200–300 μ, rigid, dark gray,
        reddish, or violet gray, underside black verrucose with infrequent
        short ball-tipped black rhizinae   *Umbilicaria cinereorufescens*

19 (9) Upper surface black to brown black, smooth but with fine network of ridges and lightly pruinose at umbo; thallus fragile; lower side dark brown or black, patchy, papillose-areolate; apothecia common, stipitate *Agyrophora rigida*

Upper surface dark gray with fine reticulation of white pruinose ridges; thallus leathery, lower side uniformly dark, smooth or furfuraceous; apothecia rare, stipitate *Agyrophora lyngei*

KEY TO THE GENERA OF THE UMBILICARIACEAE (after Llano)

1 Apothecia with flat disk, surface smooth and with continuous proper margin 2

Apothecia with flat or convex disk, surface with a central button or fissures or furrows 3

2 (1) Thallus pustulate 1. *Lasallia*

Thallus not pustulate 2. *Agyrophora*

3 (1) Apothecia with flat disk with a sterile central button, or fissure, or with secondary fissures 3. *Omphalodiscus*

Apothecia with convex disk which is furrowed and with or without a margin 4

4 (3) Furrows of concentric gyri within a continuous proper margin 4. *Umbilicaria*

Furrows of radial gyri without any margin 5. *Actinogyra*

1. LASALLIA Mérat

Thallus monophyllous, foliose, the margins entire or dissected, the upper surface pustulate with the lower surface correspondingly excavate; upper surface smooth to areolate, centrally attached by an umbilicus, lacking rhizinae, the lower surface smooth to verrucose; upper cortex paraplectenchymatous, medulla loose, lower cortex scleroplectenochymatous. Apothecia common in North American species, lecideine, sessile to pedicellate; disk flat, smooth or irregular, with proper exciple; asci 1 or 2 spores; spores hyaline to dark brown, muriform. Algae *Protococcus*.

1. **Lasallia pensylvanica** (Hoffm.) Llano Thallus large foliose, monophyllus, membranaceous to rigid, the margins often incised to with holes, upper side pustulate, smooth, with a few folds, roughened over the umbo, dull or shiny, brown to olive brown; lower surface black with coarse papillae, the margin sometimes grayish; thallus 250–300 μ thick, upper cortex paraplectenchymatous, medulla loose, containing gyrophoric acid. Apothecia to 2 mm, sessile to subpedicellate; disk flat, black,

exciple prominent; hypothecium brown; epithecium dark brown to black; hymenium 100 μ; paraphyses coherent, simple, septate, brown-capitate, asci clavate, spores 1, muriform, light to dark brown 36–60 × 15–29 μ.

Growing on acid rocks. A circumpolar temperate and boreal species common in the Appalachian Mountains but ranging into Alaska, the Northwest Territories, Greenland, and south into Mexico.

Localities: 8, 11. Llano reported this species from the North Slope at Anaktuvuk Pass, Lake Peters, Lake Schrader, the Sadlerochit River, and at Wiseman.

2. AGYROPHORA Nyl.

Thallus foliose, non-pustulate, monophyllous or polyphyllous, attached by umbilicus, upper surface partly or weakly ribbed but not strongly wrinkled, the surface smooth, powdery, or areolate-papillose, underside smooth, areolate-papillose to verrucose, lacking rhizinae or with rhizinae. Apothecia common or rare, lecideine, sessile to stipitate, the disk flat, smooth, slightly cracked or continuous with a proper exciple, black, asci 8-spored, spores simple and hyaline, or brown and muriform.

1. **Agyrophora lyngei** (Schol.) Llano Thallus to 3.5 cm broad, mono-phyllous, with irregularly torn margins, umbo elevated with low ridges which become reticulate, sometimes crystalline pruinose; the ridges fade toward the margins; upper surface dull, dark brown to brown black; lower surface dull, smooth, furfuraceous black or with orange patches, marginally with pale pruinose band, lacking rhizinae. Apothecia rare, stipitate, to 2 mm broad, leiodisk with continuous surface and marginal exciple; paraphyses simple, septate, hyaline; spores simple, hyaline, 10–14 × 5–7 μ.

A species of acid rocks. Known from Europe and North America; in the latter it occurs north of the Gulf of St Lawrence in the east, and from Oregon northwards in the west.

No specimens collected by the writer. Llano reported this species from Lake Peters and Anaktuvuk Pass on the North Slope.

2. **Agyrophora rigida** (DuRietz) Llano Thallus to 15 cm broad, mono- to polyphyllous, rigid, fragile, leathery, the margins usually split, often perforate; upper surface dark brown to black, with frosty pruina over the umbo which has a fine network of ridges; underside smooth with areolate to verrucose cortex, dove gray to dark brown, the color in patches and darkening to the outside, lacking rhizinae, the thallus to 200

μ thick. Apothecia common, to 2 mm broad, stipitate, the disk black, flat, the margins rough, subshiny; hypothecium brown; paraphyses septate with short apical branches; spores simple, hyaline, 9.6–15 × 2.4–9 μ.

A species of acid and calcareous rocks. This species is distributed in Scandinavia and in North America from Greenland across the Arctic Archipelago to Alaska and south to Washington.

No specimens were collected by the writer. Llano reported it along the North Slope at Lake Peters, Lake Schrader, Chandler Lake, Umiat, and Anaktuvuk Pass.

3. **Agyrophora scholanderi** Llano Thallus small, to 2 cm broad, mono- to polyphyllous, margin slightly and infrequently lacerate, the umbo prominent, lightly pruinose, with rounded ridges; upper side smooth, soft, with papillae, slightly reticulate veined, marginally entire or sub-perforate, marginally dark brown to black; underside ebony black, verrucose, with rhizinae of irregular length, round or flat, irregularly branched, tapering, the surface rough. Apothecia scarce or numerous, adnate; disk black, to 1 mm broad, leiodisk; the margin irregular and persistent; asci 40–45 × 14–24 μ; paraphyses simple, the apices dark and inflated to 2.5 μ; spores 8, simple, hyaline, ellipsoid, 11–14 × 5.7–7 μ.

A species of acid rocks. The species was described from Lake Peters and is also known from Mt McKinley National park and from Mt Ellinor, Mason Co., Washington. It is endemic to North America.

Locality: 8.

3. OMPHALODISCUS Schol.

Thallus foliose, monophyllous or partly polyphyllous, attached by an umbilicus; with or without rhizinae below. Apothecia lecideine, adnate or slightly stipitate, of omphalodisk type with a sterile central column sometimes becoming fissured in the center and with secondary fissures across the disk; paraphyses simple to branched, septate; spores 8, simple and hyaline in most, brown and muriform in two species which do not occur in Alaska.

1. **Omphalodiscus decussatus** (Vill.) Schol. Thallus to 8 cm broad, monophyllous, leathery, rigid, the umbo peaked, coarsely granulose, pruinose, reticulate ridged, becoming vermiform and decreasing toward the periphery, appearing areolate, occasionally with lobulae from the ridges or margins, the margins occasionally perforate; upper side dull, light yellowish buff to brown or blackish; underside dark, sooty black

with a narrow gray to light brown peripheral zone, smooth to puckered, sometimes gray blotched, the thallus 300–600 μ thick. Apothecia rare, to 3 mm broad, black, adnate, with sterile button in center, with or without fissures; the disk flat, round, becoming convex and distorted; spores 8, simple, hyaline, ellipsoid, 6.5–13 × 3.3–6.6 μ.

A species of nitrogenous rocks as near bird rookeries. It is circumpolar in both northern and southern hemispheres. In North America it is northern and western, occurring the length of the continent. The writer did not collect this species. Llano reported it from Chandler Lake and Anaktuvuk Pass.

2. **Omphalodiscus krascheninnikovii** (Savicz) Schol. Thallus to 5 cm broad, usually smaller, monophyllous, rigid, the margins rarely incised, usually rounded; upper side with a coarsely granulose umbo and sharp edged net-like ridges which extend to the margins, dull ashy, ashy black, or brown, more or less pruinose; underside dove gray peripherally and with a black spot around the umbilicus, obscurely granulose or chinky, occasionally with a few simple to branched, cylindrical rhizinae near the margins (f. *rhizinosus* Schol.). Apothecia to 3 mm broad, common, sessile to substipitate; disk black, flat with sterile button in center and many thin secondary fissures; spores 8, simple, hyaline, 5–7 × 3.8–4 μ.

This species grows on a variety of rocks. It is apparently circumpolar. In North America it is northern and western, occurring from Greenland to Alaska and south in the west to California.

Although it is common in the north it was not collected on the North Slope by the writer. Llano reported it from Anuktuvuk Pass.

3. **Omphalodiscus virginis** (Schaer.) Schol. Thallus to 10 cm broad but much smaller in arctic conditions, monophyllous, leathery, the umbo not raised, often puckered with weak network of low ridges which fade marginally; surface undulating, appearing wrinkled, finely granulose and indistinctly cracked, dull, olive brown or yellowish glaucous, the margins browner; underside with well developed umbilicus which has buttresses extending outwards into folds and creases, the surface smooth to slightly granular, shell pink to yellowish buff, covered with long cylindrical or flattened rhizinae which are dark brown; thallus about 266 μ thick. Apothecia to 4.5 mm broad, sessile; the disk round, flat, becoming convex with sterile button in center and fissures sometimes not well visible except in sections; hypothecium light brown; spores 8, simple, hyaline, 10–17 × 4–8 μ.

A species of acid rocks. It is circumpolar northern hemisphere, in North America ranging from Greenland to Alaska and southwards in the alpine regions to California. A variety replaces the species on the volcanic peaks of Mexico.

The writer did not collect this species on the North Slope. Llano reported it from Anaktuvuk Pass.

4. UMBILICARIA Hoffm.

Thallus foliose, umbilicate, monophyllous or polyphyllous, smooth to reticulate rugose above; underside smooth or with rhizinae or lamellae or raised plates called trabeculae, or with combinations of these. Apothecia lecideine, gyrose, with a common continuous margin, young apothecia with superficial central sterile column or small gyri, new gyri forming by longitudinal splitting and the old apothecia becoming more or less concentrically gyrate. The asci 8-spored, the spores simple and hyaline or becoming brown and muriform in some species.

1. **Umbilicaria angulata** Tuck. Thallus to 5 cm broad, monophyllous, thick, rigid, sometimes in overlapping lobes, the umbo raised and with broad folds the folds fading marginally; upper surface smooth to obscurely wrinkled and cracked, sometimes rough with numerous spermogonia, the cracks black, the surface dull or somewhat shiny, dark brown to with grayish olive blotches, pruinose with a purplish bloom; lower surface coarsely papillate, appearing shaggy, covered with a mat of coarse tangled, cylindrical or flattened, coarsely papillate dark rhizinae; thallus 200–425 μ thick. Apothecia usually present, adnate to partly immersed, angular or stellate with thin exciple, flat, becoming convex, weakly gyrose; spores when young hyaline simple to reticulate, becoming brown and muriform, 12–28 × 7–18 μ.

On acid rocks. A North American endemic species, western, ranging in the mountains from Alaska to California.

Locality: 11. Llano reported this species from the Aleutian Islands but not the North Slope.

2. **Umbilicaria arctica** (Ach.) Nyl. Thallus to 20 cm broad, usually much less, monophyllous, leathery, the upper surface with vermiform ridges which extend to the margins, the umbo only slightly raised and ridged, the surface dull, dark brown or brownish black, seldom a light brown or clay colored, wrinkled and bullate but not with a network of ridges, underside bare, becoming obscurely areolate papillate, black around the umbilicus and pinkish buff or gray toward the periphery; thallus about

1000 μ thick. Apothecia to 2.5 mm broad, adnate to sessile, strongly convex, gyrose; hypothecium very dark; spores 8, hyaline, simple, 8.5 μ, 15 × 4–8.5 μ.

On granitic rocks, preferably with bird perches, as it is reputed to be a nitrophile. A circumpolar arctic and alpine species occurring across the north, and southwards in the west to Mexico.

Locality: 9. Also reported by Llano from the Chandler Lake area and on several of the Aleutian Islands.

3. **Umbilicaria caroliniana** Tuck. Thallus to 6 cm broad, usually less, monophyllous to polyphyllous, membranaceous to fragile and thin, with many soft folds especially around the umbo, the margins often cleft; upper surface smooth to slightly papulose, dull or shining, brown with an olive tinge; underside black, uniformly granulose to verrucose with scattered, black, shreddy rhizinae; thallus 250–350 μ thick. Apothecia to 5 mm, black, scattered, sessile to subpedicellate but depressed in thallus hollows, gyrate with thick exciple, becoming convex with few gyri; hypothecium dark; spores 6–8 in ascus, brown, muriform, 30–50 × 20–26 μ.

On rocks. A rare species known from North Carolina, Alaska, Japan, and Siberia.

Localities: 3, 8. Llano reported this species from Lake Peters, Anaktuvuk Pass, the Kobuk River, and Cape Nome.

4. **Umbilicaria cinereorufescens** (Schaer.) Frey Thallus to 6 cm broad, monophyllous or polyphyllous, umbo raised with folds and areolate papillate; upper side very undulating, margins wavy, incised to crenate and often reflexed, upper surface dull, pale gray or mouse gray blotched with cinnamon brown; underside black, umbilicus broad with coarse plates which fade peripherally, the under surface strongly verrucose, without rhizinae or with short, occasionally ball-tipped rhizinae, peripherally lighter and more areolate; thallus 200–300 μ thick. Apothecia to 2 mm broad, rare, adnate to subimmersed, irregular, gyrose, the exciple becoming cracked; hypothecium black-granular; spores brown, muriform, 12–20 × 6.8–14 μ.

On acid rocks. A cosmopolitan species but uncommon. In North America occurring from Greenland to Alaska and south in the west to Arizona.

Localities: 8, 11. Llano also reported it as abundant at Anaktuvuk Pass, and present at Ikiakpuk Valley, Chandler Lake, Lake Peters, Lake Schrader, and Umiat.

5. **Umbilicaria cylindrica** (L.) Del. Thallus to 10 cm broad; polyphyllous, usually raised and curled; upper surface with fine reticulate pattern of dark lines or smooth, or becoming obscurely vermiform or areolate pruinose, rigid or thin, more or less perforated, sometimes with rhizine-like growths from the upper side; surface dull to subshiny, light to dark blackish brown; underside appearing smooth but finely granular, pale brown, sometimes darker toward the umbilicus or with a gray pruinose outer zone, with sparse to abundant rhizinae which are long and cylindrical or flat and branched; thallus about 450 μ thick. Apothecia to 4 mm broad, stipitate, black, flat then convex, gyrose; hypothecium dark; spores hyaline, staining with a dark central band, simple, 8.5–15 × 3.4–8 μ.

Growing on acid rocks. A cosmopolitan species which is very variable. In North America arctic alpine, occurring across Canada and Alaska, south to New England and to Colorado.

Locality: 8. Llano reported it as rare on the North Slope but present at Lake Schrader, Lake Peters, Ikiakpuk Valley, and Anaktuvuk Pass.

6. **Umbilicaria deusta** (L.) Baumg. Thallus to 9 cm. broad but usually very small, monophyllous or usually polyphyllous, paper thin and fragile when dry, curled and ragged, the lobes often reflexed, sometimes perforated, umbo not usually discernible; upper surface smooth but covered with greater or less abundance of small isidia which may be spinulose, cylindrical, coarse and lobed, or forming small lobules, dull brown to black; underside smooth to areolate-papillate, weakly or strongly veined or bullate, dark brown to blackish; thallus 200–300 μ thick. Apothecia rare, adnate to subdepressed, round, flat to subconvex, gyrose; hypothecium light brown; spores hyaline, simple, 18–27 × 7–12 μ.

Growing on rocks especially in rills and seepage surfaces. A circumpolar species which is arctic and boreal, occurring in North America from Greenland to Alaska and south to New England, Michigan, Minnesota, Colorado, and Arizona.

This species was not collected by the writer but Llano reported it on the North Slope from Ikiakpuk Valley.

7. **Umbilicaria hirsuta** (Sw.) Ach. Thallus to 8 cm. broad, monophyllous to polyphyllous, thin but rigid, with an umbo which is obscure or raised, upper cortex areolate and becoming powdery sorediate at the margins; upper surface with a fine network of fissures, the margins torn lacerate, surface dull, light brown to dark brown; lower side with lamellae developed around the umbilicus and extending radially; lower surface buff to dove gray or black, with areolate-papillate lamellae of same color, extending more or less over the surface, rhizinae poorly developed or

small, cylindrical to branched sparse to more or less dense; thallus 140–225 μ thick. Apothecia rare, to 2 mm broad, adnate and depressed in thallus to sessile, black, gyrose; hypothecium brown; spores hyaline, simple, 10–13.6 × 5–6.8 μ.

Growing on rock ledges. European and North American, a rare northern and alpine species occurring in Greenland, Alaska, Vermont, and New Hampshire and in California, Colorado, and Montana.

This species was not collected by the writer. Llano reported it from Lake Schrader and Anaktuvuk Pass on the North Slope.

8. **Umbilicaria hyperborea** (Ach.) Hoffm. Thallus to 5 cm. broad, monophyllous but sometimes with thalline growths over the umbo and appearing polyphyllous, rigid or thin and pliable; umbo raised, smooth or wrinkled to granulose; upper surface covered with raised vermiform ridges which extend to the margins, dull to subshiny, gray brown, dark brown, or olive brown, darker appearing between the ridges; underside bullate to lacunose, smooth to granulose, dark around the small umbilicus or brown, rarely with marginal pruinose peripheral zone; thallus 190–230 μ thick. Apothecia common, to 2 mm broad, sessile, gyrose, black, convex; hypothecium brown; spores hyaline, simple, with central belt staining dark 6.6–20 × 3.3–8 μ.

On acid rocks. A common arctic and boreal species, circumpolar and also in the southern hemisphere. In North America ranging southward to Pennsylvania in the east and into Mexico in the west.

Localities: 3, 8, 9, 13. The latter is var. *radiculata* Zett. which is pale below and with rhizinae. Llano reported this species from the North Slope at the Sadlerochit River, Lake Schrader, Ikiakpuk Valley, Lake Noluk, and Anaktuvuk Pass.

9. **Umbilicaria proboscidea** (L.) Schrad. Thallus to 8 cm broad, monophyllous, round, rigid to thin and fragile, the umbo peaked with a network of sharp edged ridges, usually white pruinose, the network of ridges continuing to the border of the thallus or fading toward the edges, the edges sometimes ragged or perforated, sometimes with a dead, whitish border, the center sometimes with small thalline lobes giving a polyphyllous appearance; upper side dull, brown to black brown; underside smooth to areolate, obscurely bullate, buff, dove gray or dark brown, with a gray pruinose peripheral zone, the umbilicus small, usually without rhizinae, when these are rarely present they are scattered near the periphery; thallus 120–230 μ thick. Apothecia common, sessile to subpedicellate, to 1.2 mm broad, flat to convex, gyrose; hypothecium brown; spores hyaline, simple, 10–17 × 3.5–7 μ.

Growing on acid rocks, usually in the open. A very common species in the north, arctic-alpine, ranging from Greenland to Alaska, south to the White Mts in New England, and Colorado and Oregon in the west. It is circumpolar and also in the southern hemisphere.

Localities: 3, 8. Llano reported this species on the North Slope from Anaktuvuk Pass, Lake Schrader, Chandler Lake, Lake Peters, Sadlerochit River, Noluk Lake, and the Kobuk River.

10. Umbilicaria torrefacta (Lightf.) Schrad. Thallus to 6 cm broad, monophyllous, thin to thick, rigid or very fragile, upper side crumpled to pulvinate, the upper side smooth to obscurely granular, occasionally with network of fine dark lines, the margins always perforated with small to large holes; upper surface dull to shining, brown to olive brown; underside smooth or finely granular papillose with well developed trabeculate structure covering the whole underside, often becoming fimbriate or fibrillose, buff to dark brown; thallus 250–400 μ thick. Apothecia common, adnate to subimmersed in thalline depressions, convex, black, gyrose; hypothecium dark brown; spores hyaline, simple, 7–12 × 5–7 μ.

Growing on acid rocks. A circumpolar arctic-alpine species occurring in North America from Greenland to Alaska, south to the Adirondack Mts in the east, and in the west to Nevada and California.

Locality: 9. Llano reported this species on the North Slope from Lake Schrader, Lake Peters, Noluk Lake, and Anaktuvuk Pass.

11. Umbilicaria vellea (L.) Ach. Thallus to 25 cm broad, usually smaller, monophyllous, rigid, umbo moderately raised or lacking, upper side undulating, sometimes with folds; upper surface smooth to scabrous, areolate papillate, often with cracks and holes through which the black rhizinae project, dull, pale brown, gray, smoky gray or violet gray, sometimes with reddish stains; underside black, the umbilicus broad often with trabeculae which may extend to the margins, the whole underside covered with dense nap of short thick ball-tipped rhizinae, sometimes with longer cylindrical, light brown rhizinae; thallus 300–500 μ thick. Apothecia uncommon, to 3 mm broad, sessile, black, convex, gyrose; hypothecium brown; spores simple, hyaline, 8.5–13.6 × 6.8–10 μ.

Growing on acid rocks, usually with some moisture seepage, on cliff faces, etc. A circumpolar species also found in the southern hemisphere. In North America it is wide ranging, from Greenland to Alaska, south to Georgia in the eastern mountains and into Mexico in the west, and into Illinois in the midwest.

This species was not collected by the writer; Llano reported it from Anaktuvuk Pass on the North Slope.

5. ACTINOGYRA Schol.

Thallus foliose, umbilicate, upper cortex paraplectenchymatous, medulla dense, lower cortex paraplectenchymatous to scleroplectenchymatous; under surface with well-developed rhizinae or with lamellae. Apothecia adnate, radiate without a common exciple around the gyri as in *Umbilicaria*; hypothecium poorly developed; asci with 8, simple, hyaline spores.

1. **Actinogyra mühlenbergii** (Ach.) Schol. Thallus to 20 cm broad, usually less, monophyllus, stiff, with folds or pleats centrally, the edges torn or deeply cut; upper surface smooth, dull, brown; underside olive brown, darker toward the umbilicus, with papillate lamellae and trabeculae which form plate-like layers over the underside, lower cortex areolate. Apothecia common, adnate, in indentations in the upper side, convex, of free gyri, branching dichotomously, black; spores 8.5–13 × 3.5–5 μ.

This species grows on acid rocks, including sandstones. It has a disjunct range in Asia and in eastern North America. In the latter it ranges northwestward to the Great Slave Lake region. Llano (1950) reported a specimen 'between Point Barrow and Mackenzie River, Capt. Pullen 8/1849.' This would perhaps place the species on the North Slope but it has not subsequently been collected this far to the northwest of the rest of its range in North America. Pullen was on a small boat journey from Point Barrow to the Mackenzie River thence to Fort Simpson during a Franklin search expedition of 1848–51 (Arctic Bibliography entry 2638) and further trips took him eastward to Hudson Bay. It is more likely that the specimen came from the range where the species exists just to the east of Fort Simpson rather than from the boat journey along the Alaskan coast where the species would be less likely to occur.

PERTUSARIACEAE

Crustose lichens, the thallus superficial or in some cases imbedded in the substratum, with an upper cortex and an algal layer. Apothecia immersed to adnate, the exciple thalloid, the spores 1–8, large with thick walls. Algae: *Trebouxia*.

Of the four genera known from North America only *Pertusaria* is currently known from the North Slope.

1. PERTUSARIA Lam. & DC.

Thallus crustose, over or under bark or over rocks or soil, smooth, chinky, areolate, or of papillae; upper cortex of rounded cells; many species lacking fruits, these replaced by soredial masses. Apothecia single to several or many in special verrucules, the disk either puncti-form or opening and then appearing lecanorine, the hymenium either spherical or flattened, a proper exciple lacking or poorly developed; epithecium pale or darkening; hypothecium hyaline to brownish; para-physes hyaline, slender, usually branched and loosely interwoven; asci large, clavate or cylindrical, 1–8-spored, I+ blue; spores uniseriate or irregular, 1-celled, many nucleate, hyaline, always large but the size dependent on the number in the ascus, thick walled. Algae: *Trebouxia*.

1		Thallus with dense, erect, finger-like isidia 1–10 mm tall	2
		Thallus crustose, not isidiate	3
2	(1)	Isidia thick, over 0.5 mm broad, very white or pale gray; apothecia black pruinose, immersed in the tips of the thallus	1. *P. dactylina*
		Isidia slender, less than 0.5 mm broad; dark gray; apothecia with broad, epruinose disks	2. *P. oculata*
3	(1)	On rocks	4
		On vegetation, humus, or soil	5
4	(3)	Thallus tan, areolate, fruit-like verrucae covered with coarse, almost verruculose soredia, K+ red, containing norstictic acid; lacking apothecia	3. *P. excludens*
		Thallus roseate, esorediate, areolate, K−; apothecia compounded in the fertile areolae which give the appearance of simple apothecia	
			4. *P. subplicans*
5	(3)	Forming apothecia, esorediate (except No. 5)	6
		Forming abundant capitate soredia, with or without apothecia; thallus with bitter quinine taste, KC+ violet (picrolichenic acid), soredia P- or P+ red, thallus P−	16. *P. amara*
6	(5)	Disk broadly open and pruinose, the margins more or less sorediate	7
		Apothecia not pruinose	8
7	(6)	Thallus K−, lacking hypothamnolic acid	5. *P. panyrga*
		Thallus K+ violet (slowly), hypothamnolic acid present	
			6. *P. subdactylina*
8	(6)	Disk open, lecanorine; ascus containing 1 spore	9
		Disk closed, opening by a narrow pore	10

9 (8) On humus; epithecium K+ violet; medulla K+ yellowish; disk light
       brown to black   7. *P. bryontha*
       On bark of shrubs; epithecium K-; medulla KC+ red; disk reddish
                                                          8. *P. velata*

10 (8) Base of fruiting verruca broad, the verruca broadly conical; thallus
       under bark   11
       Base of fruiting verruca narrowed, the verruca cushion-like; thallus
       over substratum   12

11 (10) Epithecium light, almost colorless K-; spores uniseriate 35–75 × 21–32 μ
                                                          9. *P. alpina*
        Epithecium dark, thick, K+ violet; spores uniseriate, 21–40 × 12–20 μ
                                                          10. *P. sommerfeltii*

12 (10) Thallus and verrucae K+ red (norstictic acid)   13
        Thallus and verrucae K+ yellow, P+ orange; containing stictic acid 16

13 (12) Spores 2–4 per ascus   14
        Spores 8 per ascus; apothecium 1 per verrucle   14. *P. octomela*

14 (13) Spores 2 per ascus; thallus P+ orange; apothecia usually 2 or more per
        verrucle   15
        Spores 4 per ascus; thallus P+ orange, containing norstictic acid;
        apothecium 1 per verrucle   12. *P. glomerata*

15 (14) Apothecia 2–5 per verrucle; K+ reaction not uniform, only in parts of
        thallus   11. *P. coriacea* var. *obducens*
        Apothecia 5–25 per verrucle; K+ reaction over entire thallus
                                                  11. *P. coriacea* var. *coriacea*

16 (12) Spores 2 per ascus   15. *P. subobducens*
        Spores 4 per ascus   13. *P. trochisea*

1. **Pertusaria dactylina** (Ach.) Nyl. Thallus thin, white or whitish with
many upright isidia from low warty protuberances to branching taller,
cylindrical forms 10 mm tall, 0.4–1 thick, the bases narrow, the apices
rounded or flattened, esorediate but sometimes becoming decorticate,
K+ yellow to red, KC+ pale yellow, C-, P+ orange red (fumarprotocetraric acid). Apothecia uncommon, at the tips of the isidia, surrounded
by a raised thalline margin, disk black, pruinose, somewhat convex;
spores single in ascus, hyaline, the walls smooth, elliptical, 125–314 ×
62–100 μ.

Growing on mosses, humus, and mineral soil. A circumpolar arctic-
alpine species known from Greenland to Alaska and common in the
Northwest Territories.

Localities: 2, 3, 8, 9, 11.

2. **Pertusaria oculata** (Dicks.) Th. Fr. Thallus thin, dispersed, whitish

gray with very slender, usually branched, cylindrical isidia, 1–3 mm tall 0.3–0.4 mm thick, the tips rounded and darkened, not sorediate; K+, KC+ yellow then red, C-, medulla P+ yellow then orange red, tips of isidia I+ violet, medulla I-, epithecium I+ violet; containing fumarprotocetraric and gyrophoric acids. Apothecia lecanorine 1–2 mm broad, round; disk black, roughened, epruinose; margin thick, raised, concolorous with the thallus; spores uniseriate, 8 in ascus, hyaline, ellipsoid, 1-celled, the walls thin, 18–30 × 10–14 μ.

Growing mainly on humus soils and mosses, rarely on thin soil over rocks. A common circumpolar arctic-alpine species ranging from Greenland to Alaska and in the Northwest Territories.

Locality: 4.

3. **Pertusaria excludens** Nyl. Thallus over rocks, of small areolae close together, 0.5–1.0 mm broad, angular, with black sides the surface brownish white, K+ red, containing norstictic acid, scattered over the thallus are higher verrucae about 1–1.5 mm broad which appear as fruiting verrucae but lack hymenia and instead have coarse soredia which resemble verrucae on their inner surface.

Growing on non-calcareous rocks. A species reported as montane in Europe and from Connecticut in North America (Erichsen 1941). The specimens reported by me as *P. monogona* Nyl. from Coppermine, NWT, have been corrected to this species by Martyn Dibben, as was my comparison specimen Sampao 163.

Locality: 11.

4. **Pertusaria subplicans** Nyl. (*Pertusaria hultenii* Erichs.) Thallus thin and smooth or becoming chinky-areolate and somewhat dispersed, lacking soredia, the margins indistinct, the areoles rounded or flat, yellowish or roseate white, with fertile areoles 4–5 mm broad appearing as if apothecia of lecanorine type, projecting above the thallus, sometimes containing a single apothecium but developing to contain a number deeply imbedded with the small pale ostioles showing in the folds of the upper side, the underside almost umbilicate; K-, C+ red, KC+ red, P- containing gyrophoric acid; hymenium I+ blue; spores 8 in ascus, hyaline, ellipsoid, 36–87 × 25–45 μ.

This species grows on calcareous and acidic rocks. It is apparently a circumpolar high arctic species. The type is from Lawrence Island. The type of *P. hultenii* is from Unalaska Island. I have seen specimens from Bering Island and Alaska determined as *P. rhodoleuca*. The type specimen of *P. determinanda* Darb. from Ellesmere Island is in Copenhagen and is this species.

Localities: 1, 8.

5. **Pertusaria panyrga** (Ach.) Mass. Thallus quite thin, whitish or bluish gray, esorediate, lacking isidia, very uneven in thickness, all usual reactions negative, fruiting verrucae numerous, thickly clustered, high convex, the base narrowed, 0.8–2 mm broad, the top flattened, usually with 1, rarely 2–3 apothecia or more; the disk opening broadly to 0.4–0.6 mm, black, rough, bare or usually whitish pruinose, surrounded by a whitish, usually sorediate, margin, irregularly and strongly crenate; spores 1 per ascus, hyaline, 1-celled, 91–185 × 50–63 μ.

Growing on humus, old moss, soil, thin soil on boulders, and occasionally on the base of shrubs. A circumpolar arctic and subarctic species which is very abundant.

Localities: 1, 2, 3, 8, 9, 10, 11, 14.

6. **Pertusaria subdactylina** Nyl. Thallus a membranaceous crust over mosses, becoming papillate-mounded in spots and these rising to short erect papillae which may have a narrow columnar base which enlarges to a capitate broadened head with a rough top; not definitely corticate, with solid medulla; K+ violet (sometimes slowly), C-, KC rose, P-; containing hypothamnolic acid (boat-shaped small crystals in GE). Apothecia immersed in the tips of the enlarged tips of the papillae, bursting through the tip; disk black, pruinose; asci clavate; spores 1 per ascus, hyaline, 1-celled, ellipsoid, 37–45 × 18 μ.

There is a strong resemblance with *P. panyrga*, especially with the apothecia, and some specimens of this species I had misdetermined as *P. panyrga*. However the reactions of the thallus, K+ violet, KC+ rose, versus all negative reactions in *P. panyrga*, and the hypothamnolic acid content versus an undefined content, make these two species separable. Smoothly papillate specimens differ from young *P. dactylina* by the K+ violet reaction instead of K+ yellow to red and P- instead of P+ red reaction.

Growing on mosses and humus. A probable member of the amphi-Beringian element in the flora, this ranges east to Coppermine, NWT.

Localities: 1, 2, 9.

7. **Pertusaria bryontha** (Ach.) Nyl. Thallus on a thin white protothallus over mosses, of rounded granules or hemispherical scattered areoles to 1 mm broad becoming thicker, or else the plant composed only of protothallus and fruiting verrucae, reactions negative. Apothecia in verrucae resembling *Lecanora epibryon*, when young with narrow mouthed opening but soon widening into a broad disk, to 3 mm broad; disk rough, epruinose, light brown or blackish or variegated; margin thick, rounded, crenulate, disappearing with age; spore 1 in ascus, hyaline, ellipsoid, the wall thick and thicker at the ends, 112–230 × 39–100 μ.

Growing over mosses, humus, and remains of plants. A circumpolar arctic-alpine species distributed in North America from Greenland to Alaska across the Northwest Territories.

Localities: 3, 9, 13.

8. Pertusaria velata (Turn.) Nyl. Thallus thin, smooth or slightly roughened, sometimes slightly chinky, not areolate, white to whitish gray, the margin sometimes somewhat radiate-chinky, paler, often zonate, sometimes showing a brownish hypothallus, lacking soredia or isidia, reactions negative except medulla K+ reddish. Fruiting verrucae numerous, 0.5–1.2 mm broad, appearing lecanorine, rounded, the base narrowed, of same color as the thallus or paler with 1 (rarely 2) apothecium; disk pale reddish, white pruinose, slightly concave or flat, surrounded by a thalloid margin which may be fine sorediate and becomes very crenulate with age; epithecium pale; hypothecium pale; spores 1 or 2, hyaline, ellipsoid, 1-celled, the walls 9–15 μ thick, smooth, 150–310 × 40–90 μ.

Growing on tree trunks and bark of shrubs. A species with an exceedingly broad range in the tropics as well as in the temperate zones. It has previously been reported from Alaska.

Locality: 2.

9. Pertusaria alpina Hepp Thallus under the bark or a thin film over the bark, smooth, whitish gray, grayish green, blue white, or yellowish white, lacking soredia or isidia, K+ yellow, KC+ yellowish, C-, P-. Fruiting verrucae with broad flaring base, to 2.5 mm broad, apothecia 1–8 per verruca, with a narrow, punctate, dark ostiole; epithecium brownish or hyaline; hypothecium hyaline; spores 8 per ascus, uniseriate, hyaline, ellipsoid, 1-celled, 34–75 × 17–32 μ.

Growing on broad-leaved trees and shrubs, seldom on conifers. A subarctic species known from Europe and North America.

Locality: 1.

10. Pertusaria sommerfeltii (Flörke) Flörke Thallus under bark, smooth, thin, uneven, continuous, seldom slightly chinky, white, grayish white, or yellowish white, lacking soredia; reactions negative except that epithecium and upper part of margin are K+ violet. Fruiting verrucae numerous, dispersed, flattened or with slightly dimpled apex, the base mainly spreading but some verrucae with narrowed base, to 1 mm broad, with 1–4(–7) black ostioles conspicuous on top; epithecium dark, forming an unusually thick layer 30–120 μ thick; hymenium and hypothecium hyaline; spores uniseriate, 8 per ascus (seldom 6 or 7), hyaline, ellipsoid, the wall thin, 1.5–3 μ, smooth, 21–40 × 12–20 μ.

Growing on the smooth bark of trees or shrubs. A circumpolar arctic-alpine species, in North America ranging from Greenland to Alaska.

Localities: 1, 2.

11. **Pertusaria coriacea** (Th. Fr.) Th. Fr. Thallus thick and knotted, continuous, rounded and piled up, white or yellowish white, to brownish or roseate, shining, K+ yellow, then red, KC+ yellow to red, C-, P+ orange containing norstictic acid. Apothecia abundant, the mouths of the black ostioles many and conspicuous along the tops of the folds and high rounded parts of the thallus; epithecium dark; hypothecium hyaline to brownish; spores 2 per ascus, hyaline, smooth, 160–230 × 50–75 μ.

There are two variants described: var. *coriacea* with many apothecia, 5 to 25, per verrucule and a K+ red reaction over the entire thallus; the other, var. *obducens* (Nyl.) Vain. has fewer apothecia and a non-uniform K reaction over parts of the thallus.

This is a very common species growing on humus, soil, mosses, and the soil at the edge of frost boils. It appears to be circumpolar arctic. It occurs commonly from Greenland across the Northwest Territories to Alaska.

Localities: var. *coriacea*: 11; var. *obducens*: 1, 3, 11.

12. **Pertusaria glomerata** (Ach.) Schaer. Thallus thick, glomerulate, white, yellowish, or brownish, smooth to warty, K+ yellow then red, C-, medulla K-, P+ red, containing norstictic acid. Fruiting verrucae numerous, almost spherical, the base narrowed, to 1.5 mm broad, yellowish or pale reddish brown; apothecia 1 or 2 immersed in each verruca, the black ostiole prominent, at first sunken, then projecting nipple-like above the verruca; epithecium dark, K+ violet as is outer part of hymenium; spores usually 4 (vary from 1 to 6), uniseriate, hyaline, the walls thick, 5–7 μ on sides, 14 μ on ends, 1-celled, elliptical, 70–126 × 25–51 μ.

Growing on mosses, humus, soil, rarely on old wood or over rocks. A widely distributed species known from the southern hemisphere as well as circumpolar in the north, boreal and alpine. In North America reported from Baffin Island, Labrador, Alaska, and the Northwest Territories.

No specimens have yet been seen from the North Slope, although it is fully expected there as it occurs in the Aleutian Islands and the Northwest Territories.

13. **Pertusaria trochiscea** Norm. Thallus a thin gray white crust over the mosses, fruiting verrucae to 1.0 mm broad, usually less, the base

strongly constricted, white, K+ yellow, becoming red, C-, P+ orange; containing stictic acid. Apothecia 1 per verrucule, rarely 2–3, the ostiole black, punctate, slightly pruinose; epithecium black, it and hymenium K+ violet; ascus cylindrico-clavate; spores uniseriate, 4 per ascus, ellipsoid, hyaline, 1-celled, the end walls scarcely thicker than the sides, 67–97 × 27–51 μ.

Growing over mosses and humus. Known from northern Europe, Siberia (Port Clarence, Kamchatka, and Artillery Lake, NWT). New to Alaska.

Locality: 2.

The very similar 4-spored *P. glomerata* (Ach.) Schaer. contains norstictic acid instead of stictic acid.

14. **Pertusaria octomela** (Norm.) Erichs. Thallus thin, whitish, K+ yellow then red, C-, KC-, P+ orange; containing norstictic acid. Fruiting verrucae to 0.8 mm broad, almost spherical, the base constricted, smooth, the ostiole slightly sunken containing 1, seldom 2–3 apothecia; ostiole black bordered; epithecium dark, K+ violet; spores uniseriate or irregularly biseriate, 8 per ascus (rarely other numbers from 4–7), hyaline, the walls thick, 4–6 μ, ellipsoid 63–94 × 26–41 μ.

Growing on mosses, humus, soil, remains of plants, and occasionally old wood. A circumpolar, arctic species known from Siberia, Scandinavia, Greenland, Northwest Territories, and Alaska. Specimens reported by me as *P. glomerata* (12558, 12970) and *P. coriacea* (12473) from Coppermine should be corrected to this species.

Localities: 1, 13.

15. **Pertusaria subobducens** Nyl. Thallus thin, soon superceded by fruiting verrucae which form a thick, knotted mass, 1 mm thick, rounded areolate, convex, shining, yellowish gray, K+ yellow, P+ orange, K̆C-, containing stictic acid. Apothecia within the thallus, with the numerous black ostioles sunken in the upper surface; epithecium dark, K+ violet; hypothecium pale; spores 2, hyaline, simple, ellipsoid, walls thick, to 15 μ, 130–200 × 55–95 μ.

Growing on humus and over plant debris. An uncommon species described from Konyam Bay, Siberia. I have seen specimens from Jenisejsk, Siberia, and Miquelon.

Localities: 1, 2, 3, 9, 10, 11, 13.

This species is very similar to *P. glomerata* but differs mainly in the thallus KOH reaction, yellow instead of red (stictic acid instead of norstictic acid), in numbers of spores 2 instead of 4, and in number of apothecia per verrucule. From *P. coriacea* which it resembles very strongly in habit, it is distinguished by the KOH+ yellow instead of red reaction, (stictic instead of norstictic acid), the epithecium K+ violet instead of K-, and the hymenium I+ blue instead of only the asci being colored.

16. **Pertusaria amara** (Ach.) Nyl. Thallus thin, dark or pale gray, slightly uneven to roughened, not chinky to slightly so, more or less paler zonate toward the margins, sorediate with round to spherical soralia, the soralia sometimes becoming confluent, the soredia whitish to pale greenish, fine in texture, cortex K- then slowly brown, medulla and soralia K+ red to brown red, C-, medulla KC+ violet red, P-, the soralia P- or P+ orange-red (fumarprotocetraric acid), taste very bitter, like quinine; containing picrolichenic acid. Apothecia rare, in fruiting verrucae resembling the soralia, at first covered with the soredia, then opening and the edges reflexing to display the disk, 1, 2, or 3 reddish or brown disks per verruca, becoming epruinose; epithecium of two layers, an outer of loose hyphae, an inner of the brownish tips of the paraphyses; hypothecium yellowish; spores 1 per ascus, simple, hyaline, ellipsoid, the wall thick, 7–9 μ, smooth 143–230 × 50–72 μ.

Growing on tree trunks of broad leaf and coniferous trees, also on rotting logs and over rocks. Circumpolar in the north temperate forests.

Locality: 11. The specimen is var. *flotowiana* (Flörke) Vain., which grows on rocks and has a thick, chinky crust with the soralia KC+ violet turning to dirty red.

ACAROSPORACEAE

Thallus crustose, squamulose, or foliose, homiomerous or heteromerous; in one species attached by an umbilicus; lacking rhizinae. Apothecia immersed in the thallus to adnate, simple or compound, with a proper exciple or a thalloid exciple; ascus containing many spores, the spores very small, hyaline, 1-celled (1–2 in Maronea). Fulcra exobasidial.

1   Proper exciple dark only on the outside, or else lacking; hymenium pale or brownish above   2
        Apothecia with completely dark exciple, immersed in a radiate thallus which is C+ red; hymenium blue green above   5. *Sporastatia*

2 (1) Apothecia immersed in the thallus   3
        Apothecia not immersed in the thallus, the thallus poorly developed   4

3 (2) Apothecia simple; thallus attached along most of the underside by the medullary layer   1. *Acarospora*
        Apothecia compound; thallus nearly umbilicate, only the central portion attached and the marginal portion free   2. *Glypholecia*

4 (2) Apothecia lecideine, the upper part of the hymenium and/or the exciple
   black; growing on rocks   3. *Sarcogyne*
   Apothecia biatorine, the upper part of the hymenium and the exciple, pale,
   not black; on soil or bark   4. *Biatorella*

1. ACAROSPORA Mass.

Thallus crustose, areolate, squamulose, or radiate lobate; heteromerous
with cortex, algal layer and medulla; the upper cortex paraplectenchymatous with isodiametric cells, the lower side ecorticate or sometimes
marginally corticate. Apothecia immersed, the margin lecanorine; spores
numerous, 1-celled, hyaline, small. Pycnidia immersed, sterigmata simple, pycnoconidia cylindrical or broadly ellipsoid. Algae: *Trebouxia*.

1   Thallus yellow   2
      Thallus pale to darker reddish brown, sometimes reddened rust colored,
         occasionally pruinose   3

2 (1) Thallus with the margins radiate lobate; disk of apothecium pale waxy to
         pale brown; on acid rocks in shade   1. *A. chlorophana*
      Thallus squamulose areolate; disk dark reddish brown; on calcareous
         earth   2. *A. schleicheri*

3 (1) Upper cortex C−   4
      Upper cortex C+ red, K−   8. *A. fuscata*

4 (3) Paraphyses 1–1.5 μ thick, hymenium more than 125 μ high   5
      Paraphyses 1.5–2.5 μ thick, hymenium usually less than 125 μ 6

5 (4) Apothecia with smooth disk; thallus K+ red   3. *A. smaragdula*
      Apothecia with rough disk; thallus K−   4. *A. scabrida*

6 (4) Thallus pale below, upper side epruinose   7
      Thallus dark brown to black below, upper side pruinose
                                                        7. *A. glaucocarpa*

7   Apothecia in the areoles; paraphyses less than 2 μ 5. *A. veronensis*
      Apothecia between the areoles; paraphyses greater than 2 μ
                                                        6. *A. badiofusca*

1. **Acarospora chlorophana** (Wahlenb.) Mass. Thallus distinct, to several cm broad, with radiate lobes to 1 mm broad and 1 mm thick, the
center warty areolate or with massed warty lobes, or numerous apothecia, bright yellow, rarely greenish yellow, reactions negative. Apothecia
usually abundant, the disk at first immersed, becoming flat or slightly
convex, smooth, often shining, with a more or less distinct thalloid
margin, the disk at the same height or strongly swollen and the margin
disappearing, concolorous with the thallus or slightly darker; epithe-

cium yellow; hypothecium yellowish; hymenium 50-60 μ; paraphyses 1.5-1.8 μ, tips slightly thicker; spores numerous, elongate with vacuoles at each end, hyaline, 3-4 × 1.5-1.7 μ.

Growing on rocks, especially under overhangs on cliffs. A circumpolar arctic and arctic-alpine species, in North America widely distributed from Greenland to Alaska.

Locality: 11.

2. **Acarospora schleicheri** (Ach.) Mass. Thallus squamulose, not radiate, the squamules 1-5 mm broad, to 0.4 mm thick, small and rounded, becoming areolate, the outer becoming overlapping dark or pale greenish yellow, reactions negative. Apothecia single or 2-3 per squamule, the disk to 1 mm broad, dark rust red, sometimes pruinose when young, flat or convex, level with the thallus or projecting; epithecium yellow brown; hypothecium gray or yellowish; hymenium 85-120 μ; paraphyses 1.8 μ; spores numerous, hyaline, 3-4 × 2-2.5 μ.

Growing on calcareous earth and especially common on steppe soils, occasionally on rocks. A circumpolar species, in North America across the north and southwards to Mexico in the west.

Locality: 1.

3. **Acarospora smaragdula** (Wahlenb.) Th. Fr. Thallus indeterminate, squamulose or areolate, the squamules dispersed or in small groups, to 2 mm broad, 0.4 mm thick, rounded or crenulate, rarely angular, dull, or shining when dark colored, flat or slightly uneven, the pale underside widely attached, yellowish or pale brown, seldom rust brown, C-, K+ red. Apothecia immersed, usually 3-7 per areole, rarely 1-2; the disk to 0.4 mm broad, dark brown or black, even, slightly roughened, sunken or level with the thallus; epithecium brownish yellow; hypothecium dense; hymenium 125-200 μ; I+ brownish or reddish; paraphyses cylindrical, 1-1.5 μ thick, the upper cells moniliform; spores very numerous, elongate or cylindrical, hyaline, 3-4.5 × 1-1.5 μ.

Growing on cliffs, boulders, and especially on boulders manured by birds. A circumpolar arctic and arctic-alpine species ranging in North America south to Mexico in the West.

Localities: 1, 3, 14.

The pale dispersed areoles, the thick hymenium, distinct proper exciple, thin paraphyses, narrow spores, red hymenial reaction with I, and the C-, K+ red reactions of the thallus aid in distinguishing this species.

4. **Acarospora scabrida** Hedl. Thallus crustose, of dispersed or contiguous areoles, sometimes imbricate, 1.5-5 mm broad, rounded, lobate,

with uneven high convex surface, dull or shining, margins free, underside pale, upper side pale or yellowish brown to dark brown, reactions negative. Apothecia usually several in an areole, soon flat and with a thin proper margin concolorous with the disk; the disk dark reddish brown, scabrid, sometimes adjacent disks confluent; epithecium yellowish brown; hypothecium very thick, I+ blue, yellowish; hymenium 120–170 μ, I+ blue or wine-red; paraphyses 1.5 μ thick; spores numerous, hyaline, some appearing 2-celled, 3.5–5.5 × 1.8–2 μ.

Growing on rocks or soil. Known from Europe as an arctic-alpine species and in the arctic from Greenland, Iceland, and Ellesmere Island.

Locality: 3.

The large, thick squamules, the scabrid open disk of the apothecia, the high hymenium, and the reactions aid in identifying this species.

5. **Acarospora veronensis** Mass. Thallus indeterminate, areolate, squamulose, usually the areoles more or less discrete, to 1.5 mm broad, 0.5 mm thick; angular or rounded, convex, seldom flattened, dark red brown, rarely pale or black; reactions negative, under side pale brown. Apothecia deeply immersed, 1–7 in each areole; the disk to 0.5 mm broad, smooth, often shining, concolorous with the thallus; epithecium yellowish brown covered by a dead layer; hypothecium brownish, I+ blue as is also a thin proper exciple; hymenium 65–80 (–100) μ, I+ bright brownish red; paraphyses 1.6–1.9 μ; spores numerous, cylindrical, hyaline, 3.5–6 × 1.7–1.9 μ.

Commonly growing on siliceous rocks, also on calcareous and bird perch rocks. A circumpolar species widely reported across the arctic and southwards.

Localities: 2, 3.

Characteristic of this species are the usually scattered high convex areoles with few impressed apothecia, the pale under surface, the negative reactions of the thallus, and the usually red I+ reaction of the hymenium.

6. **Acarospora badiofusca** (Nyl.) Th. Fr. Thallus of dispersed or contiguous areoles to 3 mm broad, sometimes rugulose, the underside margin pale brown to black, the upper side dull or shiny, grayish brown or reddish brown, reactions negative. Apothecia commencing on top of areoles, becoming large, to 1.5 mm broad; the disk pale reddish brown or darkening to black, sometimes shining, flat or becoming convex and rough; proper margin well developed, raised, blackish, thick, thin, or occasionally disappearing, I+ blue; epithecium yellow brown; hypothecium yellowish, I+ blue; hymenium 60–75 μ, I+ blue; paraphyses thick,

2-3 μ the tips to 4-5 μ; spores often poorly developed, numerous, elliptical, hyaline, 3-4 × 1.5-2 μ.

Usually on trickle surfaces on either calciferous or siliceous rocks. An arctic-alpine species known from Europe, Siberia, and North America.

Locality: 1.

7. **Acarospora glaucocarpa** (Wahlenb.) Körb. Thallus indeterminate, areolate, squamulose, or disappearing, the areoles to 4.5 mm broad and 1 mm thick, often covered by the apothecium, rounded to crenate, flat or subconvex, sometimes with raised white margined edges, underside white, upper side greenish brown, brownish gray, or darker gray brown, commonly pruinose; reactions negative. Apothecia common, mainly solitary in the areoles, finally covering most of the areole and leaving only a thin thalloid margin; disk to 2.5 mm broad, light to dark reddish brown or rarely black, commonly pruinose; epithecium dark brown; hypothecium 30-45 μ, I+ blue; hymenium 65-90 μ, I+ dark blue; paraphyses stout, 2-4 μ thick, the apices brown; spores numerous, simple, hyaline, cylindrical 3.5-5.5 × 1.6-1.8 μ.

This species is primarily found on rocks containing lime. It is widely distributed over the north temperate zone and in the arctic has been reported from Siberia, Greenland, Novaya Zemlya, Spitzbergen, Lapland, and Ellesmere Island.

Locality: 14.

Like *A. fuscata* this is a highly variable species. The commonly pruinose covering, thick paraphyses, low hymenium pale underside, I+ blue reaction of the hymenium and the negative thallus reactions should aid in identification.

8. **Acarospora fuscata** (Nyl.) Arn. Thallus of dispersed, contiguous, or heaped areoles, to 3 mm broad, 0.7 mm thick, smooth, dull or shining, flattened, angular or sublobate, sometimes distributed along cracks in the rocks, the edges quite free, sometimes the areoles almost peltate, underside brown to black, upper side pale or dark brown, red brown, or reddish yellow, C+ red, K-, P-. Apothecia usually abundant, solitary or several in an areole, immersed; the disk punctiform or opening to 1 mm, reddish brown to blackening, smooth or slightly rough, level with the thallus or impressed; epithecium yellowish; hypothecium yellowish, I+ blue as is thin proper exciple; hymenium 70-100 μ, I+ blue, becoming dirty wine red; paraphyses 1.5-2 μ, apices capitate, brownish, 3-5 μ; spores numerous, hyaline, 1-celled, narrowly ellipsoid to subcylindric, 4-6 × 1.5-1.8 μ.

On granitic rocks, cliffs, boulders, especially on bird perch rocks, occasionally on old wood. According to Magnusson this is the most common species of the genus and is widely distributed in the north temperate zone. He mentioned that it avoids the polar regions but it has been reported from Siberia, Lawrence Island, Ellesmere Island, and Spitzbergen.

Localities: 3, 11, and Wainright.

Although this is a very variable species, many forms having been described, the rather flattened squamules with very dark underside, red C+ reaction of the cortex, small spores, and slightly thickened paraphyses aid in identification.

2. GLYPHOLECIA Nyl.

Thallus umbilicate, lobate, attached centrally with free margins, lacking rhizinae. Upper cortex paraplectenchymatous, algal layer well developed, medulla thickly inspersed with calcium oxalate crystals, lower cortex lacking. Apothecia compound, forming multiple structures, the individual apothecia circular or elongate, each with a proper exciple; paraphyses simple; asci clavate, many spored; spores hyaline, simple, spherical. Algae: *Trebouxia*.

1. **Glypholecia scabra** (Pers.) Müll. Arg. Thallus foliose, umbilicate with broad attachment, rounded lobate, lacking rhizinae, the lobes broadly rounded, to 5 mm, partly ascending, concave or slightly convex, the margins rolled under, the central part of the thallus with more or less areolate appearance but all attached together, upper side chalky or bluish white when dry, pale reddish when moist; underside white or darkened, upper cortex C+ red, K-, P-. Apothecia common, immersed, at first single and small, becoming compounded in groups with a pruinose net between, the disks bare, red brown to dark brown, becoming slightly higher than the thallus surface; hypothecium and exciple I+ blue; hymenium 100 μ, I+ blue, hyaline, the upper part yellow brown; paraphyses 2–2.5 μ, the tips clavate, 3.5–4 μ; spores numerous, hyaline, spherical, 3.5–4 μ diameter.

Growing on limestone rocks. A rare species previously known in Europe from northern Norway to France, in Asia Minor, the Himalayas, and in North America from Colorado and Mt McKinley, Alaska.

Locality: 11.

3. SARCOGYNE Flot.

Thallus crustose, very little developed in most species, sub-squamulose in one, endolithic in some. Apothecia lecideine, dark brown to black with distinct exciple which is dark brown to black on the exterior; spores 100–200, cylindric, ellipsoid, or nearly spherical. Alga: *Trebouxia*.

1. **Sarcogyne regularis** Körb. (*S. pruinosa* (Sm.) Körb.) Thallus mainly endolithic and invisible, or thin pale gray or ochraceous to farinose, rarely thicker and slightly chinky, the algae mainly under the apothecia. Apothecia dispersed, sometimes immersed in the substratum, mainly sessile, 0.3–2 mm, quite regularly round, usually with a thickish smooth margin which may disappear with age and may occasionally be pruinose; disk flat or slightly convex, usually blue gray pruinose but can be bare even on the same specimen; epithecium dark reddish yellow to brown; hypothecium pale gray to brownish, I+ blue; hymenium 70–100 μ, I+ blue; paraphyses strongly gelatinous, 1.5–2.5 μ thick, the tips 3–4.5 μ and pale or dark brown; spores 100–200, hyaline, cylindrical or elongate, rarely elliptical, 3–6 × 1.5–2 μ.

This species grows on calcareous rocks, usually in full sun. It is widely distributed around the northern hemisphere and in New Zealand in the southern hemisphere; in the arctic it is known from Scandinavia, Greenland, Ellesmere Island, and the Northwest Territories, and southward in the interior of the North American continent.

Localities: 1, 2, 8, 11, 14.

4. BIATORELLA DeNot.

Thallus crustose, poorly developed, sometimes disappearing, farinose, granulose, or subsquamulose, undifferentiated. Apothecia biatorine with poorly developed pale proper exciple; spores numerous, globose or ellipsoid. Algae: *Trebouxia*.

1. **Biatorella fossarum** (Duf.) Th. Fr. Thallus very thin, arachnoid-farinose to thinly granular, ashy gray to gray greenish, homiomerous. Apothecia dispersed, 0.5–1.4 mm broad, convex and immarginate to almost globose; brown yellow to yellow red; proper exciple disappearing, pale; epithecium pale; hypothecium grayish yellow, sometimes inspersed with oil droplets; I+ blue; hymenium 150–200 μ, hyaline or the upper part yellow; I+ blue; paraphyses 1–1.5 μ, the tips branched and 2–2.5 μ; spores 200–400, hyaline, cylindric, 5–8 × 2–2.5 μ.

Growing over mosses on calcareous soil. Apparently circumpolar boreal and temperate. In the arctic it has been reported from the Bering Straits region. In North America there are reports from Alberta, Illinois, New Jersey, and Washington.

Locality: 2.

5. SPORASTATIA Mass.

Thallus crustose, radiate lobate, corticate, over a dark hypothallus, epilithic. Apothecia immersed, lecideine with a dark-carbonaceous proper exciple; spores numerous, spherical. Algae: *Trebouxia*.

1. **Sporastatia testudinea** (Ach.) Mass. Thallus bordered by a distinct black hypothallus which is K+ red violet; central part areolate, the marginal areoles elongate and distinctly radiate, shining or dull, the areoles sharply angular, 0.3–0.5(–1) mm broad, flat or more or less convex, the black underside bordering them, pale yellowish, coppery, brown, gray black, or brownish black, the outer areolae paler, K-, C+ reddish, P-, I-. Apothecia common, initiated in areolae but soon displacing the areolae and appearing to be between the areolae, immersed, level with the thallus; disk to 1 mm broad, black, smooth or with numerous punctate fissures; margin thin, the apothecia may be separated from the areoles by fissures; epithecium dark or blackish green; hypothecium grayish or purplish brown to yellowish brown, I+ blue; hymenium 85–120 μ, K+ blue green, or the dark part first red violet, HNO_3+ red violet, I+ blue; paraphyses 1.7–2 μ, the tips clavate, 2.5–3.5 μ, brownish or with a violet tinge; spores numerous, broadly ellipsoid or spherical, 3–4 × 2–3 μ.

Growing on siliceous rocks. A circumpolar arctic-alpine species in North America south to the White Mts in the east and to California in the west. Contains gyrophoric acid and a trace of lecanoric acid (Follmann and Hunek 1971).

Localities: 8, 9, 11.

LECANORACEAE

Thallus crustose, squamulose or subfoliose, undifferentiated or differentiated into cortical, algal, and medullary layers, attached by rhizoids but rhizinae lacking. Apothecia immersed to adnate, round, with thalloid exciple and rarely a poorly differentiated proper exciple; hypothecium generally pale; hymenium often inspersed above; paraphyses lax or coherent; spores usually 8(–16), hyaline. Algae: *Trebouxia* or *Trentepolia*.

1   Thallus crustose, lobed at the margins, and with large, rosette-shaped
         cephalodia containing blue green algae in the center   1. *Placopsis*
       Thallus crustose, lobed or not; if smaller cephalodia present then the
         thallus not lobed   2

2 (1) Spores not septate   3
       Spores septate   5

3 (2) Spores large, over 30 μ; paraphyses branched interwoven
                                                        2. Ochrolechia

       Spores small, less than 30 μ; paraphyses unbranched and free   4

4 (3) Alga *Trebouxia*; margin distinctly lecanorine with algae in exciple
                                                          3. Lecanora
       Alga *Trentepolia*; with distinct proper margin covered more or less by a
           thalloid margin   *4. Ionaspis*

5 (2) Spores 1(-3)-septate, ellipsoid; paraphyses lax   *5. Lecania*
       Spores (-1)3: or more-septate, fusiform to acicular; paraphyses coherent
                                                                      6

6 (5) Disk pale yellowish to rosy flesh colored; spore 1-3-septate, fusiform; on
          humus and moss   *6. Icmadophila*
       Disk scarlet; spores 3-7-septate, acicular; on rocks   *7. Haematomma*

1. PLACOPSIS Nyl.

Thallus crustose, adnate to substratum, appearing marginally lobate, heteromerous with upper paraplectenchymatous cortex, algal layer and medullary layer; with superficial centrally located cephalodia containing *Nostoc* algae, sometimes sorediate. Apothecia rare in our species, sessile, with thalloid margin of the same color as the thallus, a thin proper margin within; hypothecium hyaline; hymenium hyaline with thin granular inspersed epithecium; paraphyses slender, free; asci cylindrical; spores 8, biseriate, hyaline, simple, ellipsoid. Algae: *Trebouxia* in thallus, *Nostoc* in cephalodia.

1. Placopsis gelida (L.) Nyl. Thallus crustose, lobed at the periphery and forming round thalli which become confluent into larger patches, the marginal lobes to 2.5 mm long, 1.6 mm broad, the center of the thallus becoming areolate with irregularly angular areoles, cream or ivory colored, often brownish tinged, smooth, dull, not pruinose, soredia scattered over the center and granulose, not on all thalli; cephalodia single in center of small thalli, scattered in older thalli, radially folded, yellow brown, or reddish brown; thallus K-, C+ red, P-, containing gyrophoric acid. Apothecia rare, sessile, with thick thalloid margin; disk round, flat, smooth or minutely roughened, dull, dark flesh colored, yellow brown, or red brown, sometimes with a whitish pruina; epithecium yellowish, granular inspersed; hypothecium hyaline; hymenium 105-183 μ, yellowish; paraphyses not thickened at tips or only slightly;

spores 8, uniseriate or almost so, hyaline, ellipsoid, 1-celled, 12–20 × 6.8–13 μ.

On rocks, sometimes on soil. Circumpolar arctic and temperate, oceanic in distribution, in the northern and also southern hemisphere. In North America south to the White Mts in the east and to California in the west.

Localities: 1, 8, 11. Nylander reported this species from Cape Lisburne in 1887.

2. OCHROLECHIA Mass.

Thallus crustose, smooth, areolate-verrucose, granulose, papillose, tuberculate, spinulose, crumbling effuse, sometimes sorediate or isidiate, in some coralline, yellowish, white, orange, or ashy, sometimes under bark, lacking rhizinae; ecorticate or with a thin cortex of plectenchyma. Apothecia immersed to adnate or sessile, a prominent thalloid exciple present, sometimes the disk is compound; hypothecium yellowish; paraphyses branched and interwoven, septate; asci clavate or ovoid; spores 2–8, large, hyaline, 1-celled, the walls thin to slightly thickened, ovoid to ellipsoid, Algae: *Trebouxia*.

1	Thallus sorediate	2
	Thallus lacking soredia	3
2 (1)	Thallus and soredia P-, KC+ red, containing gyrophoric acid; apothecia rare; ascus with 2 spores	1. *O. androgyna*
	Thallus P-, soredia P+ orange, thallus KC+ red, containing gyrophoric acid; apothecia common; ascus with 6–8 spores	2. *O. inaequatula*
3 (1)	Thallus of small button-like glomerules over an exceedingly thin cobwebby prothallus over mosses	3. *O. grimmiae*
	Thallus more abundant, continuous or of massed glomerules	4
4 (3)	Thallus KC-; lacking spinules, containing 3 undetermined substances; disk pale yellowish or tan, K-, C-	4. *O. upsaliensis*
	Thallus KC+ red, often with white spinules, containing gyrophoric acid; disk pink to reddish brown, K+ red, C+ pink	5. *O. frigida*

1. **Ochrolechia androgyna** (Hoffm.) Arn. Thallus white to pale gray, diffuse, granulose to verrucose or verrucose-chinky, with large rounded soralia which are elevated and of granular soredia, the cortex with negative reactions to the usual reagents, the medulla K+ yellow, C+ red, P-, containing gyrophoric acid, the soredia P+ red. Apothecia rare, sessile, 2–4 mm broad; the margin thick, entire or sorediate; the disk reddish brown, concave, rough; K+ pale red, C+ red, P-; epithecium brownish yellow; hypothecium yellowish; hymenium hyaline, 100–140 μ; paraphyses thin, 1.5–2.5 μ; spores 8, 1-celled ellipsoid, 30–45 × 13–22 μ.

Growing over mosses, bark of trees and shrubs, sometimes on soil. It is a circumpolar arctic to temperate species. Howard reported it south to Maryland and Florida in eastern North America, and to Washington in the west.

Localities: 1, 2, 8.

The very similar *O. geminipara* (Th. Fr.) Vain. has both the thallus and the soredia P+ red orange. If with apothecia the disk will be black purple. It has been reported near Nome.

2. Ochrolechia inaequatula (Nyl.) Zahlbr. Thallus white or ashy white to bluish gray, thin, rough granular the granules developing into coarse soredia which are white or yellow and sometimes covering the thallus, K+ yellow, C-, KC+ red, P-, the soredia P+ orange; containing gyrophoric acid. Apothecia sessile, 1-4 mm broad; the margin thick, smooth, entire to granulose; disk concave, becoming flat or convex, pink, pale yellow brown or red brown, roughened, K-, C+ pink, P-; epithecium ashy yellow; hypothecium grayish; hymenium hyaline, 180-200 μ; paraphyses slender; spores 8 in ascus, 1-celled, broadly ellipsoid, 20-54 × 23-28 μ.

Growing on mosses and humus, sometimes on soil. A circumpolar arctic species in North America south to Nova Scotia and Washington.

Localities: 1, 2, 4, 5, 6, 8, 9, 11, 14.

3. Ochrolechia grimmiae Lynge Thallus of a very thin cobwebby white prothallus with many small scattered round glomerules over the mosses, K+ yellow, C+ red, P-. Apothecia adnate to sessile, 1-3 mm broad, concave to becoming flat or convex; the margin white, thin, entire or crenulate, esorediate; disk deep pink to yellowish or brownish red, smooth to roughened; epithecium brown; hypothecium brown; hymenium hyaline, 176-200 μ; paraphyses thin; spores 8 per ascus, 1-celled, ovoid, 37-40 × 14-19 μ.

Growing over *Grimmia* mosses. A circumpolar species, arctic. In North America from Alaska, Quebec, Baffin Island, Northwest Territories, and Greenland.

Localities: 2, 11.

4. Ochrolechia upsaliensis (L.) Mass. Thallus white or buff to gray, thin, becoming rough and verrucose, K-, C-, KC-, P-. Apothecia common, sessile, 0.5-3 mm broad; the margin thick, entire, concolorous with the thallus; the disk pale buff, yellowish, or grayish, flat, scabrid, white pruinose; K-, C-, P-; epithecium brown with greenish tinge; hypothecium brown; hymenium hyaline, 204-278 μ, inspersed with tiny crystals; paraphyses slender; spores 8, 1-celled, ellipsoid, 31-75 × 23-37 μ.

Growing over mosses, Selaginellas, plant bases, and soils. A circumpolar arctic-alpine species which in North America ranges south to
Quebec in the east and to New Mexico and California in the west. It is
also reported in the Andes in South America. It is abundant on the
North Slope.

Localities: 1, 2, 3, 8, 9, 10, 11, 14.

5. **Ochrolechia frigida** (Sw.) Lynge Thallus white, yellowish, pinkish,
grayish brown or gray, very variable in appearance, from a thin crust
over plants to thicker granules, verrucae, or a thick crust with few or no
spinulose processes to composed mainly of the branching spinulose
fruticose processes, occasionally forming cylindrical coralloid branches,
K-, C-, KC+ red, P-, containing gyrophoric acid. Apothecia common to
5 mm broad, sessile to adnate; the margin thin to slightly thickened,
entire or wrinkled, concolorous with the thallus; disk flat, red brown to
brown, slightly roughened, epruinose, K+ red, C+ pink, P-; epithecium
brown; hypothecium brown; hymenium hyaline, 130–180 μ; paraphyses
slender; spores 8, 1-celled, ovoid, 26–43 × 15–26 μ.

Growing on soil, mosses, humus, plant remains, and base of shrubs. A
circumpolar arctic-alpine species in North America ranging south to
New Hampshire in the east, Colorado in the west. It is very variable and
a key to the forms follows:

1   Thallus thin to granulose, not fruticose, with or without spinulose out-
        growths, these short when present   f. *frigida* (=f. *typica* Lynge)
        Thallus branching fruticose   2

2 (1) The thallus thick blunt, papillate becoming coralloid, yellowish
                                                    f. *gonatodes* (Ach.) Lynge
        The thallus of thin, branching, spinulose growths
        f. *thelephoroides* (Th. Fr.) Lynge (including *O. pterulina* (Nyl.) Howard)

Localities: f. *frigida*: 1, 2, 3, 4, 5, 9, 10, 12; f. *gonatodes*: 1, 9, 13; f.
thelephoroides: 1, 2, 3, 4, 5, 6, 12, 13.

The sterile thalli of this and of *Lecanora epibryon* are often growing
together and are very similar. *L. epibryon* contains atranorin and is K+
yellow, this species contains gyrophoric acid and is K-, KC+ red.

3. LECANORA Ach.

Thallus crustose, diffuse or effigurate, to subfoliose or umbilicate; the
upper cortex poorly developed or in the squamulose or foliose species
with a well developed palisade-like upper cortex, algal layer usually well
developed, medulla well developed and in some a poorly developed lower

cortex. Apothecia immersed in the thallus or sessile to adnate; disk concave to flat or convex; a well developed thalloid margin colored like the thallus; hypothecium hyaline, yellowish, or brownish; hymenium colored above; paraphyses unbranched, coherent or free; asci clavate; spore 2–8, rarely 16–32, hyaline, non-septate. Alga: *Trebouxia*.

KEY TO SUBGENERA OF LECANORA

1   Thallus crustose   2
      Thallus squamulose to subfoliose or umbilicate; both upper and lower
        cortices present; apothecia adnate to sessile   Subgenus *Placodium*

2 (1) Apothecia adnate to sessile on the thallus; epithecium pale or brownish,
        lacking 'aspicilia green' and not turning green with HCl or HNO_3
            Subgenus *Lecanora*
      Apothecia immersed in the thallus; epithecium greenish and containing
        'aspicilia green' which turns very green with HCl or HNO_3
            Subgenus *Aspicilia*

Subgenus *Lecanora*

1   On wood, shrubs, or over mosses or soil   2
      On rocks   12

2 (1) On wood and shrubs   3
      Over mosses and on soil   8

3 (2) Thallus sorediate, yellowish   1. *L. expallens*
      Thallus esorediate   4

4 (3) Thallus white, K+ yellow (atranorin); paraphyses coherent   5
      Thallus ashy gray to yellow brown, K–   8

5 (4) Disk pale reddish brown; large crystals present in apothecia   6
      Disk becoming black; cortex of apothecium with large crystals; margins
        P–   4. *L. coilocarpa*

6 (5) Epithecium with fine granules insoluble in concentrated HNO_3;
        margins P+ red   2. *L. chlarona*

      Epithecium with coarse granules at surfaces soluble in concentrated
      HNO_3; margins P–   3. *L. rugosella*

7 (4) Disk pale yellow to yellow brown ± pruinose; paraphyses lax; spores
        10–16 × 6–7 μ 5. *L. distans*
      Disk black, ashy to bluish pruinose; paraphyses coherent; spores 10–14
        × 6.5–8 μ 6. *L. vegae*

8 (2) Apothecia red brown, brown, or blackening   9
      Apothecia olive black; thallus yellow or greenish yellow
            11. *L. leptacina*

9 (8) Apothecia brown, dull, K+ yellow (atranorin), thallus indistinct, brown;
spores 14–17 × 5–7 μ   *7. L. castanea*
Apothecia red brown to blackening, disk shining or pruinose; thallus
white, prominent   10

10 (9) Exciple of irregular paraplectenchyma; thallus K+ yellow; spores 14–19
× 8–9 μ   *8. L. epibryon*
Cortex of reverse side of apothecium of palisadeplectenchyma; thallus
and apothecial margins K–; spores less than 6 μ broad   11

11 (10) Disk red brown, epruinose; exciple cortex 35–50 μ and only slightly
thinned to upper edge of margin; spores 9–14 × 4–6 μ
9. L. behringii
Disk blackish lead colored, pruinose; exciple cortex thinning to upper
edge; spores 11–16 × 5 μ   *10. L. palanderi*

12 (1) Thallus mainly lacking   13
Thallus well developed   20

13 (12) Apothecia with thick, deeply radiately cracked and divided margins; disk
pruinose, reddish brown   *12. L. crenulata*
Apothecial margin not radiately divided; disk bare   14

14 (13) Disk yellow to pale orange; margin thin, disappearing *13. L. polytropa*
Disk dark, olive green, red brown, or black   15

15 (14) Disk olive green, margin yellowish   ·   *14. L. intricata*
Disk dark, red brown to black   16

16 (15) Cortex of apothecium with radiate, palisade-like hyphae; disk red
brown to black   *9. L. behringii*
Cortex of irregularly arranged hyphae   17

17 (16) Margin thin, epruinose, disk brown to brown black   18
Margin thick, crenulate, pruinose, ecorticate; disk red brown
15. L. dispersa

18 (17) Apothecia with narrow base, constricted; epithecium brown; spores
9–15 × 4–6 μ   *16. L. nordenskioeldii*
Apothecia with broad base   19

19 (18) Epithecium blue green brown; cortex poorly developed; spores 8–11 ×
4–5 μ   *17. L. torrida*
Epithecium brown; cortex 30 μ thick, gelatinoid; spores 7–14 × 4–6 μ
18. L. hageni

20 (12) Thallus brown, red brown, or ashy brown   21
Thallus yellow or white   23

21 (20) Thallus of red brown squamules in an areolate crust; apothecia blood
red, immersed in the thallus   *19. L. granatina*
Thallus brown to ashy brown; apothecia black, adnate   22

22 (21) Thallus thin, inconspicuous; apothecia to 0.35 mm broad; upper part of

hymenium emerald green; spores ellipsoid, 14–15 × 7–9 μ

20. *L. microfusca*

Thallus thicker, conspicuous; apothecia to 2 mm broad, upper part of hymenium brown; spores fusiform, 10–16 × 3–6 μ 21. *L. badia*

23 (20) Thallus yellow, K+ very yellow   24

Thallus white or ashy white   28

24 (23) Thallus C+ red, pale sulphur colored; apothecia black

22. *L. atrosulphurea*

Thallus C−   25

25 (24) Apothecia blue gray pruinose, immersed; thallus sulphur colored, rugose   23. *L. caesiosulphurea*

Apothecia epruinose, adnate; thallus yellow to greenish yellow   26

26 (25) Areolae very swollen convex; apothecia red brown; spores 10–18 × 5–9 μ, ellipsoid; containing atranorin and zeorin   24. *L. frustulosa*

Areolae thinner, granulate-areolate, containing usnic acid   27

27 (26) Disk yellow to light orange   13. *L. polytropa*

Disk olive to olive black   14. *L. intricata*

28 (23) Thallus with brownish cephalodia   25. *L. pelobotrya*

Thallus lacking cephalodia   29

29 (28) Hymenium violet   26. *L. atra*

Hymenium not violet   30

30 (29) Upper part of the hymenium blue green to emerald green

17. *L. torrida*

Upper part of the hymenium brown; disk blue pruinose, C+ yellow; thallus K+ yellow containing atranorin and rocellic acid

27. *L. rupicola*

Subgenus *Aspicilia*

1   On humus and over mosses; apothecia black, sometimes pruinose, immersed; spores 3–48 × 16–32 μ 28. *L. urceolaria*

On rocks and bones   2

2 (1) Thallus effuse, neither zonate nor radiate   3

Thallus either radiate or zonate   17

3 (2) Upper part of hymenium blue to green; thallus thin, ashy orange, sub-granular; spores 12–18 × 7–11 μ 29. *L. flavida*

Upper part of hymenium yellowish brown to olive   4

4 (3) Thallus I+ blue, K−; disk red brown; thallus dark gray, areolate

30. *L. cinereorufescens*

Thallus I−   5

5 (4) Thallus K+ red (norstictic acid), ashy gray, often with black hypothallus

31. *L. cinerea*

Thallus K−   6

6 (5) Apothecial disk yellowish to pale brown or reddish; spores 11–20 × 5–8
       μ; on rocks by lakes, streams, tarns   32. *L. lacustris*
       Apothecial disks black; growing in dry sites   7

7 (6) Disk white pruinose; margin thick   33. *L. nikrapensis*
       Disk bare   8

8 (7) Spores large, 20–35 μ, medulla transparent   9
       Spores smaller; thallus with medulla gray with granules which dissolve
       in HCl   10

9 (8) Growing in dry places, cortical cells large, not perpendicular; spores
       20–30 × 13–16 μ; thallus blue gray, verruculose areolate
                                                        34. *L. caesiocinerea*
       Growing in wet places, frequently submerged, cortical cells in distinctly
          perpendicular rows; spores 22–25 × 14–18 μ; thallus blue gray,
          chinky-areolate   35. *L. aquatica*

10 (9) Thallus white or orange white   11
       Thallus ashy gray, blackish gray, to brownish gray   13

11 (10) Thallus white   12
        Thallus orange and white, the orange part lower, white areoles stand-
           ing higher and scattered over the rest, fertile areoles also higher and
           prominent; apothecia partly confluent; spores 15–21 × 11–12 μ
                                                           36. *L. heteroplaca*

12 (11) Thallus of discrete verrucae; apothecia compound   37. *L. composita*
        Thallus chinky-continuous; apothecia simple   38. *L. anseris*

13 (10) Thallus very thin, of discrete, scattered areoles, blackish gray, to 0.5
           mm broad; apothecia solitary; hymenium I+ wine red; spores 11–17 ×
           7–10 μ   39. *L. ryrkaipiae*
        Thallus thin or thicker, contiguous, ashy gray or brownish gray   14

14 (13) Apothecia prominently adnate above the thallus, the margins thick; the
           thallus thin   15
        Apothecia at same level as thallus; thallus thick, areolate; spores greater
           than 17 μ   16

15 (14) Thallus membranaceous, clay colored, fanning out at the edges; spores
           small, 12–17 × 8–10 μ   40. *L. annulata*
        Thallus slightly thicker, chinky-areolate, pale ashy to ashy brown;
           spores broadly oval, 14–23 × 10–12.5 μ   41. *L. elevata*

16 (14) Thallus very verrucose, whitish gray; margin of apothecium separated
           from thallus by a fissure; exciple I+ blue; spores 17–25 × 12–16 μ
                                                           42. *L. pergibbosa*
        Thallus brownish gray; apothecial margin not fissured; exciple I–;
           spores 15–25 × 10–14 μ   43. *L. supertegens*

17 (2) Thallus zonate   18
       Thallus distinctly radiate, lobate, grayish to white   24

18 (17) Thallus K+ (either yellow or red)   19
        Thallus K-   22

19 (18) Thallus K+ yellow (atranorin)   20
        Thallus K+ red (salazinic acid)   21

20 (19) Thallus lobes separate, narrow, brown, radiate; exciple I+ blue; spores 12–17 × 7–8.5 μ   44. *L. rosulata*
        Thallus lobes contiguous, low, zonate toward the margins, gray, radiate splaying over a black hypothallus; exciple I+ blue; spores 15–18 × 8.5–11 μ   45. *L. perradiata*

21 (19) Thallus blackish, shiny; hypothallus black, lobes indistinct; spores 17–25 × 10–15 μ   46. *L. stygioplaca*
        Thallus brownish gray, lobes discrete, narrow; paraphyses non-moniliform; spores 14–16 × 8.5 μ   47. *L. fimbriata*

22 (18) Thallus yellowish gray, thin, indistinctly zonate   23
        Thallus brownish gray, chinky, medulla filled with granules; paraphyses submoniliform; spores 17–22 × 10–12 μ   43. *L. supertegens*

23 (22) Medulla filled with granules; paraphyses distinctly moniliform; spores 13–20 × 8–13 μ   48. *L. polychroma*
        Medulla transparent; paraphyses not moniliform; spores 13–17 × 8–10 μ
                                                  49. *L. cingulata*

24 (17) Surface of exciple with K+ yellow diffusion   25
        Surface of exciple K-   26

25 (24) Thallus broad, lobes not distinct; apothecia pruinose, exciple I-; spores 14–24 × 10–16 μ   50. *L. candida*
        Thallus small, lobes short, distinct; apothecia epruinose; exciple I+ blue; spores 9–16 × 7–10 μ   51. *L. lesleyana*

26 (24) Thallus brownish gray; cortical hyphae perpendicular   27
        Thallus white; cortical hyphae not perpendicular   28

27 (26) Larger apothecia over 0.5 mm broad; thallus lobes of contiguous lobes separated by cracks; spores 14–26 × 8–10 μ   52. *L. sublapponica*
        Apothecia less than 0.5 mm broad; thallus of very thin radiately plicate lobing not separated by cracks but by thallus portions; spores 10–15 × 5–8 μ   53. *L. plicigera*

28 (26) Thallus bluish white; lobes separate, paraphyses moniliform
                                                  54. *L. disserpens*
        Thallus with yellowish cream tinge; lobes contiguous; paraphyses non-moniliform; spores 18–23 × 10–12.5 μ   55. *L. concinnum*

Subgenus *Placodium*

1   Thallus with cephalodia   2
        Thallus lacking cephalodia   3

2 (1) Cephalodia rosette-shaped, central on thallus; thallus chinky centrally, lobate marginally, often sorediate; medulla C+ red   *Placopsis gelida*
    Cephalodia rounded capitate, scattered over the thallus; margins not effigurate, esorediate; medulla C-   25. *L. pelobotrya*

3 (1) Thallus umbilicate   4
    Thallus not umbilicate   6

4 (3) Disk red brown, often pale yellowish pruinose; thallus whitish or yellowish green, dull; medulla P+ or P-   56. *L. chrysoleuca*
    Disk brown or black   5

5 (4) Disk black, greenish pruinose; apothecia broadly attached; thallus bright yellow green, smooth, often shining; paraphyses dark capitate; P+ yellow (psoromic acid)   57. *L. melanopthalma*
    Disk yellowish brown to brown, epruinose; apothecia soon adnate, thallus dull; paraphyses not capitate; P+ or P-   58. *L. peltata*

6 (3) Thallus of white flat lobes forming rosettes on soil, pruinose above, P-, containing usnic acid and atranorin; disk flat or convex, brown
    59. *L. lentigera*
    Thallus darker green, not white   7

7 (6) Thallus grayish glaucous to dark olive, lobes convex, epruinose, cortex 12 μ; medulla P+ red (norstictic acid); apothecia red brown to black
    60. *L. melanaspis*
    Thallus yellow green, olive to brownish, areolate squamulose, lobes flat, crenulate, cortex 50 μ; medulla P+ or P-, containing usnic acid; apothecia pale olive to brown red   61. *L. muralis*

1. **Lecanora expallens** Ach. Thallus thin, chinky areolate, yellowish, sorediate. Apothecia adnate to sessile, to 0.8 mm broad, the margin thick and crenulate or disappearing; disk flat to convex, pale brownish to yellowish flesh colored; sometimes pruinose; epithecium pale yellowish; hypothecium pale; spores 8, simple, hyaline, ellipsoid, 10–17 × 6–7.5 μ.

Growing on wood and bark, especially of conifers. Circumpolar boreal and arctic, in North America ranging south to New England in the east, California in the west.

Locality: 13.

2. **Lecanora chlarona** (Ach.) Nyl. Thallus thin crustose, areolate, corticate, darker yellowish gray. Apothecia small, to 1 mm broad, margin crenulate, P+ red (fumarprotocetraric acid); disk pale yellow, red brown, to dark red brown; epithecium with fine granules on the surface and between the paraphyses tips; upper hymenium pigmented red brown to brown black between the paraphyses tips; amphithecium usually with large crystals; hypothecium brownish; spores 8, hyaline, simple, ellipsoid, 9–15 × 7–10 μ.

Growing on bark of conifers and deciduous woody plants. Circumpolar, north temperate, and boreal. In North America south to Maine, Minnesota, Oregon.

Localities: 3, 14.

3. **Lecanora rugosella** Zahlbr. Thallus continuous verrucose to dispersed verrucose, K+ yellow. Apothecia pale brown to pink brown, rarely black, with very thick crenulate to verrucose margins, the margins with large, coarse, oxalate crystals; epithecium with coarse granules at the surface, the granules soluble in HNO_3, P-; disk C-; apothecial cortex over 15 μ thick, gelatinous; hymenium brown above; paraphyses capitate; spores with walls more than 1 μ thick, 11–18 × 6–11.5 μ.

Growing on barks and old wood. Pending monographic work on *Lecanora* by Dr I.M. Brodo the distribution cannot yet be stated accurately. It may be circumpolar temperate to subarctic.

Locality: 8.

4. **Lecanora coilocarpa** (Ach.) Nyl. Thallus over a blue black hypothallus, thin, white, K+ yellow. Apothecia small, to 1 mm broad, margin thin, usually entire to slightly crenulate, P-, disk dark red brown to usually black; epithecium with fine granules on the surface and between the tips of the paraphyses; upper hymenium pigmented; red brown to dark red black; amphithecium with large crystals; hypothecium brownish; spores 8, hyaline, ellipsoid, simple, 10–18 × 8–10 μ.

Growing on conifers and on broad leaved shrubs and trees. A boreal and alpine species, circumpolar in the boreal forest and in the Rocky Mountains.

Localities: 3, 8, 13, 14.

5. **Lecanora distans** (Pers.) Nyl. Thallus with more or less zonate edge, whitish, thin, K-. Apothecia small, margin entire, thin, gelatinous, the cortex 50–100 μ thick; disk waxy red to roseate, bare or pruinose; epithecium with fine granules; upper part of the hymenium pale yellow brown; medulla filled with small octahedral crystals; hypothecium hyaline; paraphyses free; spores 8, simple, ellipsoid, 10–16 × 6–7.5 μ.

Growing especially upon aspen. A circumpolar boreal species.

Locality: on *Salix*, 3.

6. **Lecanora vegae** Malme. Thallus crustose, of dispersed or contiguous semiglobose to irregular 0.5 mm broad verrucules, ashy, yellowish brown or brownish ashy, dull, K-, C-, I-; cortex irregular, 20 μ thick. Apothecia dense, at first innate and with a thick, entire margin, becoming adnate and well elevated, to 0.8 mm broad, the margin becoming

excluded, with a thin, black, proper margin; disk flat, black, ashy or bluish pruinose; hypothecium hyaline; epithecium brown red; hymenium 75 μ, the upper part colored red brown, I+ blue or turning reddish; paraphyses 1 μ, coherent, not capitate; spores 8, ovoid, 10–14 × 6.5–8 μ.

This species was described on old wood in Siberia. It is new to North America. It resembles *L. caesiosulphurea* but the different cortical reactions and the disappearing margins will separate these two species.

Locality: 8.

7. **Lecanora castanea** (Hepp) Th. Fr. Thallus thin, granulose, brownish, ashy, often disappearing, K+ yellow (atranorin), dull. Apothecia flat to slightly convex, to 3 mm broad, disk brown, to liver brown, shining, margin thick and inflexed at first, becoming reflexed and disappearing; epithecium brown; upper part of hymenium brown; hypothecium hyaline; spores 8, simple, oblong, fusiform, to cylindrico-elongate, 14–17 × 5–7 μ.

Growing on mosses, sometimes soil. A circumpolar arctic-alpine species.

Localities: 1, 3, 6, 8, 13.

8. **Lecanora epibryon** (Ach.) Ach. Thallus white, verrucose, shining, K+ yellow (atranorin). Apothecia large, to 3 mm, margins entire to crenulate with a distinct cortex 12–35 μ thick; containing many small crystals in the medulla of the amphithecium; disk dark red brown, often shining, smooth; darkening in severe conditions; epithecium not inspersed, with hyaline gelatinous surface over the hymenium; upper part of the hymenium a clear red brown; hypothecium brownish; spores 8, hyaline, simple, ellipsoid, 14–19 × 8–9 μ.

Growing over mosses and other lichens and on soil, especially clay soils and frost boils, pioneering in such places. A circumpolar arctic-alpine species which in North America ranges south into New England in the east, to New Mexico in the west. Abundant on the North Slope.

Localities: 1, 2, 3, 4, 7, 8, 9, 10, 11, 12, 13, 14.

9. **Lecanora behringii** Nyl. Thallus crustose, of verrucules or areoles, thin to disappearing, bluish white or ashy, dull or shining, K-. Apothecia to 2 mm broad, the base well constricted, almost stipitate; the margin thick, pale, entire, with a thick cortex of palisade-like hyphae; disk flat, red, red brown, or blackening; epithecium reddish or brownish to violet reddish; hymenium 40–65 μ, upper part colored; paraphyses coherent, slender, the apices slightly thickened; spores 8, hyaline, simple, oblong to ellipsoid, 9–14 ×4–6 μ.

Growing on a wide variety of substrata, wood, humus, dead mosses, rotting bones, and rocks. An arctic species known from Siberia, Greenland, Northwest Territories, and Alaska.

Localities: 11, 13, 14.

10. **Lecanora palanderi** Vain. Thallus crustose, thin, continuous, smooth, gray to bluish white, dull, K-; hypothallus indistinct. Apothecia 0.6–1.6 mm broad, the base constricted; disk flat, reddish brown to lead brown or blackish, dull, pruinose; margin lecanorine, thin, white or whitish, dull, entire to slightly flexuous, persistent; cortex of palisade-plectenchyma, thick walled cells, hyaline, not granulose; hypothecium thin, 15–30 μ, pale, of irregular conglutinate hyphae; hymenium 60–70 μ, I+ blue; epithecium reddish, K-, HNO$_3$+ reddish; paraphyses coherent, slender, 1–1.5 μ septate, the apices clavate, sparsely branched; asci clavate; spores 8, biseriate, simple, hyaline, ellipsoid, 11–16 × 5 μ.

Growing on old wood and vegetational detritus. A species described from Siberia and reported by Ahti et al. (1973) from the Reindeer Preserve, NWT, area. Additional Canadian records in WIS include the Keele River region, NWT, Scotter 15533, 15604, 15645; Ellesmere Island; Marvin Peninsula, Brassard 4066; and Lineham Lake area, Waterton Lakes National Park, Alberta, Scotter 4926. It is probably to be classed as a member of the Beringian element in the flora.

Locality: 11.

11. **Lecanora leptacina** Sommerf. Thallus scant, granulose-squamulose with the squamule margins crenulate, yellow, yellowish green, K-. Apothecia to 1 mm broad, margin thick persistent; disk olive green to blackish; epithecium yellowish brown; hypothecium pale; hymenium yellowish brown above; paraphyses coherent; spores 8, hyaline, simple, ellipsoid, 10–12 × 4.5–6 μ.

Growing over mosses and humus. Known from northern Europe, Greenland, Bear Island, Alaska.

Localities: 2, 3, 5.

12. **Lecanora crenulata** (Dicks.) Nyl. Thallus lacking or of a few scattered granules, K-, the algae mainly under the apothecia. Apothecia to 1 mm, concave with thick white, chinky radiately divided margin; disk yellowish or reddish brown with gray pruina; epithecium inspersed with granules; hypothecium pale; paraphyses thick with the tips brown and clavate, slightly moniliform; hymenium hyaline below, yellow brown above; spores 8, simple, hyaline, ellipsoid, 10–15 × 4–7 μ.

Growing on calcareous rocks. A circumpolar, arctic-alpine species.
Localities: 1, 2, 3, 14.

13. **Lecanora polytropa** (Ehrh.) Rabenh. Thallus crustose, granulose or chinky areolate to subsquamulose or more commonly disappearing, pale yellowish, KC+ yellow (usnic acid). Apothecia waxy yellow to pale fleshy yellow, small, to 1.5 mm, margin at first thick, yellow, becoming excluded; disk flat to convex, bare; epithecium pale; hypothecium yellowish; paraphyses slender; spores 8, hyaline, simple, ellipsoid, 10–15 × 5–7 μ.

On rocks, both calcareous and acid. A circumpolar, arctic-alpine and boreal species, ranging south in North America across the northern states and into California.
Localities: 2, 3, 8, 9.

14. **Lecanora intricata** (Schrad.) Ach. Thallus of areoles which are partly separated, slightly lobed, greenish gray, whitish, or greenish yellow, sometimes on black hypothallus. Apothecia small, to 0.8 mm, adnate, separate or clustered on an areole; flat to convex; margin prominent and colored like the thallus; disk olive black; epithecium dark; hypothecium pale; spores 8, hyaline, simple, ellipsoid, 10–13 × 4.5–6.5 μ.

Growing on calcareous or acid rocks. A circumpolar arctic-alpine species reported south in North America to New England mountains.
Localities: 2, 11.

15. **Lecanora dispersa** (Pers.) Somm. Thallus lacking or of small scattered greenish or whitish granules, K-. Apothecia scattered or grouped, sometimes angular in the groups; margins persistent, thick, entire or becoming crenulate but not radiately cracked as in *L. crenulata*, white or yellowish, pruinose; disk flat, red brown, yellowish brown or olive, bare, to 1.2 mm broad; epithecium granular inspersed, yellowish brown; hypothecium brownish; hymenium yellowish brown above; paraphyses coherent, slender, the tips clavate; spores 8, simple, hyaline, ellipsoid, 8–13 × 3.5–6 μ.

On rocks, calcareous or acid. Circumpolar, arctic, boreal, and temperate.
Localities: 1, 2, 3, 8, 14.

16. **Lecanora nordenskioeldii** Vain. Thallus crustose, of more or less dispersed small granules, whitish or ashy, K-. Apothecia to 1.2 mm, the base constricted, dispersed or aggregated; margin thin, colored like the thallus, entire or crenulate; disk flat or becoming convex, black or brownish black; exciple with thick cortex of irregular paraplectenchyma,

not palisade-like; epithecium red, red brown, or brown, HNO_3 purple; hypothecium pale; hymenium 80 μ, pale; paraphyses coherent, thin, tips clavate, upper part slightly branched and interconnected; spores 8, hyaline, simple, ellipsoid, 9–15 × 4–6 μ.

On calcareous rocks. A species of the arctic, known from Siberia and Greenland as well as Alaska.

Localities: 2, 3, 8, 9, 11.

17. **Lecanora torrida** Vain. Thallus of small sparse verrucules, white, dull, K-. Apothecia to 0.4 mm broad, scattered, adnate, flat or slightly convex; margin thin, white, entire, persistent, with very poorly developed cortex; disk black, dull, bare; epithecium bluish or bluish brown, HNO_3+ violet; hypothecium hyaline; hymenium I+ blue; paraphyses coherent, septate, unbranched or slightly branched, 1.5–2 μ thick; spores 8, simple, hyaline, ellipsoid, 8–11 × 4–5 μ.

Growing on calcareous rocks. An arctic species known from Siberia, Bear Island, Greenland, and Alaska. New to North America.

Locality: 1.

18. **Lecanora hageni** (Ach.) Ach. Thallus usually lacking. Apothecia usually dispersed, to 0.5 mm broad, the margin white, thin, smooth or crenulate, often slightly powdery, level with the disk; disk flat, dull, flesh colored to brownish or blackish, bare or more or lees blue gray pruinose; epithecium brown to olive brown, HNO_3-; hypothecium hyaline; hymenium 60–66 μ; paraphyses slender, unbranched, not capitate; spores oval, 7–14 × 4–6 μ.

Usually on barks and old wood, this species may occur on rocks and then be confused with L. *dispersa* which has an ecorticate, crenulate, usually pruinose margin. It is circumpolar, temperate to subarctic.

Locality: 8.

19. **Lecanora granatina** Somm. Thallus of granular, chinky, umbilicate verrucules, black below, red brown above. Apothecia minute to small, to 0.3 mm, adnate; margin thin, crenulate, disappearing; disk flat to convex, red brown; epithecium brown; hypothecium brownish; hymenium hyaline, 125–150 μ; paraphyses slender, unbranched, septate, not thickened at tips, 2 μ, gelatinized; spores hyaline, simple, oblong, 9–13 × 4–5.5 μ.

Growing on acid rocks. A circumpolar arctic alpine species, rather rarely collected. Known from Europe, Siberia, Greenland, Ellesmere Island, Novaya Zemlya, Maine, New Hampshire, and recently collected in the Olympic Mountains, Washington.

Locality: 11.

20. **Lecanora microfusca** Lynge Thallus very inconspicuous, areolate, the areoles 0.3 mm broad, thin, scattered or contiguous, irregular, angular, flat, rough, brown, paler margins slightly raised. Apothecia small, to 0.35 mm, the base constricted, margins brown, entire, persistent, disk black, scabrid, flat to convex; exciple with thin cortex, 15 μ thick, the exterior brown, interior pale, epithecium blue; hypothecium brownish; hymenium 65 μ, upper part emerald green; paraphyses coherent, thin, the tips not thickened; spores 8, simple, hyaline, ellipsoid, 14–15 × 7–9 μ.

On calcareous rocks. An arctic species known from Greenland. New to North America.

Locality: 2.

21. **Lecanora badia** (Pers.) Ach. Thallus granulose, warty, to areolate or subsquamulose, olive green, ashy, or brown, K-, C-, containing lobaric and usnic acids. Apothecia to 1.5 mm broad, adnate to sessile; the margin thick, colored as the thallus, entire to crenulate; disk chestnut brown to brown or brownish black, shining; epithecium brown; hypothecium hyaline or yellowish; hymenium brownish above; paraphyses coherent, septate, the tips thickened, brown; spores 8, hyaline, simple, fusiform, the tips pointed, 10–16 × 3–6 μ.

Preferring acid rocks. A circumpolar arctic alpine species in North America south to New England mountains and west to New Mexico.

Localities: 3, 8, 9, 11.

22. **Lecanora atrosulphurea** (Wahlenb.) Ach. Thallus chinky areolate to globulose heaped, pale sulphur colored, C+ red, containing thiophanic acid. Apothecia at first immersed, then adnate, to 2 mm broad, margin thin, sometimes appearing lecideine but with algae still under the hypothecium; disk black, olive or greenish when wet; epithecium black or bluish brown; hypothecium pale; hymenium 65–75 μ, upper part black brown or blue brown; paraphyses coherent, thin, apices black brown to blue brown; spores 8, hyaline, simple, ellipsoid 8–13 × 4–6 μ.

Growing on acid rocks. An arctic circumpolar species known from Greenland, Alaska, Labrador, Northwest Territories in North America.

Locality: 3.

23. **Lecanora caesiosulphurea** Vain. Thallus dispersed or areolate to chinky diffract, uneven, from thin to 2.5 mm thick, sulphur yellow, the medulla white, slightly shining, K+ yellow, cortex C+ red, KC+ red. Apothecia to 2 mm broad, at first immersed, becoming adnate with the

base more or less constricted; margins persistent, colored as the thallus with a distinct bluish pruinose proper margin inside; disk convex, bluish lead colored, KC+ red; epithecium granulose inspersed, olivaceous brown, HNO_3 violet; hypothecium pale, of vertical hyphae, I+ blue; upper part of hymenium bluish, hymenium 70–80 μ; paraphyses coherent, slender, to 2 μ, tips slightly thickened and rarely branched; spores 8, simple, hyaline, ellipsoid to ovoid, 9–12 × 5–7 μ.

On acid rocks. A species described from Siberia. New to North America.

Locality: 11. The specimen was compared directly with the type specimen and agreed.

24. **Lecanora frustulosa** (Dicks.) Ach. Thallus of spherical verrucules, the margins of the verrucules slightly crenate when large, yellow, greenish yellow, or whitish, containing atranorin, zeorin, epanorin, norstictic acid all + or -. Apothecia to 2 mm broad, immersed to sessile, margin thick, colored as the thallus, becoming crenate or flexuous; disk convex, red brown to black; epithecium brown; hypothecium pale; hymenium hyaline below, brown above, 75 μ; paraphyses coherent, septate and capitate to 0.3 μ at tips, upper part brown; spores 8, hyaline, simple, ellipsoid 10–18 × 5–9 μ.

Growing on acid rocks. A circumpolar arctic, boreal, and north temperate species. Reported across the northern United States and south to California as well as in Canada and Alaska.

Locality: 3.

25. **Lecanora pelobotrya** (Wahlenb. ex Ach.) Sommerf. (*Lecidea pelobotrya* (Wahlenb. ex Ach.) Leight.). Thallus moderately thick, areolate chinky or areolate, the areolae flat or rarely convex, smooth or roughened, dirty white or pale ashy to somewhat pinkish ashy, lacking soredia, cephalodia present. Apothecia immersed in the areolae, to 1.5 mm broad, disk black, bare, dull; exciple brownish black, at first immersed in the thallus and inconspicuous, becoming distinct and forming margin later; hypothecium brownish black, the upper part of somewhat erect hyphae, the lower part irregular; hymenium 180–240 μ, I+ blue; epithecium brownish; paraphyses conglutinate, slightly branched, tips little thickened; spores 4 or 8, hyaline, 1-celled, ellipsoid, 20–30 × 12–16 μ.

On acid rocks. A high arctic species, possibly circumpolar, known from northern Scandinavia, Greenland, Jan Mayen Island, Baffin Island, and Alaska.

Locality: 11.

This species is very similar to *Lecidea panaeloa*. For the differences consult that species.

26. Lecanora atra (Huds.) Ach. Thallus crustose, verrucose or in a smooth or warty crust, becoming areolate, greenish gray or ashy white, K+ yellow, containing atranorin and α-collatollic acid. Apothecia to 2 mm broad, adnate; disk flat to becoming slightly convex, black; margin thick, entire or crenulate, commonly flexuose, of same color as thallus; epithecium dark; hypothecium black; hymenium dark violet, 180 μ; paraphyses coherent, septate, not clavate, violet; spores 8, simple, ovoid to ellipsoid, 10–15 × 5–8 μ.

On rocks, rarely trees. A circumpolar, boreal and temperate species. Locality: 1.

27. Lecanora rupicola (L.) Zahlbr. Thallus crustose, thin or thicker, smooth or becoming chinky or areolate, ashy or brownish or greenish gray, K+ yellow, containing atranorin, thiophanic acid, roccellic acid. Apothecia to 1.5 mm, partly immersed to adnate; disk flat to convex, waxy flesh colored to livid black, bluish pruinose; C+ lemon yellow; margin persistent, thin, entire, or disappearing; epithecium pale to darker olive brown; hypothecium pale; hymenium hyaline, paraphyses coherent, slender, slightly thickened at tips; spores 8, simple, hyaline, ellipsoid, 8–15 × 5.5–7.5 μ.

Growing on rocks, usually granitic but on many other types. A circumpolar boreal and alpine species, in North America in the northern states and southward in the west.

Localities: 1, 3, 8, 11, 14.

28. Lecanora urceolaria (Fr.) Wetm. (*Pachyospora verrucosa* (Ach.) Mass.) Thallus continuous or dispersed, chinky or more frequently composed of rounded warts, dark gray, greenish gray or ashy gray, K-. The apothecia innate in the warts, 1 per wart, dark around the mouth which may be slightly fimbriated; disk concave, black, sometimes grayish pruinose, often nearly closed; margin rather thick, entire, usually inflexed, of same color as thallus except around the mouth of the apothecium; epithecium greenish, very green with HCl; hypothecium thin, brownish; hymenium thick, to 300 μ, hyaline; paraphyses very slender, unbranched, the tips not thickened, not interwoven; spores (4–)8, simple, hyaline, ellipsoid, 30–62 × 16–37 μ.

Growing over mosses and soil in the arctic, also on tree trunks toward the southern edge of the range. Mainly arctic-alpine and boreal, circumpolar, in North America south to North Carolina, Minnesota, New Mexico, and California.

Localities: 1, 2, 3, 8, 9, 11, 13, 14.

29. **Lecanora flavida** Hepp Thallus thin, effuse, chinky-areolate, pale yellowish to orange; K-, I-. Apothecia minute, to 0.3 mm broad, innate; the disk concave, becoming flat, black; margin thin, entire; epithecium blue, HCl+ green; hypothecium hyaline; hymenium very blue, 65–70 μ; paraphyses coherent, gelatinized, the tips blue; spores 8, hyaline, simple, ellipsoid 12–18 × 7–11 μ.

On rocks. A rare circumpolar arctic lichen known from Scandinavia, Siberia, Iceland, Greenland, Northwest Territories, and Alaska.

Locality: 9.

30. **Lecanora cinereorufescens** (Ach.) Hepp Thallus rough, verrucose to chinky-areolate, non-radiate; dark gray to ashy gray, usually on a distinct black hypothallus. Apothecia innate to raised, to 1 mm broad; margin prominent; disk red to red brown, dark red when wet, bare, concave to flat; epithecium brown, HCl+ green; hypothecium hyaline; paraphyses thin, the tips thickened and brown; spores 8, hyaline, simple, ellipsoid 12–24 × 7–10 μ.

Growing on rocks, acid or calcareous. A circumpolar arctic and boreal species, ranging south in the western mountains to Colorado.

Locality: 1.

31. **Lecanora cinerea** (L.) Sommerf. Thallus usually with a more or less prominent black hypothallus, chinky-areolate, indistinctly radiate marginally, the areoles flat to slightly convex, ashy gray, K+ red, containing norstictic acid. Apothecia 1 or 2–5 per areole, innate in the thallus with a darker margin around the opening, flat or concave, disk dark brown black; epithecium brown, HCl+ green; hypothecium brownish-yellow; hymenium 100–115 μ, upper part brownish; paraphyses 2 μ, slender, the upper part distinctly moniliform (of bead-like cells); spores 8, simple, hyaline, ellipsoid, 14–21 × 8–12 μ.

Growing on a wide variety of rocks. A circumpolar arctic, boreal, and temperate species common in much of North America.

Localities: 2, 3, 8, 11.

32. **Lecanora lacustris** (With.) Nyl. Thallus crumbling, thin, smooth, chinky divided, pale reddish or orangish, K-, I-. Apothecia small, to 0.6 mm broad, margin thin to thicker and like the thallus, disk concave,

yellowish or reddish brown or paler reddish; epithecium pale brownish; hypothecium hyaline to brownish; hymenium 125 μ, hyaline; paraphyses coherent, the tips thickened and brownish or yellowish; spores 8, hyaline, simple, ellipsoid, 11–20 × 5–8 μ.

Growing on acid rocks in places which are frequently inundated as along the edges of tarns and seepage areas on outcrops as well as the edges of lakes. A circumpolar species, arctic, boreal, and temperate, with a wide range reported in North America.

Locality: 9.

33. **Lecanora nikrapensis** (Darb.) Zahlbr. Thallus thick, white, chalky, sometimes slightly radiate with very indistinct lobes, K-. Apothecia in prominent verrucae, 1–5 in a verruca; the disk innate, concave, black, often pruinose, the margin prominent, often farinose; epithecium brown, HCl+ green; hypothecium with vertical hyphae, brownish; hymenium 75–85 μ, upper part brownish; paraphyses coherent, the upper part only slightly moniliform, the tip cells globular; spores 8, simple, hyaline, broadly ellipsoid 14 –24 × 10–16 μ.

On calcareous rocks. A high arctic species known from Greenland, Spitzbergen, Novaya Zemlya, and Ellesmere Island, and new to Alaska. Additional specimens in WIS include: Yukon, St Elias Mts, Divide Camp nunataks, Murray 2378; and NWT Canoe Lake, Richardson Mts, Thomson & Larson 16487.

Localities: 8, 14.

34. **Lecanora caesiocinerea** Nyl. Thallus crustose, without distinct hypothallus; areolate with angular uneven areoles, whitish gray to bluish gray, K-. Apothecia usually solitary in the areoles, innate; with a dark edge around the disk which is concave to flat, black, dull; epithecium dark olive, HCl+ green; hypothecium brownish; hymenium 80–120 μ, upper part blue or greenish olive; paraphyses coherent, only slightly moniliform, the tip cells globular; spores 8, hyaline, simple, ellipsoid, 15–18 × 10–11 μ.

Usually on acid rocks, dry. A probably circumpolar arctic and alpine species.

Localities: 1, 8.

35. **Lecanora aquatica** (Körb.) Hepp (*Aspicilia aquatica* Körb.) Thallus gray or blue gray with yellowish tinge, limited with a thin margin, chinky-areolate, the areolae flat to slightly convex, sometimes with a thin dark hypothallus; upper cortex 30–50 μ with the cells in distinctly perpendicular rows, K-. Apothecia 1–3 per areole, immersed; disk to 0.5 mm

broad, black, epruinose; margin thick, gray or sometimes with thick dark proper margin; epithecium dark olive; hypothecium obscure, I+ blue; hymenium 110–200 μ, I+ blue, upper part dark olive; paraphyses sometimes much branched toward the apices, not moniliform; spores 8, hyaline, simple, broadly ellipsoid, 22–35 × 14–18 μ.

On rocks and boulders in or near brooks, often submerged for long periods. A boreal alpine species distributed in the Alps and in northern Europe and Siberia. New to North America.

Locality: 8.

36. **Lecanora heteroplaca** Zahlbr. Thallus bimorphic, for the greater part thin, to 1.5 mm thick, yellowish ashy or paler, K-, C-, continuous or rarely minutely and thinly chinky-areolate, the areoles to 0.2 mm broad, flat to convex, in portions thicker, to 0.9 mm in dispersed white, dull, areoles separated by chinks, flat to 1 mm broad, medulla I-. Apothecia in the raised or thin portions of the thallus, sessile, constricted at base, the margin white to smoky, smooth; disk concave, black, dull, bare, the margin entire; epithecium olive brown, HNO$_3$+ bluish green; hypothecium narrow, hyaline; hymenium 90–100 μ, upper part olive; paraphyses slender, coherent, simple or slightly branched, non-capitate; spores 8, simple, hyaline, containing large oil droplets, ellipsoid, 15–21 × 11–12 μ.

Growing on rocks. A high arctic species previously known from Novaya Zemlya, and Bear Island. New to North America.

Locality: 2.

37. **Lecanora composita** Lynge Thallus dispersed, growing over a large area, soft, non-lobate, epruinose, white, of verrucules to 1.5 mm broad, convex, the surface irregularly reticulately chinky rough, K-; the orange, more or less chinky primary thallus visible between the verrucules. Apothecia single or 2 in elevated, almost columnar verrucules, with proper dark margin as well as with the white thalloid margin; the disk to 1.5 mm broad, compounded of a number of smaller units, black, subpruinose, occasionally with a raised umbo in the center; epithecium olive brown, inspersed, HCl+ green; hypothecium pale; hymenium 85–100 μ, upper part olive brown, inspersed with granules and oil droplets; paraphyses distinct, coherent, conspicuously septate but not moniliform, slender, slightly branched; spores 8, simple, hyaline, ellipsoid, 17–22 × 10–12.5 μ.

Growing on acid rocks. A species described from northeast Greenland. New to Alaska.

Locality: 11.

38. **Lecanora anseris** Lynge Thallus thin effuse, not orbicular nor lobate, irregularly chinky-areolate, soft, toward the margins becoming less chinky, more continuous, merging into a thin, indistinct ashy hypothallus, the sterile parts low and ashy yellowish, the fertile parts elevated, convex but not verrucose, white or ashy white, K-. Apothecia in the raised warts, single or several per wart; the disk to 0.5 mm, concave, black, lightly pruinose; margin raised, pale banded, sides of the apothecium dark ashy; epithecium brownish, HCl+ green; hypothecium pale; hymenium 80–100 μ, upper part yellowish brown; paraphyses slender, easily separable in KOH, not moniliform, becoming branched and interwoven; spores 8, hyaline, simple, ellipsoid 16–20 μ.

On calcareous and non-calcareous rocks. A species described from Greenland, and new to Alaska.

Localities: 1, 2, 3, 9.

39. **Lecanora ryrkaipiae** Magn. Thallus of dispersed verrucae, small, to 0.5 mm broad with broadened bases, blackish gray, cortex thin, 25 μ, transparent, K-. Apothecia 1 or 2 in areole, with very thin proper margin; disk concave, black, bare, small; epithecium olive, HCl+ green; hypothecium thin, pale; hymenium 85–90 μ, upper part olive green to brownish; paraphyses coherent, simple, moniliform, upper cells globose; spores 8, simple, hyaline, subglobose, 11–17 × 7–10 μ.

Growing on calcareous rocks. A species described from Siberia and possibly belonging to the Beringian element in the Alaskan flora. New to North America.

Localities: 1, 2, 3.

40. **Lecanora annulata** Lynge Thallus exceedingly thin, not radiate but partly fimbriate, chinky but not areolate, gray, clay ashy, not pruinose, the cortex transparent, thin, ca. 25 μ, K-, I-. Apothecia in prominent verrucae with vertical sides or slightly narrowed at base, to 0.5 mm broad, with a prominent wall formed by the verruca and of the same color as the thallus, giving a ring-like appearance; disk concave, black, or brownish black when wet; epithecium brown, HCl+ green; hypothecium hyaline; hymenium 70–100 μ, upper part brownish yellow; paraphyses fairly free in KOH, slender, rarely branched, upper part submoniliform, the cells constricted at septum but elongate; spores 8, simple, hyaline, ellipsoid, 12–17 × 8–10 μ.

On acid or calcareous rocks. Described from Greenland, new to Alaska.

Locality: 2.

41. **Lecanora elevata** Lynge Thallus pale ashy to ashy brownish, thin, non radiate, areolate in small areolae to 0.35 mm broad, convex, angular to rounded, K-, I-. Apothecia dispersed and well raised over the thallus; margins thick and with flabelliform radiating hyphae constricted at the septa; disk black, epruinose, urceolate and appearing divided but without excipular columns; epithecium brownish; hymenium 90–100 μ, the paraphyses distinctly moniliform, the tips thickened; spores broadly ovate, 14–23 × 10–12.5 μ.

Growing on rocks. This species is known from west Greenland and the west coast of Hudson Bay. A report from Bear Island in Magnusson's monograph of Aspicilia (1939) appears more like *Lecanora (Aspicilia) ryrkaipiae*. New to Alaska.

Locality: 8.

42. **Lecanora pergibbosa** Magn. Thallus crustose, non-radiate, marginally verruculose, centrally thickened verrucose, uneven, the verrucules occasionally constricted at the base, on a scarcely visible ashy hypothallus, K-, C-, P-; the cortex transparent, not inspersed. Apothecia abundant, in most verrucules, the disks single to 2–3 per verrucule, to 1.6 mm broad, concave, black, with thin fissure around the margin which is thin or thick; epithecium yellowish brown, HCl+ green; hypothecium pale; hymenium 100–120 μ, upper part yellowish brown; paraphyses distinct, with the upper 4–6 cells moniliform and globose to 3.6 μ; spores 8, simple, hyaline, broadly ellipsoid to subglobose, 17–25 × 12–16 μ.

Growing on acid or calcareous rocks in the open. A probably circumpolar species known from Scandinavia, Greenland, the Northwest Territories. New to Alaska.

Locality: 2.

43. **Lecanora supertegens** (Arn.) Magn. Thallus determinate, limited by a distinct blackish hypothallus, zonate to radiate at times, indistinctly chinky areolate or smooth, dull, brownish gray with a violet tinge, K-, C-, P-, the cortex granular inspersed. Apothecia toward the center, innate, disk to 1 mm broad, concave, black; margin indistinct, concolorous with the thallus; epithecium olive brownish, HCl+ green; hypothecium cloudy and oil inspersed; hymenium 90–135 μ, upper part dark olive brownish; paraphyses coherent, more or less branched and anastomosing, the tips globular, the cells below the tip submoniliform; spores 8, simple, hyaline, ellipsoid, 15–25 × 10–14 μ.

On acid rocks. A circumpolar arctic-alpine species reported from Siberia, Novaya Zemlya, Bear Island, Scandinavia, central Europe, Greenland, and Alaska.

Localities: 8, 9.

44. **Lecanora rosulata** (Körb.) Stiz. Lacking distinct hypothallus, the thallus with the margin of radiate discrete lobes which are not contiguous, of serially formed, convex, chinkily separated or contiguous areoles, dark gray, to olive or dark brown, centrally the areoles more contiguous, the cortex transparent but the upper part dark brown, medulla yellowish with KOH. Apothecia innate, to 0.5 mm broad; disk concave, black; the margin thick prominent and dark or of same color as thallus; epithecium dark olive brown, HCl+ yellowish mist; hypothecium dense, darkened; hymenium 70–85 μ, upper part dark olive to brown; paraphyses coherent, branched, submoniliform, the uppermost cell globular; spores 8, simple, hyaline, ellipsoid, 12–17 × 7–8.5 μ.

Growing on noncalciferous rocks. A circumpolar arctic species previously reported from Siberia, Novaya Zemlya, Greenland, and Alaska.

Locality: 9.

45. **Lecanora perradiata** Nyl. Thallus limited by a distinct black, 1–2 mm broad thin hypothallus, the marginal lobes radiate, dense and contiguous, tapering down toward the exterior, 3–5 mm long and to 0.4 mm broad, ashy gray to dark gray, the center verrucose-areolate, the areolae convex, separated by broad cracks which narrow to the base, cortex transparent, the upper part dark brown; K+ yellow. Apothecia to 1.5 mm broad, prominent to almost sessile, disks 1 or 2–4 per areole, to 0.7 mm broad, concave or flat, black, sometimes with thin pruina; the margin pale gray and prominent, sometimes inflexed; epithecium greenish to brown; hypothecium pale slightly darkened; hymenium 80–100 μ, the upper part greenish yellow to reddish brown; paraphyses indistinct, coherent, submoniliform, upper cells subglobose; spores 8, simple, hyaline, broadly ellipsoid to subglobose, 13–18 × 8.5–11 μ.

Growing on acid rocks. A circumpolar arctic species reported from Siberia, Novaya Zemlya, Spitzbergen, Greenland, Northwest Territories. New to Alaska.

Localities: 9, and Sahligvik Ridge, Kukpuk River, Melchior, 739.

46. **Lecanora stygioplaca** Nyl. Thallus with more or less distinct black hypothallus; marginal part not radiate or only indistinctly so, centrally areolate, the areoles convex or flat, shining or dull, olive black to black, medulla K+ yellow, becoming red, containing salazinic acid; cortex transparent, the upper part dark. Apothecia sparse, the margins black, shining, the disk concave, black; epithecium dark olive brown, HNO$_3$ green; hypothecium pale; hymenium 100–150 μ, the upper part olive brown; paraphyses distinct, non-moniliform, sparsely branched anastomosing, the cells elongate and constricted between; spores 8, hyaline, simple, 17–25 × 10–15 μ.

Growing on granitic rocks. A member of the Beringian element in the arctic, reported from Siberia and the Northwest Territories. New to Alaska.

Locality: 9.

47. Lecanora fimbriata Magn. Hypothallus lacking; thallus thin, marginally with discrete narrow lobes which are radiating, narrow, convex, ashy brown, the cortex transparent, the upper part olive brown, the medulla K+ red; the center part of the thallus of dense verrucules containing apothecia. The apothecia to 0.5 mm broad, the margin thick, dark brown, the outer part paler, the disk concave, black, solitary in the verrucules; epithecium dark; hypothecium darkened; hymenium 85 μ, upper part olive brown; paraphyses coherent, simple or rarely branched, non-moniliform; spores 8, hyaline, simple, 14–16 × 8.5 μ.

Growing on rocks, calciferous or not. Apparently a member of the Beringian element in the arctic, reported from Siberia. New to Alaska and North America.

Locality: 9.

48. Lecanora polychroma (Anzi) Nyl. Hypothallus not visible; thallus verruculose-areolate, thin at the edges and partly indistinctly radiate, the areoles flat or slightly convex, the fertile areoles verruciform, pale gray or whitish gray with a yellowish tinge, cortex opaque, medulla K-. Apothecia to 0.8 mm broad, solitary or 2–3 per verruca, margin prominent, not darkened; disk concave, black, more or less pruinose; epithecium olive brown; hypothecium pale; hymenium 85–110 μ, upper part olive brown; paraphyses unbranched, with oil drops, moniliform with globose end cells and those below swollen ellipsoid; spores 8, simple, hyaline, ellipsoid, 13–20 × 8–13 μ.

Growing on calcareous rocks. A species known from Italy, Novaya Zemlya, and Ellesmere Island, rather rare. New to Alaska.

Locality: 1.

49. Lecanora cingulata Zahlbr. Hypothallus indistinct, of same color as the thallus, yellowish gray, thallus circular, zonate in bands, the areoles also more or less radiate but contiguous, thinning down toward the periphery, the center chinky-areolate, the areoles angular, cortex opaque, medulla K-. Apothecia abundant centrally, to 0.5 mm broad; margin not prominent, not darkened; disk concave to flat, black; epithecium olive brown; hypothecium pale; hymenium 85–110 μ, the upper part olive brown paraphyses coherent, in much gelatin, indistincly moniliform; spores 8, hyaline, simple, broadly ellipsoid to subglobose, 13–17 × 8–10 μ.

On non-calciferous rocks. A species known from Novaya Zemlya. Rare, arctic. New to North America and Alaska.

This species resembles *L. plicigera* but has shorter conidia, 13–17 instead of 25–30 μ, and is less radiate and not so yellowish.

Locality: 9.

50. **Lecanora candida** (Anzi) Nyl. Thallus on indistinct or dark hypothallus; continuous, to slightly chinky but not areolate, to disappearing in f. *evanescens*, chalky white to cream colored, the margin sometimes determinate and radiate with short, indistinct lobes; cortex opaque with granules; medulla K-. Apothecia at first immersed, becoming prominent; the disk to 0.6 mm broad, concave, black, pruinose; the margin thick, farinose, K+ yellow, I-; epithecium olive brown; hypothecium pale; hymenium 75–85 μ, upper part darkened olive brown; paraphyses coherent, submoniliform, the apical cells globular to subellipsoid; spores 8, hyaline, simple, broadly ellipsoid, 14–24 × 10–16 μ.

Growing on rocks. A species reported from the European Alps, Novaya Zemlya, Ellesmere Island. New to Alaska. Specimens from Colorado and Alberta are in University of Wisconsin herbarium.

Locality: 1 (the specimen is f. *evanescens* Magn.).

51. **Lecanora lesleyana** (Darb.) Pauls. Hypothallus blackish and more or less visible; thallus in small rosettes to 2 cm broad, the outer areoles lobate and radiate, about 1.5 mm long and 0.5 mm broad, convex, separated by cracks, the central areoles angular, chalky white with a bluish tinge, the cortex and medulla filled with crystals, K-, C-. Apothecia in prominent verrucules, solitary or 2–3 per areole, surrounded by a thick margin which is black around the disk, the outer part of same color as the thallus; the margin K+ yellow, I+ blue; the disk concave, black, bare, to 0.3 mm broad; epithecium olive brown; hypothecium pale; hymenium 75–120 μ, upper part olive brown; paraphyses coherent, in thick gelatin, containing oil drops, unbranched or sparingly so, submoniliform, the tips ellipsoid; spores 8, hyaline, simple, ellipsoid, 9–16 × 7–10 μ.

Growing on calciferous rocks. A circumpolar arctic species known from Siberia, Novaya Zemlya, Spitzbergen, Greenland, Ellesmere Island, and Devon Island. New to Alaska. Specimens from Utah and Colorado are in University of Wisconsin herbarium.

Localities: 1, 9, 11.

52. **Lecanora sublapponica** Zahlbr. Thallus brown gray to lead gray, sometimes olive, the margins distinctly radiating with lobes to 0.5 mm

broad, 2–3 mm long, convex, the center areolate with the areolae separated by cracks, K–; the cortex 30–50 μ thick, transparent, the cells oriented perpendicularly; medulla K–, I–. Apothecia single in the areolae, projecting or sessile, the margins thick; disk impressed, black; epithecium olive blackish; hypothecium pale, I+ blue; hymenium 100–110 μ, I+ blue, non-inspersed; paraphyses coherent, only slightly moniliform; spores 8, hyaline, simple, oval, 14–26 × 8–10 μ.

Growing on rocks. This arctic species is known from Novaya Zemlya, Siberia, northern Sweden, and Greenland. In North America it has been collected on Baffin Island and on mountains in Maine, Colorado, and Alberta. New to Alaska.

Locality: 8.

53. **Lecanora plicigera** Zahlbr. Thallus gray with very yellowish or orange cast, very continuous and thin with slightly raised radiating marginal lobe-like ridges giving a plicate appearance, the thallus interior more areolate appearing; the hyphae of the upper cortex distinctly perpendicular in orientation, 30–40 μ thick, K–; medulla opaque. Apothecia 1(–2) per verruca, 0.2–0.4 mm broad, impressed, the margin darkened toward the disk; disk black; epithecium dark greenish brown; hypothecium dense, I+ blue; hymenium 70–100 μ, I+ blue, upper part dark; paraphyses coherent, moniliform with globose cells; spores hyaline, broadly ellipsoid, 10–15 × 5–8 μ.

On non-calciferous rocks, shaly or slaty. This species was previously known from Novaya Zemlya and is new to North America.

Locality: 8.

54. **Lecanora disserpens** (Zahlbr.) Magn. Hypothallus not visible between the lobes; thallus of distinctly radiate discrete lobes, rarely contiguous, the lobes forking and chinkily divided into sections transversely, the center of the thallus of still radiate but more convex areoles, bluish white to bluish gray, the tips darkening; cortex opaque, medulla K–. Apothecia immersed, solitary; the margin thick and prominent, of same color as thallus; disk flat or concave, black, rough, bare; epithecium olive brown; hypothecium pale; hymenium 80–90 μ, the upper part olive brown; paraphyses branched toward the tips, moniliform, the upper cells globose; spores 8, hyaline, simple, ellipsoid, 16–18 × 8.5–10 μ.

On calciferous or acid rocks. A circumpolar arctic species reported from Novaya Zemlya, Spitzbergen, Greenland, and Northwest Territories. New to Alaska.

Localities: 9, 14.

55. **Lecanora concinnum** Thoms. Hypothallus thin, black, bordering the thallus; thallus 2–3 cm broad, yellowish cream colored, areolate, radiate, the areolate lines contiguous or discrete, the marginal sterile lobes elongate and slightly convex, the central areoles angular and convex, contiguous, the fertile areoles verruciform, containing 1, occasionally 2 or 3 apothecia, cortex indistinct, medulla K-. Apothecia to 0.5 mm broad, with thick thalloid margin which darkens toward the disk and is K-; disk concave, black, dull, roughened, frequently pruinose; epithecium olive brown; hypothecium pale; hymenium 135 μ, upper part olive brown; paraphyses thin, non-moniliform, tips thickening to 2.5 μ; spores 8, simple, hyaline, ellipsoid, 18–23 × 10–12.5 μ.

On siliceous boulders. New to North America.

Locality: Known only from the type locality (9), Franklin Bluffs on the Sagavanirktok River, and probably eliminated there by construction of the pipeline.

56. **Lecanora chrysoleuca** (Sm.) Ach. (*L. rubina* (Vill.) Ach.) Thallus squamulose, becoming umbilicate, the squamules thick, cartilaginous, flat, lobulate, rounded to crenulate, greenish yellow to yellow above, brown to black below, corticate above and below, K+ or K-, P+ or P-, containing usnic acid, ± atranorin, + psoromic acid. Apothecia quite large, reaching 5 mm broad, sessile on the thallus, the margin colored as the thallus, thick or thin, sometimes disappearing; disk flat to convex, pale waxy to reddish or brownish, bare or slightly pruinose; cortex of the margin of vertical hyphae; epithecium pale yellowish; hypothecium pale yellowish; hymenium yellowish, 50 μ thick; paraphyses coherent, slender; spores 8, simple, hyaline, ellipsoid, 8–15 × 4.5–8 μ.

Growing on acid rocks. A circumpolar arctic, boreal, and temperate species with very broad range.

Localities: 3, 8, 11, 14.

57. **Lecanora melanopthalma** (Ram.) Ram. Thallus squamulose to umbilicate in massed squamules, the squamules thick and cartilaginous, flat, lobed or crenate, greenish yellow, shining; corticate above and below, underside brown to black, K-, P+ yellow, containing usnic and psoromic acids. Apothecia sessile on the upper surface, the margin thick and concolorous with the thallus, disk black or dark olive; epithecium dark; hypothecium yellowish; hymenium 50 μ, pale below, darkening above; paraphyses coherent, blackish and capitate at the tips; spores 8, simple, hyaline, ellipsoid, 9–15 × 5–7.5 μ.

Growing on rocks, usually acid. A circumpolar arctic, boreal, and alpine species, in North America south in the Appalachian Mts and in the west in the mountains to Texas and California.
Localities: 3, 11.

58. **Lecanora peltata** (Ram.) Steud. Thallus umbilicate, lobate, the lobes cartilaginous, thick, dull, yellowish green above, bluish black below, rarely pale below, corticate above and below, K-, P+ or P-, containing usnic acid ± psoromic acid and zeorin which is diagnostic. Apothecia adnate, brown to reddish brown or yellowish brown, to 4 mm broad; margin concolorous with the thallus; epithecium pale brownish; hypothecium yellowish; hymenium pale; paraphyses slender, coherent, non-capitate; spores 8, simple, hyaline, ellipsoid, 9–15 × 5–9 μ.
Growing on acid rocks. A circumpolar arctic, boreal, and alpine species.
Locality: 11.

59. **Lecanora lentigera** (G. Web.) Ach. (*Squamarina lentigera* (G. Web.) Poelt) Thallus squamulose, close to the substratum, the center areolate but the periphery distinctly lobed, the lobes rounded to crenulate, elongate, greenish or yellowish white and covered with white pruina, corticate above and below, K+ yellow, P-, containing usnic acid and atranorin. Apothecia sessile, to 2 mm broad, with pale margin; disk red brown to buff, concave to flat, bare; epithecium yellowish; hypothecium yellowish; hymenium pale yellowish 60–75 μ; paraphyses coherent, slender, unbranched, non-capitate; spores 8, simple, hyaline, ellipsoid 9–14 × 4.5–5.5 μ.
Growing on calcareous, argillaceous soils. Arctic and also in the western plains. A circumpolar species.
Locality: 14.

60. **Lecanora melanaspis** (Ach.) Ach. Thallus lobate, close to the substratum, the center more or less areolate, the margins distinctly radiate lobate, the lobes very convex to almost cylindrical, upper side grayish olive to dark olive, or glaucous, not pruinose, corticate above and below, K+ red, P+ orange, containing norstictic acid. Apothecia to 2 mm, adnate, with entire margin concolorous with the thallus; disk brown or black brown, sometimes pruinose; epithecium pale or brown; hypothecium hyaline; hymenium 70–75 μ, pale; paraphyses coherent, thick 4.5–6 μ, the upper part moniliform with the uppermost cell globular; spores 8, simple, hyaline, ellipsoid, 8–14 × 5–9 μ.

On rocks, usually acid types. A circumpolar arctic, boreal, and temperate species.

Locality: 11.

61. Lecanora muralis (Schreb.) Rabenh. Thallus lobate at the periphery, areolate toward the center, the lobes crenate to rounded, flattish, sometimes pruinose above especially along the centers of the lobes, corticate above and below, K-, P+ or P-, containing usnic acid + or - fumarprotocetraric acid. Apothecia to 2 mm, sessile or adnate, common; the margin thick, entire or crenulate, concolorous with the thallus; the disk yellowish to brownish, light colored; epithecium hyaline; hypothecium hyaline; hymenium 65-75 μ, pale; paraphyses slender, coherent, non-capitate; spores 8, simple, hyaline, ellipsoid, 9-15 × 4.5-7 μ.

Growing on rocks, often calcareous, sometimes perch rocks of birds which provide lime. A very broad ranging species, boreal, temperate, circumpolar.

Localities: 1, 11.

4. IONASPIS Th. Fr.

Thallus crustose, endolithic or epilithic, effuse or with definite borders, not layered, the entire lower surface attached to the rocks, pale whitish or orangish to dark olive. Apothecia immersed in the thallus, with a proper exciple of the same color as the disk; disk flat or concave, pale pinkish to dark blackish; paraphyses simple or slightly branched, septate; asci clavate; spores 8, simple, hyaline, ellipsoid to ovoid spores with thin walls. Pycnidia immersed, simple, sterigmata simple, pycnoconidia short, cylindric. Algae: *Trentepohlia*.

1	Apothecia pale to brownish; hymenium pale to olive, not bluish green	
		1. *I euplotica*
	Apothecia black; hymenium bluish green	2
2 (1)	Colored parts of apothecia not violet in HNO_3	2. *I. suaveolens*
	Colored parts of apothecium violet in HNO_3	3
3 (2)	Apices of paraphyses thickened	4
	Apices of paraphyses not thickened	5
4 (3)	Thallus scant, pale dirty yellow; margin of apothecium blackish	
		3. *I. reducta*
	Thallus thick, whitish; margin of apothecia inconspicuous	
		4. *I. heteromorpha*

5 (3) Thallus orange yellow, even; apothecia 0.3–0.4 mm, margin inconspicuous; paraphyses constrictedly septate   5. *I. ochraceella*
Thallus white or pale yellowish, uneven; apothecia 0.5–0.7 mm, margin
conspicuous; paraphyses septate but not constricted   6. *I. schismatopsis*

1. Ionaspis euplotica (Ach.) Th. Fr. var. **arctica** (Lynge) Magn. Thallus
crustose, thin, disappearing, ashy, ashy yellowish, to brownish,
indistinctly chinky-areolate, reactions negative. Apothecia 0.3–0.5 mm
broad; thalloid margin thick and prominent; disk concave, pale flesh
colored, surrounded by a pale proper margin which is separated from
the thallus by a crack; epithecium hyaline; hypothecium hyaline;
hymenium 75–135 μ, pale; paraphyses more or less separating in water,
2.5–4 μ, septate, not clavate; spores 8, uniseriate, simple, hyaline,
ellipsoid, 7–16 × 5–8 μ.

Growing on rocks, both acid and calcareous. An arctic-alpine species
known from Novaya Zemlya, Spitzbergen, Siberia, Bear Island, and
Northwest Territories. New to Alaska.

Localities: 8, 11.

2. Ionaspis suaveolens (Schaer.) Th. Fr. Thallus thin, effuse, yellow
brown to olive brown, indistinctly chinky and areolate. Apothecia
immersed, thalloid margin indistinct; disk 0.2–0.4 mm broad, concave to
flat, black; proper exciple in upper part of margin 17–25 μ, blue green,
darker than hymenium, becoming colorless below; hypothecium
hyaline, with vertical hyphae; hymenium 50–55 μ, the upper half
emerald green to blue green, of same color in HNO_3, olive yellow in
KOH; paraphyses indistinct, 2–2.5 μ thick, shortly and constrictedly
septate; spores 6–8 biseriate, globose or broadly ellipsoid, 6–8 μ or
10 × 6–7 μ.

On rocks in stream beds. A boreal alpine species known from
Greenland and Europe. New to Alaska.

Locality: 11.

3. Ionaspis reducta Magn. Thallus poorly developed, thin, smooth or
slightly chinky around the apothecia, dirty yellow, sometimes reduced to
a thin film around the apothecia. Apothecia 0.1–0.125 mm broad,
immersed in the thallus or the substratum; disk concave, black; upper
part of proper exciple dark brown green, K+ violet, lower part hyaline;
epithecium dark blue green; hypothecium granular, partly bluish; hymenium 85 μ, upper half dark bluish green, fading below; paraphyses in
much gelatin, little branched, 3–3.5 μ at apices, thicker toward tops,
septate; spores 8, biseriate, simple, hyaline, ellipsoid, 8–13 × 6–8 μ.

Growing on calcareous rocks. Described from Novaya Zemlya. New to North America.

Locality: 2.

4. **Ionaspis heteromorpha** (Krempelh.) Arn. Thallus crustose, thin, with thin dark hypothallus showing, minutely chinky-areolate, mainly smooth, whitish or yellowish gray. Apothecia immersed, 0.2–0.3 mm broad; disk concave or flat, black, with thin proper exciple and sometimes a crack between this and the thallus; the exciple bluish green above, fading below into a thin brownish line, brownish green in KOH; hypothecium grayish; hymenium 65–100 μ, the upper half bright bluish green, HNO_3 + red-violet, K+ paler green; paraphyses coherent, indistinct, thickened above to 2–3.5 μ; spores 8, hyaline, simple, globose to broadly ellipsoid, 7–8 μ or 8–12 × 6–9 μ.

Growing on calcareous rocks. Known from Scandinavia and the Alps in Europe, in Spitzbergen, and in Siberia. New to Alaska and North America.

Locality: 11.

5. **Ionaspis ochraceella** (Nyl.) Magn. Thallus small, thick for the genus, minutely chinky-areolate, the areolae flat or slightly concave, pale orange yellow. Apothecia immersed, 0.3–0.4 mm broad; disk concave, black; proper exciple surrounded by a distinct fissure, pale yellowish brown, the upper part dark bluish green, HNO_3 + red-violet; hypothecium gray; hymenium 85–100 μ, the upper part dark emerald green; paraphyses coherent, in gelatin, 2–2.5 μ thick, septate with constricted septa but not moniliform; spores 8, simple, hyaline, ellipsoid, 10–11 × 7 μ.

Growing on calcareous rocks. Known from Scandinavia. New to Alaska and North America.

Localities: 2, 8.

6. **Ionaspis schismatopsis** (Nyl.) Hue. Thallus effuse, continuous, smooth to becoming chinky-areolate, the edges smooth, white or roseate white, lacking hypothallus. Apothecia 0.3–0.7 μ, sometimes 2–3 confluent, immersed, the thalloid margin slightly prominent, a fissure between it and the proper margin; proper exciple dark bluish green above, tapering off and fading below; disk concave, black; hypothecium pale yellow; hymenium 85–120 μ, upper part dark emerald to bluish green, HNO_3 + violet, hyaline below; paraphyses slender, 1.7–2 μ, imbedded in gelatin, unbranched, densely septate but the septa not constricted, the apices not thickened; spores 8, simple, hyaline, subglobose to broadly ellipsoid, 10–15 × 6–10 μ.

Growing on calcareous rocks. Known from Siberia, Bear Island, Norway, and the Arctic islands of Canada. New to Alaska.

Localities: 2, 11.

5. LECANIA Mass.

Thallus crustose, granulose, heteromerous or poorly developed. Apothecia sessile, lecanorine, small, the disk concave, flat or convex, the hypothecium hyaline, the hymenium hyaline or brownish above, I+ blue turning wine red; paraphyses unbranched, slender, lax; asci clavate; spores 8–16, hyaline, ellipsoid, 1–3-septate transversely. Algae: *Trebouxia*.

1. **Lecania alpivaga** Th. Fr. Thallus thin, of dispersed or heaped granules, gray, K-, I-. Apothecia adnate, to 1 mm broad, with thick thalloid, inflexed, crenulate, persistent, margin; disk blackish brown; hypothecium hyaline, I+ blue; hymenium brown above, I+ blue, the asci turning wine red; paraphyses lax, the apices brown, capitate; spores 8, hyaline, blunt ellipsoid, 1-septate, 15–21 × 5–6 μ.

Growing on calcareous rocks. Previously known from Scandinavia, Novaya Zemlya, and Baffin Island. New to Alaska.

Locality: 1.

6. ICMADOPHILA Trevis

Thallus crustose, a thin upper cortex present; medulla loose and loosely attached to the substratum. Apothecia short stipitate or adnate, lecanorine, with a hyaline proper exciple inside; paraphyses slender, simple or little branched; asci narrowly clavate to cylindrical; spores 8, hyaline, spindle-shaped, 2–4-celled. Sterigmata endobasidial; pycnoconidia straight, cylindrical with somewhat thickened ends. Algae: *Coccomyxa*.

1. **Icmadophila ericetorum** (L.) Zahlbr. Thallus crustose, smooth or granular, when thick becoming chinky-areolate, in part becoming massed soredia, dull, gray or gray green or blue green, over a more or less visible white prothallus, K+ yellow, becoming brown, C-, I-, P+ yellow, then orange (thamnolic acid). Apothecia common, dispersed, on short stipes or sessile, round, to 4 mm broad, with thalloid margin which is smooth or powdery sorediate; disk pale flesh colored or roseate, becoming reddish brown in severe conditions, flat or slightly swollen, dull or with a white pruina; epithecium yellow brown, inspersed with granules; hypothecium yellowish; hymenium hyaline or yellowish, 145–150 μ; paraphyses slender, 1 μ thick; spores 8, hyaline,

spindle-shaped, mainly 4-celled but simple spores or 1-2-septate also present, 16–28 × 5.5–7 μ.

Growing in shady spots over mosses, humus and well-rotted wood, circumpolar, boreal, but occurring in the arctic north of the tree line. It occurs south in the east to New England, to Michigan and Wisconsin in the center, and to Colorado and Washington in the west.

Localities: 2, 3, 14.

7. HAEMATOMMA Mass.

Thallus crustose, smooth to wrinkled or chinky to areolate with upper cortex, other layers poorly developed, underside attached to the substratum over the surface. Apothecia adnate to sessile with thin thalloid margin, sometimes disappearing, hypothecium hyaline; hymenium brown above, hyaline below, I+ blue, K+ violet; paraphyses coherent, thickened and darkened at tips; asci clavate; spores 8, often twisted, hyaline, needle-shaped, 3–7-septate, the cells cylindrical. Pycnidia projecting, pycnoconidia cylindrical, straight. Algae: *Trebouxia*.

1. **Haematomma lapponicum** Räs. Thallus becoming thick, verrucose rugose or areolate with convex areoles, warty, greenish yellow to very bright yellow, K-, P-, containing usnic acid ± barbatic acid, ± zeorin. Apothecia to 3 mm, adnate, with thin thalloid margin concolorous with the thallus; disk blood red, flat to convex, violet with KOH; hypothecium hyaline; hymenium pale below, the upper part reddish brown, K+ blue violet; paraphyses very coherent, the tips thickening and darkened reddish; asci clavate; spores 8, 3–7-septate, the cells cylindrical, hyaline, commonly curved or twisted, one end more tapered than the other, 35–55 × 3–5 μ.

Growing on rocks, mainly non-calcareous. Circumpolar, arctic-alpine, in North America southward in the eastern mountains in New England, in the west to Alberta and Washington, occurring also on Volcan de Toluca, Mexico, at 4160 meters (Iltis, 2178, WIS).

Localities: 3, 8, 9, 11.

CANDELARIACEAE

Crustose or foliose lichens with upper cortex poorly or well developed, a lower cortex in one genus (not arctic); characterized by producing calycin and pulvinic dilactone, bright yellow in color. Apothecia adnate to sessile, with thalloid exciple colored as the thallus; hypothecium hyaline;

hymenium hyaline, darkening brownish above, paraphyses unbranched or slightly branched toward the apices; asci clavate; spores 8, 16, 32, hyaline, ellipsoid, oblong, sometimes one end thinner than the other, sometimes curved, non-septate or 1-septate. Algae: *Trebouxia* or *Cystococcus*.

1. CANDELARIELLA Müll. Arg.

Crustose thallus granulose, warty, areolate, slightly coralloid, to somewhat lobed at margins, sometimes lacking and the algae only in the base of the apothecium, yellow, olive green or yellowish orange, K-, C-, P-, I-, producing calycin and pulvinic dilactone. Apothecia sessile to adnate, disk flat to convex; thalloid margin entire to crenate, sometimes disappearing; epithecium usually yellowish, granular inspersed; hypothecium hyaline; hymenium hyaline, the upper part usually becoming brownish; paraphyses lax, capitate or not, the upper part slightly branched; asci clavate or saccate; spores 8 to numerous (32), hyaline, simple to 1-septate, often curved, ellipsoid to oblong. Algae: *Cystococcus*.

1	On humus or over moss	2
	On rocks and bones or on bark	3

2 (1) Spores 8 per ascus, 14–18 × 6–8 μ; thallus of very small round granules or becoming lobate, sometimes forming glomerules   1. *C. terrigena*
Spores 16–32 per ascus, 6.5–15 × 5–6 μ; thallus of rounded granules, occasionally glomerate, dull and becoming sorediate   2. *C. placodizans*

3 (1) Asci with 8 spores; thallus granulose, sometimes nearly lacking; spores 12–30 × 5–7.5 μ 3. *C. aurella*
Asci with numerous spores, spores 9–15 × 4.5–6.5 μ 4

4 (3) Thallus lacking or present only as granules which are precursors of apothecia   5
Thallus granular or verrucose areolate   6

5 (4) On barks; apothecia high convex, margin thin to disappearing; epithecium yolk colored   4. *C. lutella*
On rocks, apothecia flat to slightly convex; margin entire to subcrenate; epithecium lemon yellow   5. *C. athallina*

6 (4) Thallus coarsely crenulate edged, granulose to areolate heaped; on rocks and old wood   6. *C. vitellina*
Thallus powdery granulose; on broad leaved trees or shrubs, sometimes on wood or humus   7

7 (6) Thallus forming a thick granulose crust; epithecium pale yellow to orange yellow; apothecia flat   7. *C. xanthostigma*
Thallus of dispersed granules or lacking; epithecium yolk colored; apothecia high convex   4. *C. lutella*

1. **Candelariella terrigena** Räs. Thallus crustose, of round granules or becoming slightly lobate granules, the granules often dispersed, dull, lemon yellow to yellow, flattened or convex. Apothecia adnate, flat to convex; margin of the same color as the thallus, entire or becoming crenate, occasionally disappearing; disk yellow to olive yellow or grayish; epithecium granulose, yellow to yellow orange or greenish; hypothecium hyaline; hymenium 50–60 μ, hyaline, upper part darker, I+ blue; paraphyses seldom branched, little thicker at tips; asci clavate to saccate; spores 8, hyaline, slightly curved, oblong-ellipsoid, simple or appearing 1-septate, 10–17 × 4.5–5.5 μ.

Growing on soil and humus. An American arctic and alpine species in the west growing south to New Mexico, and in South America.

Localities: 3, 8, 14.

2. **Candelariella placodizans** (Nyl.) Magn. Thallus dispersed to granulose verrucose, the margins becoming effigurate-crenate, subsorediate, lemon yellow to yolk colored. Apothecia to 1.5 mm broad; disk flat, granulose, darker than the margin; margin flexuose to crenulate, very golden pulverulent; epithecium yellow; hymenium 60–120 μ, I+ blue; paraphyses simple or sparingly branched, septate; asci broadly clavate; spores numerous, hyaline, oblong to ellipsoid, simple or 1-septate, 6.5–15 × 5–6 μ.

Growing on soils and humus. It is circumpolar arctic and alpine. It is known from the Northwest Territories and Colorado.

Locality: 8.

3. **Candelariella aurella** (Hoffm.) Zahlbr. Thallus granulose, effuse to dispersed, usually obsolete to lacking, lemon yellow to yolk colored or greenish ashy. Apothecia to 1.2 mm broad, often grouped, flat to slightly convex; the thalloid margin entire to slightly crenulate, thin to disappearing; disk darker than the margin; epithecium granulose, yellow to orange yellow; hypothecium hyaline; hymenium 60–75 μ, I+ blue; paraphyses unbranched, slightly capitate to 3 μ, lax; spores 8, hyaline, oblong, ellipsoid, straight or curved, usually simple, 12–30 × 5–7.5 μ.

Growing on rocks, and on old bones. A circumpolar species, arctic and boreal.

Localities: 1, 2, 3, 8, 9, 11.

4. **Candelariella lutella** (Vain.) Räs. Thallus of minute dispersed granules or lacking, yolk colored. Apothecia dispersed, to 0.4 mm broad, convex to subglobose; margin entire, thin to disappearing; epithecium granulose, yolk colored, dull to subshining; hymenium 60–65 μ, I+ blue;

paraphyses unbranched, capitate, partly septate, 1.5–2 μ, the tips 2.5–3 μ; asci broadly clavate, spores numerous, hyaline, oblong or ovoid, slightly curved, 9–12 × 4–5 μ.

Growing on the barks of *Alnus, Populus, Salix, Picea* and other woody plants. Probably circumpolar arctic and boreal, known from Northwest Territories.

Locality: 8.

5. **Candelariella athallina** (Wedd) DR. Thallus lacking or of minute dispersed granules, yolk colored. Apothecia flat and considerably smaller than in *C. vitellina*, to 0.7 mm; margin entire, rarely crenate; epithecium clear yellow; hypothecium hyaline; hymenium hyaline, 50–60 μ, I+ blue; paraphyses slightly capitate; asci broadly clavate; spores 32–36 per ascus, hyaline, ellipsoid, often poorly formed, with one or sometimes both ends pointed, filled with oil drops which sometimes coalesce to give the appearance of 2-celled spores, straight or slightly curved, 9–15 × 4–6 μ.

This is a species of fissures and holes in acidic rocks, particularly along seashores. It is known from France and Scandinavia. New to North America but specimens in WIS are from Dubawnt Lake, NWT; Trout Lake on the Babbage River, Yukon; Whirlpool Canyon of the Liard River, BC; and Arizona.

Locality: 8.

6. **Candelariella vitellina** (Ehrh.) Müll. Arg. Thallus granulose to verrucose-areolate, dull, lemon yellow to yolk yellow, or brownish yellow, the granules to 0.2 mm broad but usually flattened and forming masses to 5 mm broad, in groups, the color very variable. Apothecia to 1.5 mm, adnate, flat to slightly convex; margin of the same color as the thallus, entire or becoming granulose to partially disappearing, flexuous or subcrenate; epithecium granulose, yellow, olivaceous, or brownish; hypothecium hyaline; hymenium 60–90 μ, yellowish to brownish above, I+ blue; paraphyses unbranched, non-capitate, non-septate, 2–3 μ broad; asci saccate; spores numerous, hyaline, oblong to ellipsoid, slightly curved, the apices rounded, 9–15 × 4.5–6.5 μ.

Usually growing on stones of many types, also on wood, rarely on bark or soil. A species of world-wide distribution, common in North America.

Localities: 8, 14.

7. **Candelariella xanthostigma** (Pers.) Lettau Thallus granulose or farinose leprose, the granules 30–60 μ in diameter, often widely dispersed, lemon yellow to yolk colored. Apothecia not common, to 0.9

mm broad, adnate between granules, flat to slightly convex; the margin thin and entire to becoming leprose, of same color as the thallus; epithecium granulose, yellow to orange yellow; hypothecium hyaline, hymenium 65–75 μ, I+ blue; paraphyses slender, only partially slightly capitate, lax, rarely branched; asci saccate; spores numerous, hyaline, oblong to ovoid, partly septate, 9–12(–15) × 4–5 μ.

Growing on humus or bases of broad leaved shrubs as *Salix, Betula*, or *Alnus*. A common circumpolar species.

Localities: 9, 10, 14.

PARMELIACEAE

Thallus foliose to fruticose, usually dorsiventrally flattened, from closely appressed to erect, with or without rhizinae, in the latter case fixed to the substratum by the underside, heteromerous, the upper cortex prosoplectenchymatous (the hyphal strands densely coherent, very thick walled and with minute lumina) or paraplectenchymatous (with the hyphae also densely coherent but with large lumina and appearing cellular); a medulla usually present; a lower cortex present and similar to the upper but thinner. Apothecia laminal or marginal, sessile to short pedicillate, with thalloid margin; hymenium and hypothecium usually hyaline; paraphyses simple or branched, usually coherent; asci clavate; spores 6–8 (more in some non arctic species), hyaline, 1-celled. Pycnidia with endobasidial or exobasidial fulcra. Algae: Protococcaceae, *Trebouxia*.

1   Fulcra exobasidial; medulla poorly developed   1. *Parmeliopsis*
     Fulcra endobasidial; medulla well developed or hollow   2

2 (1) Medulla hollow (except in *Hypogymnia oroarctica* which is terete, lacks
        rhizines, and contains atranorin and physodic acid)   2. *Hypogymnia*
     Medulla well developed   3

3 (2) Pycnidia marginal or lacking   4
     Pycnidia laminal; rhizinae present; cortex prosoplectenchymatous
                                                                    7. *Parmelia*

4 (3) Cortex paraplectenchymatous   3. *Cetraria*
     Cortex prosenplectenchymatous   5

5 (4) Rhizinae absent; purple pigment usually produced in lower part of medul-
        la especially in dying parts   4. *Asahinea*
     Rhizinae present; no purple pigment produced in medulla   6

6 (5) Caperatic acid present; upper cortex I-; upper surface usually pseudocyphellate; soredia common; pycnoconidia without inflated ends;
spores small, subspherical, 5-8 × 3-5 μ 5. *Platismatia*

Caperatic acid absent; upper cortex often I+ blue; upper surface always
pseudocyphellate; soredia rare; pycnoconidia with inflated ends; spores
larger, ellipsoid, 11-22 × 6-12 μ 6. *Cetrelia*

1. PARMELIOPSIS Nyl.

Thallus foliose, small, growing in rosettes, closely appressed, underside
with rhizinae, lobes dorsiventral, upper cortex paraplectenchymatous
the cells vertically oriented, medulla poorly developed, lower cortex
paraplectenchymatous. Apothecia laminal, sessile, margin thalloid,
concolorous with the thallus; hypothecium hyaline; hymenium hyaline;
paraphyses unbranched or rarely so; asci clavate; spores 8, hyaline;
subspherical to ellipsoid, sometimes curved, simple. Pycnidia laminal or
marginal, pycnoconidia short and straight or longer (12 μ or more) and
curved. Algae: *Chlorococcus*.

1 Thallus yellow, K-; containing usnic and divaricatic acids   1. *P. ambigua*

2 Thallus gray, K+ yellow; containing divaricatic acid plus atranorin
    2. *P. hyperopta*

1. **Parmeliopsis ambigua** (Wulf.) Nyl. Thallus small, round, of narrow
(1 mm) lobes soft, thin, linear, flat or slightly convex, the tips slightly
broadened and crenate, upper side yellow green, dull, wrinkled toward
the center, sorediate with yellowish green soredia, underside black or
brown black, pale brown at the tips, shiny, smooth, with dark rhizinae
which are distributed over the entire underside, containing usnic acid
and divaricatic acid. Apothecia sessile, margin entire and concolorous
with the thallus; disk to 5 mm, slightly concave to convex, chestnut
brown or darker; hypothecium pale; hymenium hyaline; paraphyses
slightly branched, coherent, the tips darkened; spores 8, hyaline, oblong-
ovoid, slightly curved, 8-13 × 2-3.5 μ. Pycnoconidia curved, 12 μ or
longer.

Growing on trees, shrubs, old wood. Rarely on rocks. A circumpolar
arctic, boreal species.

Localities: 8, 14.

2. **Parmeliopsis hyperopta** (Ach.) Arn. Thallus of small 1-2 mm broad
lobes, irregularly dichotomously divided, the tips rounded or crenulate;

upper side whitish or ashy gray, brownish toward the tips, dull or shining, with capitate soralia containing blue gray soredia on the upper surface and the tips of short lobes; underside black, brown toward the margins, with dark rhizinae. Apothecia on surface or margins, slightly pedicellate, to 2 mm broad, the margin whitish or brown, crenulate, sometimes sorediate; disk pale or dark brown, flat or becoming convex; epithecium brown or yellow brown; hymenium very gelatinous, hyaline, 52–62 μ; hypothecium hyaline; paraphyses indistinct; asci clavate; spores hyaline 1-celled, short elliptical or rod-shaped with the ends rounded, 8.5–12 × 2–4 μ.

Growing on *Betula, Ledum, Salix,* and other woody plants, occasionally on rocks. It is circumpolar boreal to subarctic.

Locality: 8.

2. HYPOGYMNIA Nyl.

Thallus foliose with elongate lobes, gray or brown, corticate above and below with the cortex prosenplectenchymatous (with tiny lumina), lacking rhizinae, hollow or solid, often with perforations into the interior. Apothecia laminal, sessile or short pedicillate, margin thalloid, concolorous with the thallus, disk brown; hypothecium hyaline; epithecium brownish, paraphyses unbranched; spores 8, hyaline, ellipsoid, simple. Pycnidia laminal, the mouth black, pycnoconidia cylindrical with the center narrowed, straight. Thallus usually containing atranorin. Algae: *Trebouxia.*

1	Thallus hollow	2
	Thallus with solid medulla; containing atranorin and physodic acid	
		6. *H. oroarctica*
2 (1)	Sorediate or isidiate	3
	Lacking soredia or isidia (one species with papilliform lobes)	6
3 (2)	Thallus diffusely sorediate over the upper surface, the soredia isidioid, the lobes mottled brown and gray with black margins, K+ yellow, P-, containing atranorin and physodic acid	1. *H. austerodes*
	Thallus with the soralia at the tips of the lobes	4
4 (3)	Soralia labriform; thallus pale gray to blue gray, P+ red or P-	5
	Soralia capitate; thallus blue gray to brown, P-, containing atranorin and physodic acid	4. *H. bitteri*
5 (4)	P+ red, containing atranorin, physodic, and physodalic acids	2. *H. physodes*
	P-, containing atranorin and physodic acid	3. *H. vittata*

6 (2) With small papilliform lobes toward the center of the thallus; lobes brown or blackish, P-, K-, containing only physodic acid, not atranorin
5. *H. subobscura*

Lacking such small papilliform lobes; lobes gray or blue gray   7

7 (6) P-, K+, containing atranorin and physodic acid   3. *H. vittata*

P+, K+, containing atranorin, physodic and physodalic acids
2. *H. physodes*

1. **Hypogymnia austerodes** (Nyl.) Räs. Thallus forming rosettes, the lobes close together, upper side gray or more commonly brown, shining, the margins black, as is the underside, hollow, convex above, cracks frequently across the lobes, with isidoid warts breaking into soredia on the upper side, the lobes sometimes perforate at the ends, containing atranorin and physodic acid, lacking rhizinae. Apothecia rare, adnate, to 6 mm broad; the margins crenulate to sorediate; epithecium pale brownish; hypothecium hyaline; hymenium hyaline; spores ellipsoid, 8; hyaline, 6.5–7.5 × 4.5–5 μ.

Growing over mosses, on the bases of trees, on old wood, rarely on soil. A circumpolar arctic alpine species, in North America ranging south to Montana and Washington, and collected on Volcan de Toluca, Mexico, at 4160 m by H.H. Iltis 2179 (WIS).

Localities: 1, 8, 11, 14.

2. **Hypogymnia physodes** (L.) Nyl. Thallus forming rosettes, pinnately or irregularly branched, sometimes with smaller lobes irregularly placed, hollow, the tips usually raised, sometimes perforated, gray to blue gray with the margins more or less black, the underside brown and shining at the tips, mainly black, strongly wrinkled, lacking rhizinae, containing atranorin, physodic and physodalic acids, the tips of the lobes with terminal labriform soralia, the soredia whitish gray green. Young thalli may lack the soralia. Apothecia rare, laminal, sessile or becoming pedicellate, cup-shaped; the margin thalloid, inflexed, concolorous with the thallus; the disk yellow brown to red brown, shiny; epithecium yellowish brown; hypothecium hyaline; hymenium hyaline; spores 8, hyaline, ellipsoid 6–9 × 4–5 μ.

Growing among mosses, on barks, on twigs of trees and shrubs, on old wood, occasionally on soil or rocks. A circumpolar arctic and temperate species, south in the Appalachian Mountains in the east, and to Colorado, New Mexico, and California in the west.

Localities: 2, 3, 6, 8, 10, 11.

3. **Hypogymnia vittata** (Ach.) Gas. Forming small rosettes or imbricated lobes which spread irregularly and loosely over the substratum, the lobes slender, discrete, with or without small branches on the sides, upper side gray green or brownish green, shiny, flat or convex, often edged with a black border which is a continuation of the black underside, lacking rhizinae, some of the lobes with terminal labriform soralia and more or less curved upwards, underside brown at the tips, black and reticulately wrinkled centrally, often perforate at the tips, containing atranorin and physodic acid. Apothecia not seen in North American material.

Growing on bark or mosses. A controversial species in the literature, probably circumpolar temperate. In North America material from the eastern states appears to differ from western material. The obvious criterion to distinguish this species from H. *physodes* is the lack of the physodalic acid (monoacetylprotocetraric acid) which gives that species its P+ red reaction.

Locality: 14.

The specimen is P- and tests for atranorin in GAo-T, physodic acid in GE. Similar specimens are from along the Tanana River near Fairbanks collected by Leslie Viereck. Krog (1968) has reported collections of this species from Alaska, including from near Kotzebue.

4. **Hypogymnia bitteri** (Lynge) Ahti Thallus rosette-like of deeply cut irregular lobes; upper side gray green to brownish, shining at the tips, black spotted or in segments separated by black lines; underside black without rhizinae; with lateral branches which rise up and are tipped with blue gray or whitish gray capitate soralia; medulla P-, K+ yellow, containing atranorin and physodic acid. Apothecia rare, not seen in American material.

Growing over mosses and on bark of spruce, occasionally on rocks and soil. A circumpolar arctic-alpine and boreal species ranging into Quebec in the east, to New Mexico and Mexico in the west. A specimen from Nevada de Toluca near the edge of the crater (Iltis T-6) is in WIS.

Localities: 11, 14.

5. **Hypogymnia subobscura** (Vain.) Poelt Thallus forming more or less rosettes, of pinnate or dichotomously divided elongate lobes, hollow, the tips more or less raised, upper side gray to more commonly brownish, the tips shining; underside black and the black rising up and showing on the sides, lacking rhizinae; without soredia or isidia but toward the center of the thallus with small papilliform lobes irregularly produced on

the edges or upper surface; medulla P-, K-, containing only physodic acid. Apothecia not seen in North American material.

Growing on mosses in hummocky tundras, occasionally on boulders. A circumpolar arctic species in North America ranging south to northern Quebec, Alberta, and Alaska. Common on the North Slope of Alaska.

Localities: 1, 2, 3, 8, 9, 10, 11, 12, 13, 14.

6. **Hypogymnia oroarctica** Krog Thallus foliose, round in section, irregularly swollen, more or less dichotomous with elongate lobes, the tips pointed, upper side bluish gray to brownish gray; underside black; lacking rhizinae; medulla solid; cortex K+ yellow, medulla K-, C-, P- or P+ orange, containing atranorin and physodic acid with accessory protocetraric acid. Apothecia concave to flat, small, on upper side; margin concolorous with the thallus, crenate to flexuous; the disk blackish brown; spores 8, hyaline, ellipsoid, 8–10 × 5-7 μ.

Usually growing over rocks. A circumpolar arctic-alpine species reported from Alaska by Krog (1962, 1968), ranging south to New Mexico in the west.

Localities: 8, 11.

3. CETRARIA Ach.

Thallus foliose to fruticose and erect, more or less dorsiventral, attached to the substratum by the base or by scattered rhizinae, upper and lower sides corticate with paraplectenchymatous cortex several cells thick, marginal cilia present in some species along with the pycnidia; the medulla dense, white or yellow; pseudocyphellae along the margins or on lamellae in some species. Apothecia on the margins, sometimes appearing terminal, usually oriented toward the upper side of the thallus; margin thalloid; hypothecium hyaline; paraphyses unbranched to slightly branched, tips slightly thickened; spores 8, hyaline, simple, ellipsoid or spherical, thin walled. Pycnidia marginal and immersed in papillae or spinules, brown or black; fulcra endobasidial; pycnoconidia straight, ellipsoid, sometimes constricted in the middle. Algae: *Chlorococcus*.

1	Thallus erect, fruticose, not strongly dorsiventral	2
	Thallus foliose, spreading, obviously dorsiventral	15
2	(1) Thallus yellow, containing usnic acid	3
	Thallus brown, not containing usnic acid	6

3 (2) Thallus leathery, not brittle, straw yellow, on acid soils   4

Thallus brittle or leathery, golden yellow, on twigs or calcareous soils

5

4 (3) Lobes cucullate, smooth above, base turning red when dying; containing usnic and protolichesterinic acids   1. *C. cucullata*

Lobes flattened, more reticulate above, base turning yellow when
dying, containing usnic acid   2. *C. nivalis*

5 (3) Thallus brittle, esorediate, on calcareous soils; containing usnic, rangiformic, and pinastric acids   3. *C. tilesii*

Thallus leathery, not brittle, margins sorediate, or twigs; containing
usnic, vulpinic, and pinastric acids   4. *C. pinastri*

6 (2) Medulla UV+ white; thallus tough, horny in texture, dark brown, becoming green when moistened, clearly dorsiventral but curled up
when dry; underside with large pruinose patches, medulla KC+ pink
or red; containing alectoronic acid   5. *C. richardsonii*

Medulla UV-; thallus thinner, not horny in texture, dark brown, not
becoming green when moistened, not clearly dorsiventral; underside
not pruinose   7

7 (6) Medulla I+ blue; thallus pale to dark brown, base reddish or blackish

8

Medulla I-   11

8 (7) Thallus margins with cilia but no pycnidia, pseudocypellae poorly
developed; containing rangiformic and norrangiformic acids

6. *C. elenkinii*

Thallus not ciliate, margins with pycnidia on spinules, pseudocyphellae
present   9

9 (8) Medulla K-; thallus dark brown   10

Medulla K+ yellow or red (substance not known), P+ red (fumarprotocetraric acid); thallus pale brown with white mottling

9. *C. laevigata*

10 (9) Thallus P+ red (fumarprotocetraric acid ± rangiformic and protolichesterinic acids), lobes usually broad, pseudocyphellae broad, both
marginal and laminal   7. *C. islandica*

Thallus P- (fumarprotocetraric acid absent, protolichesterinic acid
present), lobes narrow, cucullate; pseudocyphellae mainly marginal,
narrow   8. *C. ericetorum*

11 (7) Medulla C+ rose; lobes flattish, pale brown with yellowish base, dull,
branching bimorphic with broad lower lobes and much tufting of the
upper flat and markedly smaller branches   10. *C. delisei*

Medulla C-; lobes not bimorphic except sometimes in No 12   12

12 (11) Thallus dorsiventral although not markedly so, paler below, margin not
pseudocyphellate; epithecium K+ violet; containing accessory rangiformic acid   11. *C. sibirica*

Thallus erect, cucullate (channeled above); epithecium K-; marginally
pseudocyphellate   13

13 (12) Color dark, dark brown or black brown, shining   14
        Color pale brown and olive brown mottled, dull, thallus very irregular;
        containing protolichesterinic acid   14. *C. subalpina*

14 (13) Thallus tips broadly rounded, not tufted, branching sympodial, occa-
        sionally bimorphic and with the lower parts broad, upper narrow;
        containing rangiformic and norrangiformic acids   12. *C. andrejevii*
        Thallus tips narrow, branching dichotomous, spinules rare, no laminal
        pseudocyphellae, those on margins few, narrow, inconspicuous; con-
        taining protolichesterinic acid, and rangiformic acid
                                    13. *C. kamczatica*

15 (1) Thallus bright yellow, sorediate, on twigs and bark, containing usnic,
       vulpinic, and pinastric acids   4. *C. pinastri*
       Thallus gray, brown, or black   16

16 (15) On bark or twigs, thallus brown, lobes short and broad   17
        On soil or rocks; thallus deep brown to black   19

17 (16) Medulla KC+ red, UV+, containing alectoronic acid (GE test) and some-
        times *a*-collatolic acid; thallus gray brown   (*C. halei*)
        Medulla KC-, UV-, lacking alectoronic acid   18

18 (17) Thallus deep brown, short, epithecium K-; containing protolichesteri-
        nic acid   15. *C. sepincola*
        Paler brown, slightly taller and broader, epithecium K+ violet, contain-
        ing accessory rangiformic acid   11. *C. sibirica*

19 (16) Blowing loose over the tundra, tough and horny in texture, dark brown
        when dry, greenish when moist; underside with large pruinose
        patches; medulla KC+ red or pink; containing alectoronic acid
                                    5. *C. richardsonii*
        Attached to soil or rocks, softer in texture, not markedly greenish when
        moist; not pruinose below; medulla KC-, lacking alectoronic acid but
        with other constituents   20

20 (19) On soil; lobes marginally ciliate and spinulose, I+ blue; containing proto-
        lichesterinic and rangiformic acids   18. *C. nigricans*
        On rocks; lobes not marginally ciliate   21

21 (20) Pycnoconidiospores with the center narrow, the ends swollen; under-
        side of thallus dark; medulla K+ yellow, containing stictic acid
                                16. *C. hepatizon*
        Pycnoconidiospores cylindrical, the ends not swollen; underside of
        thallus pale; medulla K-, containing *a*-colatollic acid
                                17. *C. commixta*

1. **Cetraria cucullata** (Bell.) Ach. Thallus erect, fruticose, to 8 cm tall, scattered or in masses, the lobes to 4 mm broad, cucullate (channeled) in shape, the edges crisped, the tips somewhat reflexed, smooth, yellow or straw yellow, sometimes greenish yellow, the base turning red upon dying; the edges with dark short-spinulose pycnidia; underside with

narrow whitish pseudocyphellae along the edges, containing usnic and protolichesterinic acids, both sides corticate. Apothecia infrequent, mainly formed in late snow areas, circular to elliptical, to 8 mm broad; the margins concolorous with the thallus, sometimes slightly crenulate, mainly entire; disk reddish brown, pale brownish or yellowish brown, epruinose, smooth; epithecium pale brownish; hymenium 35–45 μ, hyaline; hypothecium hyaline; spores hyaline, simple, ellipsoid, 7.5–8.5 × 3–4 μ.

Growing in many tundra habitats, scattered in the windier places, forming tufts in sheltered spots, growing well in late snow spots and fruiting there. A circumpolar arctic and boreal species, ranging south New England in the east and to New Mexico in the west in mountains. Localities: 1, 2, 3, 4, 6, 7, 8, 9, 11, 14.

2. **Cetraria nivalis** (L.) Ach. Thallus erect to spreading erect, single lobes or forming tussocks, irregularly branched with crisped tips, upper side straw yellow, reticulate sulcate, dull, underside of the same color with scattered whitish pseudocyphellae; base turning yellow to brown when dying; medulla white; containing usnic acid. Apothecia infrequent, at the sides or tips of the lobes, circular; the margins crenulate, concolorous with the thallus, the reverse netted sulcate; disk yellow brown to brown; epithecium pale brownish; hymenium pale; hypothecium hyaline; spores 8, hyaline, simple, ellipsoid or with one end thicker, 5–8 × 3–5 μ.

Growing on soil or between mosses on soil, rarely on twigs where snow covers the lower parts of plants. Arctic-alpine and boreal with a wide range in northern North America, south in mountains in the east into New England, in the west to New Mexico. Very common on the North Slope. Localities: 1, 2, 3, 5, 6, 7, 8, 9, 11, 12, 14.

3. **Cetraria tilesii** Ach. Thallus erect, brittle, golden yellow, on calcareous soils, narrow lobes to 3–4 mm broad with irregular branching, the tips somewhat dentate; upper side smooth to slightly sulcate; underside of same color, smooth; medulla yellow; containing usnic, rangiformic, and pinastric acids. Apothecia not seen.

Growing on calcareous soils and between calcareous gravels. A circumpolar arctic-alpine species originally described from Kamchatka. Localities: 1, 8, 9, 10, 11, 14.

4. **Cetraria pinastri** (Scop.) S. Gray Thallus foliose, crisped-lobate, the lobes rounded, overlapping; upper side yellow to grayish green, smooth,

dull; underside pale yellow to brownish yellow with scattered whitish rhizinae; margins with elongate soralia forming masses of yellow soredia; medulla yellow, containing usnic, vulpinic and pinastric acids. Apothecia rare, marginal toward tips of lobes; the margin confluent with the upper side of the thallus, sorediate; disk 2.5 mm broad, red brown, flat, epruinose; epithecium brownish; hypothecium hyaline; hymenium 70 μ, hyaline; spores 8 hyaline, simple, ellipsoid, 6–8.5 × 6 μ.

Growing on a wide variety of trees and shrubs, *Picea, Pinus, Betula, Ledum, Salix,* etc. and on old wood, occasionally on humus soils and on rocks. A circumpolar boreal and temperate species ranging south in New England, the Great Lakes states, and to Colorado and New Mexico.

Localities: 3, 8, 14.

5. **Cetraria richardsonii** Hook. (*Masonhalea richardsonii* (Hook.) Karnef.) Thallus foliose, dichotomous to irregularly divided, when dry rolled up into ball-like masses, horny in texture, the lobes sharp pointed and antler-like; upper side chestnut brown when dry, becoming olive green to greenish brown when moist; underside paler brown with conspicuous pruinose patches, lacking rhizinae; medulla KC+ pink or red. Apothecia rare, at the tips of lobes, the margin crenate, concolorous with the thallus, inflexed; disk 4–8 mm, brown, almost same color as the thallus, shining; epithecium brownish; upper part of hymenium brownish, lower part pale brown; hypothecium pale brownish; hymenium 125 μ; spores 8, hyaline, thin walled, simple, ellipsoid, 7.5–12.5 × 5–6 μ.

A species which grows unattached and blows across the tundras, settling in lower spots in numbers, the extra moisture causing it to flatten out and cease moving with the wind. A Beringian element in the North American lichen flora, ranging east to Chesterfield Inlet, south to British Columbia.

Localities: 1, 2, 3, 4, 6, 7, 8, 9, 10, 11, 13, 14.

6. **Cetraria elenkinii** Krog Thallus dorsiventral but loose on the substratum and rising almost fruticose, lobes to 2 mm broad and 4 cm long, slightly channeled above; upper side dark olive brown to blackish brown, smooth; underside nearly the same color, slightly dented in spots, pseudocyphellae few or absent; margins with cilia which are simple or sparingly branched; tips not rounded, pointed; pycnidia nearly lacking. Apothecia not seen.

A species of sandy soils, sometimes among mosses. Circumpolar and high arctic.

Specimens seen: No specimens have been collected yet on the North Slope although fully to be expected there. The species is included here for the convenience of workers in this region. Krog reported it from Ogotoruk Creek area collected by Hultén.

7. **Cetraria islandica** (L.) Ach. Thallus fruticose, erect, to 10 cm tall, more or less dichotomously branched or irregularly branched, the margins inrolled to make the lobes more or less channeled, the tips partly reflexed; upper side olive brown to dark brown or blackish brown, smooth; the underside paler than the upper side, with white pseudocyphellae scattered over the surface and more elongate ones common along the edges; the margins with cilia and also with abundant pycnidia on projecting thorn-like processes; the dying base often turns red; medulla white; containing fumarprotocetraric acid and accessory protolichesterinic and rangiformic acids. Apothecia rare, on broadened thallus tips, to 18 mm broad; margin poorly developed, partly inflexed, the outer reverse side roughened; disk light or dark reddish brown, dull, epruinose; epithecium yellowish brown; hypothecium yellowish; hymenium pale yellowish; spores 8, hyaline, ellipsoid to ovoid, simple, 8–9.5 × 3–5 μ.

Growing on moist or dry tundras among mosses or in the open, occasionally on old wood or on twigs of spruce in the part protected by snow cover in winter. A circumpolar arctic and boreal lichen with many modifications of size, branching, and shape. Although in Europe it is usually distinguished from *C. ericetorum*, Kristinsson in Iceland has shown it to be intergrading with that species, possibly forming only a large variable population of one species, yet he shows different areas of the island occupied by different dominants. Kristinsson disagreed with the often-used distinguishing characteristic of P+ or P- (fumarprotocetraric acid + or -); yet this test seems a more reliable way of separating the populations than the morphology (which he demonstrated to be intergrading), and hence this chemical character is still followed here. The specimens listed here are thus all P+.

Localities: 1, 2, 3, 4, 7, 8, 9, 10, 12, 13, 14.

8. **Cetraria ericetorum** Opiz Thallus fruticose, erect, to 9 cm tall, more or less dichotomously or irregularly branched, the margins inrolled to make the lobes more or less channeled, usually narrower than in *C. islandica*, the tips partly reflexed, the upper side from dark olive brown to brown or reddish or blackish brown, smooth, shining, the underside with elongate whitish pseudocyphellae along the margin, rarely over the midsurface which is of the same color as the upper surface or slightly

paler, the margins with projecting thorn-like processes containing the pycnidia and also with occasional cilia; contents lacking fumarprotoce-traric acid hence P-, with protolichesterinic acid and rangiformic acid usually present. Apothecia rare, at the tips of enlarged lobes; the margin inflexed; the disk red brown, smooth; epithecium brown; hymenium 40-60 μ; hypothecium pale; asci clavate; spores 8, hyaline, simple, ellipsoid, 7.5-10 × 3.5-5 μ.

This species appears to occupy much the same habitats as *C. islandica*, among mosses in either moist or dry tundras. Kristinsson (1969) has shown that the P- plants of his study were more common in the parts of Iceland with a more continental climate, while the P+ plants were correlated with *Rhacomitrium* moss mats and more oceanic climate. It is circumpolar boreal and arctic-alpine with North American variants, which are possibly undescribed species, in the Great Lakes area and the eastern states.

Localities: 1, 9, 13, 14.

9. **Cetraria laevigata** Rass. Thallus fruticose, erect, irregularly branched, somewhat sympodial, the margins inrolled to make the branches canaliculate as in *C. ericetorum* and *C. islandica*, the tips slightly reflexed, the upper side pale brown, the underside paler and the base usually very pale, sometimes pinkish but not red, the underside with marginal pseudocyphellae which are wider and more continuous than in *C. ericetorum*, P+ red, K+ yellow becoming red, containing fumarprotoce-traric acid plus an unknown substance. Apothecia not seen.

This species occupies somewhat moister habitats than the preceding two. It ranges widely circumpolar, and in North America is high arctic, in Canada and Alaska.

Locality: 7.

10. **Cetraria delisei** (Bory) Th. Fr. Thallus fruticose, erect, forming tufts or mats in low places, the branches markedly bimorphic with the basal part of broad lobes and the outer tip parts very narrow and densely branched, the branches pale to dark brown on both surfaces, the tip parts usually paler, flattened, lacking pseudocyphellae on upper side, pseudocyphellae scattered on lower side, and lacking spinules along the margins; the dying base yellowish; medulla C+ rose, containing hiascinic and gyrophoric acids. Apothecia not seen.

This species has a marked preference for low wet swales where the water flows late from snowbanks or in spring overflow areas along streams. It is often associated with *C. andrejevii* and may also be associated with *Lecidea ramulosa* in such places. It is circumpolar high arctic in distribution and is distributed across northern Canada and Alaska in alpine as well as lowland tundras.

Localities: 5, 7, 8, 9, 10, 13.

11. **Cetraria sibirica** Magn. Thallus in small rosettes with the dorsiventral lobes becoming more or less erect, to 1 cm tall, attached by the base, the lobes slightly concaave, upper side dark brown to blackish brown, lower side pale, not pseudocyphellate, the margins with few spinules, K-, C-, P-, containing rangiformic acid or no contents. Apothecia terminal on the lobes; the margin entire to crenulate; disk flat or concave, brown or concolorous with the thallus; epithecium brownish, K+ violet; hypothecium yellowish, I+ blue; hymenium 35–45 μ, yellowish, I+ blue; paraphyses 2–3 μ, thickened at tips; spores 8, hyaline, simple, ellipsoid, 6.5–8.5 × 4–5 μ.

This species has the appearance of a larger paler *C. sepincola*, but has thicker exciple, paraphyses and hyphae, a K+ violet reaction of the epithecium, and lacks the protolichesterinic acid of that species.

Growing on twigs of *Betula*. A species described from Siberia, Krog (1968) has reported it from the Ogotorok Creek area in the Bering Strait District, and Llano (1951) had previously reported it from Umiat. Specimens from Trout Lake, Babbage River, Yukon Territory, Thomson 14754 and 14781 are in WIS. A report of *C. magnussonii* from the Thelon River, NWT (Scotter and Thomson 1966) should be corrected to this species as the epithecium of the material is K+ violet. This is a Beringian species in the flora.

Locality: 3.

12. **Cetraria andrejevii** Oxn. Thallus fruticose, forming small mats and tufts, erect, intertwined and fusing in the groups, irregularly branched, the branch tips markedly enlarged and rounded, dilating upwards, the margins partly inrolled and the thalli therefore partly canaliculate, both sides deep shining brown to chestnut brown, the dying parts paler yellowish brown; with weak pseudocyphellae, the margins lacking isidia or cilia, the small spinules with pycnidia very scant, reactions of the medulla negative with the usual reagents including I; containing rangiformic and norrangiformic acids. Apothecia rare, at tips of lobes, the reverse wrinkled; epithecium yellowish brown; hypothecium pale; spores 8, simple, hyaline, ellipsoid, 8–10 × 4–5 μ.

Growing on the ground in very wet places, especially with cold seepage as below late snow banks and along stream banks where there is springtime overflow, sometimes in very moist places in bogs. This species was described from Siberia and ranges across northern Alaska and Canada to Greenland.

Localities: 7, 8.

This species is somewhat variable. More wrinkled, spreading and anastomosing plants with tan bases instead of gray white bases and shorter pycnidial projections but with the same chemical contents have been separated by Krog (1968) as *C. simmonsii* Krog, including the specimen cited above and including variants almost indistinguishable from *C. andrejevii* under the name *C. simmonsii* var. *intermedia* Krog. She also included similar plants with very broad lobes, to 30 mm broad, and margins covered with lobules which developed into new lobes as *C. simmonsii* var. *lobulata* Krog. However the writer feels that these may be but expressions of environmental pressure on the major species *C. andrejevii* causing some variation in the morphology and has therefore not accepted *C. simmonsii*. The similarity of the chemistry also causes doubt concerning the validity of the proposed species. Lich. Arct. 32 specimens were obtained in an area occasionally overrun in times of storm by the salt waters of the Chukchi Sea, and the specimens lie around low pools on the gravelly point of Point Barrow, a region of considerable stress.

13. **Cetraria kamczatica** Savicz Thallus fruticose, forming small mats with intertangled lobes, dichotomously branched, the angles of branching wider than in *C. islandica*, the tips becoming turned back, the lobes narrow, the edges inrolled and not only canaliculate but occasionally the edges fusing to become tubular, the surfaces chestnut brown, shining, the margins wavy, occasionally dentate, with rare spinules which are poorly developed, lacking pseudocyphellae or with few, narrow and tiny pseudocyphellae on lower margins; medulla white with reactions P-, K-, C-, KC-, I- and containing protolichesterinic and rangiformic acids. Apothecia not seen.

Growing among mosses in dry tundra. A species described from Siberia and reported by Krog (1962, 1968) from Alaska, St Matthew Island, Nunivak Island, King Mt on the Seward Peninsula, and near Cantwell in the Alaska Range. It ranges east to the vicinity of Canoe Lake west of Inuvik, NWT.

Locality: 4.

14. **Cetraria subalpina** Imsh. Thallus of densely tangled, twisted, irregularly branched, prostrate or erect lobes, the lobes flat or with a central groove but not canaliculate, shining, pale greenish olive to olive brown or chestnut brown and irregularly mottled, lower surface paler with

inconspicuous marginal pseudocyphellae, margins more or less spinulose as well as with pycnidia, reactions of medulla all negative, containing protolichesterinic acid. Apothecia marginal; margin inflexed, disappearing; disk buff to light brown, to 5 mm broad; epithecium yellowish brown; hypothecium pale; hymenium 50–80 μ; spores 8, hyaline, simple, subspherical or irregular 4.3–6.4 μ.

Usually on branches of small shrubs as *Betula, Rhododendron, Vaccinium*, occasionally on shaded mossy soil. Ranging from the Olympic Mts in Washington north to Alaska. Reported from Point Barrow by Imshaug (1950) from a collection by Lehnert (MICH). The extremely similar *Cetraria inermis* (Nyl.) Krog (differing only in more conspicuous lateral pseudocyphellae and smaller size) has been recorded by Krog (1973) from the Meade River and Voth Creek.

15. **Cetraria sepincola** (Ehrh.) Ach. Thallus small, of radiating short, to 1 cm long lobes, scantily branched, foliose; upper side olive brown to chestnut brown, shining, smooth, underside pale brownish or whitish with pale scattered rhizinae; margins lacking rhizinae, with the pycnidia more or less embedded or projecting; medulla white, reactions all negative, containing protolichesterinic acid. Apothecia at the tips of the lobes, usually abundant, to 3 mm broad; margins entire or crenulate; disk reddish brown, shining, epruinose; epithecium pale brown; hypothecium pale; hymenium 40–50 μ; spores 8, hyaline, simple, ellipsoid, 7–10 × 4.5–6.5 μ.

Growing on the twigs of shrubs in the tundra, especially common on *Betula*. A circumpolar, arctic and boreal to northern temperate species, ranging south to New York, Michigan, Minnesota, and California.

Localities: 1, 3, 8, 9.

16. **Cetraria hepatizon** (Ach.) Vain. Thallus foliose, loosely attached to rocks or gravel by rhizoids, forming large rosettes of irregularly branched lobes, upper side brown to blackish brown, shining, the margins thickened and forming a channeled upper side, with embedded dark pycnidia (in the similar *Parmelia stygia* the pycnidia are distributed over the surface); underside dark brown to black, rhizinate; medulla white, P+ orange, K+ yellow becoming reddish, containing stictic acid. Apothecia at the tips of lobes but appearing toward the center of the thallus, to 5 mm broad; margin entire to much crenulate, brown; disk dark brown, almost the same color as the thallus, shining or dull; epithecium brown; hypothecium yellowish; hymenium 45–50 μ; spores 8, hyaline, simple, ellipsoid, 6.5–10 × 5–7 μ; pycnoconidiospores with the ends thickened, 4–6 × 1 μ.

Very common on exposed dry rocks and gravel eskers in arctic and boreal regions, ranging southward to New York in the eastern mountains, to the north shore of Lake Superior, and to Arizona in the western mountains.

Localities: 3, 8, 9, 11.

17. Cetraria commixta (Nyl.) Th. Fr. Thallus foliose, forming rosettes on rock surfaces, the lobes narrow, to 2 mm broad, slightly channeled as in the previous species, upper side bronzy brown to black brown, shining or dull, underside pale or becoming brown, with scattered dark rhizinae; medulla white, K-, containing α-collatolic acid. Apothecia at the tips of short lobes, more or less central on the thallus, to 7 mm broad; margin thin, inflexed, irregularly thickened; disk of the same color as the thallus, slightly shining, smooth, concave; epithecium brownish; hypothecium pale; hymenium 45–55 μ; spores 8, hyaline simple, ellipsoid, 8–10 × 4.5 μ; pycnoconidiospores cylindrical or the center thickest, not the ends, 3–4 × 1 μ.

Growing on rocks and gravel in open dry places. A circumpolar arctic-alpine species, ranging across Canada, south to New York, Colorado, and Washington.

Localities: 8, 11.

18. Cetraria nigricans (Retz.) Nyl. Thallus foliose, forming tufts and small mats, the lobes to 1 mm broad, slightly channeled, the margins with cilia which may be simple or branched; upper side dark brown or olive brown; lower side pale brown or tan with sparse, short and inconspicuous pseudocyphellae; medulla K-, C-, KC-, P-, I+ blue. Apothecia unknown in North America. The thallus contains protolichesterinic and rangiformic acids.

This species grows on rocks and soil. It is circumpolar low arctic to alpine.

Locality: 8.

4. ASAHINEA W. L. & C. F. Culb.

Thallus foliose, the lobes broad, yellowish or grayish, upper side pseudo-cyphellate in one species, isidiate in another, the margin not ciliate, commonly with blackish spots and edges; underside black to the center, brown marginally, lacking any rhizinae; upper cortex prosoplecten-chymatous (with thick walls and tiny lumina). Apothecia rare, marginal, imperforate, the spores hyaline, simple, ellipsoid, 8–13 × 5–8 μ; pycnidia rare, marginal, immersed; pycnoconidiospores straight, 5–10 μ. Atranorin and alectoronic acid present, usnic acid in one species $\pm$ α-collatolic acid.

1 Thallus yellow mottled or edged with black, lacking isidia but pseudocyphellate above, usnic acid present   1. *A. chrysantha*
  Thallus whitish to tan, mottled with black and becoming mostly black, upper surface isidiate, non-pseudocyphellate (but scars of isidia sometimes resembling pseudocyphellae), usnic acid absent.   2. *A. scholanderi*

1. Asahinea chrysantha (Tuck.) W. Culb. & C. Culb. Thallus medium to large, the lobes to 3 cm broad, rounded, upper surface with large and small reticulate pitting, pseudocyphellate, the pores few or many, mainly on the ridges, yellow to yellowish tan with the edges black; lower surface jet black, smooth or wrinkled, brown toward the margins; upper cortex with minute crystals; medulla white or partially lavender in old parts; K-, C-, P-, KC+ red or pink; containing usnic acid and atranorin in the upper cortex, alectoronic acid $\pm$ α-collatolic acid in the medulla. Apothecia rare, described from Japanese material seen by the Culbersons, submarginal to laminal, to 2.4 cm; disk brownish black; hypothecium I+ lavender; hymenium 52–70 μ; spores 8, ellipsoid, 8–13 × 5–8 μ; pycnidia rare, marginal, pycnoconidiospores 5–6.5 × 1 μ.

Growing on boulders, among mosses, and on bases of plants. A Beringian species known from Siberia across North America to Baffin Island, high arctic.

Localities: 1, 7, 8, 9, 11, 14.

2. Asahinea scholanderi (Llano) W. Culb. & C. Culb. Thallus of medium size, the lobes to 1.5 cm broad, forming rosettes on the substratum, the lobes imbricated and massed; upper side whitish or bluish gray to tan, mottled with black and sometimes mainly black, lacking pseudocyphellae, with numerous erect unbranched or rarely slightly branched cylindrical isidia, colored like the surface or tipped with black or all black; lower surface jet black, brown toward the margins, smooth to minutely pitted, lacking rhizines; medulla white or lavender in the older parts; K-, C-, P-, KC+ pink or red; containing atranorin in the upper cortex, alectoronic and α-collatolic acids in the medulla. Apothecia very rare, described by the Culbersons as 3 mm broad with thalline margin isidiate, the isidia bearing pycnidia; spores 8, ovoid to subglobose, 10–13 × 5–8 μ; pycnoconidiospores straight or slightly curved, 5–10 μ.

On boulders, plant remains, and soil in the tundra. A Beringian element in the flora, known from Siberia and east in North America to the region of Baker Lake and the Back River in Canada.

Localities: 8, 11.

The Culbersons also list North Slope specimens from Anaktuvuk Pass, Lake Peters, and from Cape Dyer and Ogotoruk Creek near the Bering Straits.

5. PLATISMATIA W.L. & C.F. Culb.

Thallus of medium size, lobes to 2.5 cm broad; upper side usually rugose, commonly pseudocyphellate or isidiate, ashy or blue gray; underside with rhizines, black, punctate; upper cortex prosoplectenchymatous; medulla commonly containing atranorin and caperatic acid. Apothecia large, to 4 cm broad, marginal or submarginal, commonly perforate; epithecium pale; subhymenium I+ blue; spores 8, hyaline, simple, ellipsoid or subglobose.

Only one species of this genus reaches the North Slope.

1. **Platismatia glauca** (L.) W. Culb. & C. Culb. Thallus medium sized, the lobes to 2 cm broad, margins with soredia, simple or coralloid isidia or both, or branched and divided into fruticose lobes which in turn may have soredia or isidia or both; upper surface whitish or whitish green, or tinged with yellow, smooth or becoming reticulately wrinkled, not pseudocyphellate, smooth or with coralloid or sorediose isidia in patches or scattered; lower surface jet black, margins brown, colored like the upper surface or white, smooth or reticulately wrinkled and pitted, rhizines few to many, brown or black, simple or branched; medulla white I+ lavender; cortex K+ yellow, medulla K-, C-, KC-, P-; containing atranorin and caperatic acid. Apothecia very rare, to 1 cm broad, marginal, perforate or not; hymenium 34–56 μ; spores 8, hyaline, ellipsoid to ovoid, 3.5–8.5 × 3–5 μ.

Growing on rocks, soil, and on vegetation. An almost circumpolar boreal and temperate species also found in Africa and South America but apparently replaced by other related species in eastern Asia. In North America in the east in Labrador to North Carolina, in the west from Alaska to Colorado and California.

Locality: 9.

6. CETRELIA W.L Culb. & C.F. Culb.

Thallus large to medium, the lobes to 2.5 cm broad; upper surface ashy white to greenish white, pseudocyphellate, in extraterritorial species can be isidiate or sorediate; under surface black, brown at the margins, rhizinate; upper cortex prosoplectenchymatous. Apothecia not known in the species of our area.

1. **Cetrelia alaskana** (C. Culb. & W. Culb.) W. Culb. & C. Culb. Thallus medium sized, to 7 cm broad, the lobes to 1.5 cm broad, pale yellowish tan or brownish in the herbarium, greenish gray in the field, often mottled with black, smooth, esorediate, pseudocyphellate with minute pores; lower surface jet black centrally, the margins chestnut brown, rhizines few, short papilliform or longer; upper cortex 16–29 μ, medulla 86–135 μ, lower cortex 16–23 μ. Apothecia and pycnidia unknown; reactions: medulla K-, C-, KC- or + pale pink, P-; cortex K+ yellow; containing atranorin (in cortex) and imbricaric acid.

Growing between tussocks in tussock-tundra (*Eriophorum*) and also on the tussocks. A Beringian species known only from Alaska near the Bering Sea and the Chukchi Sea.

Localities: 1, 9.

Reported also from the Ogotoruk Creek area; 40 mi east of Cape Lisborne; Kuskakwin; and near Nome by the Culbersons.

7. PARMELIA Ach.

Thallus foliose, dorsiventral, flat or nearly terete, upper cortex prosoplectenchymatous as is lower cortex, attached to the substratum by rhizinae; medulla dense or loose, white. Apothecia laminal, round, adnate to sessile; margin thalloid; epithecium pale; hypothecium hyaline; asci clavate; spores 8, hyaline, simple, ellipsoid. Pycnidia laminal, innate or partly projecting, often on margin of apothecia; fulcra endobasidial; pycnoconidiospores straight, cylindrical or spindle-shaped. Algae: *Trebouxia*.

1	Thallus yellow, containing usnic acid (if lacking rhizinae and pseudocyphellate above check *Asahinea chrysantha*)	2
	Thallus either gray or brown, lacking usnic acid	5
2	(1) Thallus esorediate	3
	Thallus with capitate soralia above; medulla UV+, KC+ red, K+ yellow, containing alectoronic acid and atranorin	4. *P. incurva*
3	(2) Medulla UV-, K+ red, containing salazinic acid; underside black	1. *P. tasmanica*
	Medulla UV+, KC+ red, K+ yellow, containing alectoronic acid and atranorin; underside pale or dark	4
4	(3) Underside white, medulla I-, lobes short and wide	2. *P. centrifuga*
	Underside dark, mouse gray to black; medulla I+ blue; lobes elongate	3. *P. separata*
5	(1) Thallus lacking rhizinae, grayish to brown mottled, P- containing atranorin plus physodic acid	*Hypogymnia oroarctica*
	Thallus with rhizinae, lacking physodic acid	6

6 (5) Thallus gray   7
       Thallus brown   10

7 (6) Thallus sorediate or isidiate   8
       Thallus lacking soredia or isidia; containing salazinic and lobaric acids
           and atranorin   7. *P. omphalodes*

8 (7) Thallus sorediate   9
       Thallus isidiate, the isidia scattered over the upper surface; containing
           salazinic acid and atranorin   6. *P. saxatilis*

9 (8) Thallus sorediate with the soredia on a raised network on upper side of
           lobes, gray; containing salazinic acid and atranorin   5. *P. sulcata*
       Thallus with soredia along the edge of the lobes; yellowish gray; con-
           taining salazinic, usnic, and protolichesterinic acids and atranorin
                                   8. *P. fraudans*

10 (6) Thallus sorediate or isidiate   11
        Thallus neither isidiate nor sorediate   16

11 (10) Thallus sorediate   12
        Thallus isidiate   13

12 (11) Soredia in large capitate soralia; non-pseudocyphellate above; medulla
            C+ red; containing olivetoric and possibly imbricaric acids
                                   9. *P. disjuncta*
        Soredia in smaller groups on the thallus; upper side pseudocyphellate;
            C+ rose, containing gyrophoric acid   10. *P. substygia*

13 (11) Isidia granular, medulla P+ red, containing fumarprotocetraric acid
                                   11. *P. olivaceoides*
        Isidia cylindrical or spatulate; medulla P-   14

14 (13) Isidia spatulate with narrow bases expanding upwards
                                   12. *P. exasperatula*
        Isidia cylindrical   15

15 (14) Medulla C-, K+ yellow, containing atranorin; tips of lobes often
            pruinose   13. *P. infumata*
        Medulla C+ red, K-, containing lecanoric acid; tips of lobes shiny,
            epruinose   14. *P. glabratula*

16 (10) Lobes convex   17
        Lobes flattened   19

17 (16) Lobes with tiny white pseudocyphellae on the upper side; medulla P+
            red, K+ yellow, or P-, K-, containing accessory fumarprotocetraric
            acid, or lacking acids   15. *P. stygia*
        Lobes lacking tiny white pseudocyphellae   18

18 (17) Medulla P+ yellow, KC+ red, containing alectorialic and protoliches-
            terinic acids   16. *P. alpicola*
        Medulla P-, K-, C+ red, KC+ red, containing olivetoric acid
                                   17. *P. almquistii*

19 (19) On bark of trees and shrubs; lobes 1–3 mm broad; medulla P+ red, containing fumarprotocetraric acid   18. *P. septentrionalis*
On stones; lobes 0.5 mm broad, very imbricated; medulla P-, K+ yellow, containing atranorin   19. *P. panniformis*

1. **Parmelia tasmanica** Hook. f. & Tayl. (*Xanthoparmelia tasmanica* (Hook. f. & Tayl.) Hale) Thallus foliose, irregularly branched, the lobes loosely attached to the substratum, 1–6 mm broad, overlapping, flat to slightly rounded; the upper side yellow, smooth, lacking isidia or soredia; underside brown at the margins, black toward the center; medulla white; K+ yellow turning red, P+ orange, KC+ yellow; containing usnic and salazinic acids. Apothecia concave to deeply concave, to 8 mm; the margin concolorous with the thallus, irregularly wavy to lobulate; disk chestnut brown to dark brown, shiny or dull; epithecium brown; hypothecium and hymenium hyaline; spores 8, hyaline, ellipsoid to oval or slightly curved, 8–11 × 5–6 μ.

Growing over acid rocks or outcrops. A circumpolar, arctic to temperate species of wide distribution, ranging over most of North America.
Locality: 11.

2. **Parmelia centrifuga** (L.) Ach. (*Xanthoparmelia centrifuga* (L.) Hale) Thallus foliose, closely adnate to the substrate, commonly dying in the center and forming concentric rings, the lobes irregularly branching, tips pointed, flat to slightly convex, 2 mm broad; upper side greenish yellow, partly whitish green, darkening toward the center, dull, sometimes chinky, lacking soredia or isidia; underside whitish or pale brownish with brown to blackish short rhizines; medulla white, I-; K+ yellow, C-, KC+ red, P-, containing atranorin, usnic and alectoronic acids. Apothecia adnate, to 5 mm broad; the margin entire to crenulate to inflexed, concolorous with the thallus; disk reddish to dark brown, dull; epithecium brownish; hypothecium and hymenium hyaline; hymenium I+ blue; spores ellipsoid, 8–14 × 4.5–6 μ.

Growing on rocks and outcrops. A circumpolar arctic-alpine species ranging in North America south to the Appalachian Mountains and the northern shores of the Great Lakes.
Localities: 8, 9, 11.

3. **Parmelia separata** Th. Fr. (*Xanthoparmelia separata* (Th. Fr.) Hale) Thallus foliose, loosely attached to the substrate, dichotomously to irregularly branched, the lobes elongate as compared with *P. centrifuga*; upper side yellow, brownish in spots, often chinky separated crosswise, smooth to wrinkled; the underside mouse gray or dove gray to black,

smooth with sparse rhizinae; medulla white and I+ blue, K+ yellow, KC+ red, P-; containing usnic and alectoronic acids and atranorin. Apothecia adnate on surface of thallus, slightly pedicillate; margin entire to lobulate, concolorous with the thallus; disk concave, to 8 mm broad, bright chestnut brown, shining, epruinose; cortex of amphithecium poorly differentiated, heavily inspersed with yellow granules and opaque, the surface rough; epithecium and upper 12 μ of hymenium clear brown; lower part of hymenium hyaline, hymenium 65 μ, I+ blue; hypothecium 90 μ, hyaline to yellowish or grayish yellow, the upper 25 μ I+ violet, no reaction in the lower part; paraphyses conglutinate, thick walled, little branched, septate, the tip cells enlarged, brown, to 4 μ; asci clavate, 12 × 30 μ; spores 8 hyaline, elongate ellipsoid, simple, 10–12.5 × 4.5–6.5 μ.

Growing over rocks, occasionally over humus. A species ranging from Novaya Zemlya across Siberia into Alaska, across Canada in the arctic; not in Greenland.

Localities: 3, 8, 9, 11, 14.

The apothecia have apparently not hitherto been described. They seem to be fairly common in North American material. In the WIS herbarium are fruiting specimens from Umiat (5691), Okpilak (5900, 6364, 9592); Artillery Lake, NWT, Thomson 12220, 12337; Lake Ennadai, NWT, Thomson 11519; Father Lake, northern Saskatchewan, Scotter 289; Higginson Lake, northern Saskatchewan, Scotter 84; Fort Hall Lake, Manitoba, Scotter 2893, 2926, 2963; Pate Lake, north of Yellowknife, NWT, Thomson 10894.

The I+ reaction in the apothecia is quite spectacular with the upper hypothecium or subhymenium turning violet in contrast with the blue reaction of the hymenium. A paler blue I+ reaction slowly develops among the hyphae of the medullary layer of the amphithecium.

4. **Parmelia incurva** (Pers.) Fr. (*Xanthoparmelia incurva* (Pers.) Hale) Thallus forming small rosettes, closely attached to the substrate, the lobes narrow, to 1 mm broad, convex, closely set to each other, with capitate soralia scattered over the surface of the thallus; upper side grayish green, yellow green or grayish yellow, darkening to the center, dull, the soralia to the tips of short lobes and to 4 mm broad, usually less; underside brown at the edges, black centrally, with blackish simple rhizinae; medulla white, K+ yellow, KC+ red, P-; containing usnic and alectoronic acids and atranorin. Apothecia uncommon, small 2(-4) mm broad, adnate; margin inflexed, entire to sorediate or crenulate, concolorous with the thallus; disk chestnut brown, concave or flat, dull to shining; epithecium yellowish; hypothecium hyaline; spores ellipsoid, thick walled, 9–10.5 × 6–8 μ.

Growing on rocks and outcrops of acidic rocks. A circumpolar arctic-alpine species, ranging south to the Adirondack Mountains in the east, but only to Alaska and the Yukon in the west.

Localities: 8, 11.

5. **Parmelia sulcata** Tayl. Thallus foliose, loosely attached to the substrate, irregularly branching, the lobes elongate, to 6 mm broad, the edges entire or notched; upper side blue gray, ashy gray or brownish, with a raised network of fine ridges, sulcate between, pruinose occasionally in patches or toward the tips, the ridges with round or elongate soralia with lighter soredia than the general color; underside with the edges brown, the greater part black, with numerous simple or branched black rhizinae; medulla white; the cortex K+ yellow, medulla K+ red; containing atranorin and salazinic acid. Apothecia uncommon, adnate or on short pedicels, to 9 mm broad; the margin concolorous with the thallus, entire or often sorediate; disk flat, dull, brown, epruinose; epithecium brownish; hypothecium yellowish; spores 8, hyaline, poorly developed, ellipsoid, $12-18 \times 5-8 \ \mu$.

Growing on rocks, bark, humus, or soil. A circumpolar lichen with broad arctic to temperate range.

Localities: 3, 8, 9, 11, 14.

6. **Parmelia saxatilis** (L.) Ach. Thallus foliose, loosely attached, the lobes radiating, elongate, usually to 2-6(-10) mm broad, the sinuses rounded between lobes, sometimes the lobes very narrow and much imbricated in heaps, irregularly branching; upper side ashy gray or bluish gray, with slight netting and sulcation, lacking soredia but with short to cylindrical and sometimes branching isidia scattered over the surface, the tips of the isidia browned; underside black, the margins brown and shining, with simple or slightly branched but not squarrose rhizinae abundant and loosely attaching the plant to the substrate; K+ yellow, turning red, C-, KC-, P+ red; containing atranorin, salazinic and lobaric acids. Apothecia adnate to short pedicillate on the upper surface; margin entire to isidiate, concolorous with the thallus or browning; disk concave, brown or reddish brown, dull epruinose; epithecium brown, smooth; hymenium 75-100 μ, hyaline; paraphyses conglutinate; spores 8, simple, hyaline, ellipsoid, $13-19 \times 8-12 \ \mu$.

Growing on rocks, usually acid in type. A circumpolar arctic and boreal species which ranges south in the Appalachians and the Rocky Mountains.

Localities: 3, 8, 11.

7. **Parmelia omphalodes** (L.) Ach. Thallus foliose, of elongate lobes 5–10 mm broad, irregularly branched, the axils rounded; upper side bluish gray to brownish or blackening over much of the thallus; with slightly raised network of whitish markings, sulcate between, lacking soredia or isidia; underside black with a narrow peripheral brown zone, with black rhizinae; medulla white; K+ yellow, medulla K+ yellow, becoming red; containing atranorin, salazinic and lobaric acids. Apothecia adnate on upper surface; margin concolorous with the thallus; disk to 8 mm broad, concave to flat, dark brown, dull; epithecium brown; hypothecium hyaline; hymenium hyaline; spores 8, simple, hyaline, ellipsoid 14–20 × 8–12 μ.

On boulders and rock outcrops. A circumpolar arctic and boreal species ranging south in the Appalachians and to Washington and Montana.

Localities: 1, 2, 3, 7, 8, 9, 10, 11, 12, 13.

8. **Parmelia fraudans** Nyl. Thallus dichotomously to irregularly branched, slightly channeled above, pale brownish or yellowish gray, with coarse yellowish or gray soredia partly scattered in rounded groups on the upper surface but mostly along the margins in abundance; underside black, with abundant black, branched rhizinae; medulla K−, P+ red orange, containing usnic acid, protolichesterinic acid, and salazinic acid, cortex K+ yellow containing atranorin. Apothecia not seen.

A species of rocks. Circumpolar arctic and alpine.

Locality: 8.

9. **Parmelia disjuncta** Erichs. Thallus foliose, small, appressed to the substratum; the lobes flat, to 0.6 mm broad, irregularly branched, the margins entire; upper side dark olive brown to blackish brown, shiny, smooth, isidiate and sorediate, the isidia on the surface, verrucose, short, bursting at the tips and there forming gray or whitish soredia which may unite to form larger patches; underside black centrally, brown peripherally, slightly shining, wrinkled to slightly veiny, with black simple or branched rhizinae; medulla white; K−, KC+ red; containing olivetoric acid and possibly also imbricaric acid. Apothecia not seen.

Growing on rocks. A circumpolar arctic and boreal species; ranging south in North America in Quebec, Michigan, Minnesota, Montana, Idaho, and Washington.

Localities: 3, 8, 11, 14.

10. **Parmelia substygia** Nyl. Thallus foliose, small, appressed to the substratum, the lobes convex, to 4 mm broad, irregularly branched; upper surface olive brown to blackish brown with white pseudocyphellae, shiny, smooth, with small groups of capitate soralia scattered over the thallus; underside black, with black rhizinae; medulla white; K-, C+ rose, KC+ red; containing gyrophoric acid. Apothecia adnate, to 7 mm broad, concave to flat; margin concolorous with the thallus, pseudocyphellate and becoming sorediate; disk red brown to brown black; epithecium brown; hypothecium hyaline; hymenium I+ blue, hyaline; paraphyses little branched, the tips capitate to 7 μ; spores 8, hyaline, simple, ellipsoidal, 8–11 × 4–6.5 μ.

Growing on acid rocks. A seldom collected, probably circumpolar species which ranges southward to Arizona and New Mexico in the Rockies and into Wisconsin in the center of the continent.

Localities: 3, 8, 9, 11.

11. **Parmelia olivaceoides** Krog Thallus small, foliose, irregularly branched, 1–3 mm broad, upper side dull or shiny, brown, slightly pitted, with rounded areas like soralia containing dark brown, granular, shiny isidia; underside brown, shiny, ridged and wrinkled, with few short, simple rhizinae; medulla K-, P+ red, C-, KC-, containing fumarprotocetraric acid. Apothecia not seen.

Growing on barks. Described from Alaska (Circle Hot Springs) and also reported from Hunker Creek, Yukon (Krog 1968).

Locality: 14.

12. **Parmelia exasperatula** Nyl. Thallus foliose, small, thin, closely appressed to the substratum, lobes 2–5 mm broad, sometimes with the margins rolled under, upper side with many isidia or lobules which have narrowed bases expanding upwards into club-shaped isidia, usually simple but sometimes slightly branched, sometimes bearing ciliate outgrowths, olive green to dark brown, greener when moist; underside dark gray or dark brown to black, shining, slightly veiny and with numerous black or brown rhizinae which are distributed to the edge; medulla white; K+ yellow or K-, C-, KC-, P-; sometimes containing atranorin. Apothecia not seen in American material, in Europe described as rare, small, to 2 mm broad, adnate on the upper side of the thallus or slightly pedicillate; the margin of the same color as the thallus, smooth or crenulate; the disk concave to flat, dull or shining, concolorous with the thallus; epithecium brownish; hypothecium hyaline; spores 8, biseriate in the ascus, simple, hyaline, broadly ellipsoid, 8.5–9.5 × 6.5–7 μ.

Growing on rocks or on the bark of shrubs and trees, both broad-leaved and coniferous. A circumpolar boreal and temperate species widely distributed in North America.

Locality: 8.

13. Parmelia infumata Nyl. Thallus foliose, adnate, moderately thin, irregularly branched, with rounded sinuses; upper side dark olive green or brown, peripherally blue gray pruinose, with warty to cylindrical or coralline isidia which may form small cushion-like agglomerations in the center of the thallus; the underside pale brown gray or dark brown to black with simple or branched black rhizinae; medulla pale; K+, C-, KC-, P-; sometimes with small amounts of atranorin in Alaska material (Krog 1968). Apothecia unknown.

Growing on rocks. A circumpolar, arctic and boreal species; ranging south to Washington in the west.

Localities: 3, 8, 11.

14. Parmelia glabratula Lamy Thallus foliose, adnate, forming small rosettes, the lobes to 2 mm broad, irregularly branching with rounded or pointed tips; upper side olive green to brown or blackish brown, dull but with the ends sometimes shiny, with simple to branched cylindrical isidia scattered over the surface or lacking; underside with the edges brown, the center black and dull, somewhat wrinkled, with black rhizinae which do not quite reach the margins; medulla white; C+ red; containing lecanoric acid. Apothecia not seen.

Growing on barks. A circumpolar boreal and temperate species; ranging south to Wisconsin and Washington.

Locality: 14.

15. Parmelia stygia (L.) Ach. Thallus foliose and forming more or less large rosettes, loosely attached to the rocks; the lobes narrow, to 1.5 mm broad, irregularly branching; dark brown black to blackish, more olive brown when moist; partly dull, partly shining, smooth; underside dull black, the edges brown, the entire underside with rhizinae; medulla white; P+ red or -, K+ yellow or K-; strains of this species may contain fumarprotocetraric acid, or caperatic acid, or no acid, and atranorin is accessory (Krog 1968). Apothecia are adnate or on short pedicels, to 7 mm broad; the margin concolorous with the thallus, smooth or pitted or crenulate; disk concave to flat, shining or dull, brown or brownish black; epithecium brown; hypothecium hyaline or yellowish; hymenium to 70 μ, hyaline; spores 8, hyaline, simple, ellipsoid, 7.5–11.5 × 5–6 μ.

Growing on rocks, pebbles in gravels, rock outcrops. A circumpolar arctic-alpine to temperate species.

Localities: 3, 8, 9.

A variant of this species which causes difficulty in determinations is var. *conturbata* (Arn.) Dalla Torre & Sarnth. which is characterized by the lobes becoming erect, branching, to several mm tall and 0.5 mm in diameter. It seems to grow on windswept gravels and may be an ecotype. One such specimen was collected at the Sagavanirktok R.

16. **Parmelia alpicola** Th. Fr. Thallus of elongate narrow lobes placed close together, forming rosettes, lobes to 1 mm broad, convex, lumpy; upper side brownish green, brown to blackish brown, dull; underside black or brown black to the edges, lacking rhizinae; medulla solid, white, C+ red, KC+ red, P+ yellow; containing alectorialic acid, ± protolichesterinic acid (Krog 1968). Apothecia 4 mm broad, adnate to short pedicillate; margin concolorous with the thallus, smooth, inflexed; disk concave to becoming flat, dull, smooth or rough, epruinose, black or brown black; epithecium olive black with clearer superficial layer; hypothecium hyaline or yellowish; hymenium 80 μ, spores 8, hyaline, simple, spherical or ellipsoid, 5–7.5 μ in diameter or 6–7 × 6.5–9.5 μ.

On rocks and outcrops. A circumpolar, arctic and alpine species, ranging south to Labrador, Quebec, and Washington.

Localities: 8, 9.

17. **Parmelia almquistii** Vain. Thallus small, foliose to subfruticose, with narrow, to 0.2 mm broad, convex, tortuous to sublinear lobes, erect in the center of the thallus, appressed at the periphery, shining, brown to black above, paler below; medulla K-, C+ red, KC+ red, P-, containing olivetoric acid. Apothecia not seen, described by Vainio as sessile, to 1.3 mm broad, the disk flat to concave, shining or dull; margin thin, entire to crenulate; hymenium 60 μ, I+ blue; epithecium brown; asci broadly clavate; spores 8, simple, hyaline, ellipsoid to subglobose, 7–10 × 5–7 μ pycnoconidia straight, oblong, the apices slightly thickened, 4–5 × 1 μ.

Growing on granitic rocks in strong light. An arctic-alpine probably Beringian radiant species known from northern Siberia, Tajmyr Island, through arctic North America to northern Quebec and Baffin Island.

Locality: 11. Det. T. Esslinger.

18. **Parmelia septentrionalis** (Lynge) Ahti Foliose, tightly appressed and forming rosettes, the lobes to 3 mm broad, edges irregularly crenate, branching irregular; upper surface dark brown, rarely pruinose, wavy, peripherally shiny, small laminal pseudocyphellae present, mainly mar-

ginal and elongate; under surface light brown to black, smooth or slightly trabeculate, densely rhizinate with black rhizines; medulla white, P+ red, K-, C-, KC-; containing fumarprotocetraric acid. Apothecia numerous, covering most of the thallus, to 1.5 mm broad; margin thin to disappearing, smooth or slightly crenate, pseudocyphellate; disk brown, shining, flat to convex; epithecium brownish; hypothecium hyaline; hymenium pale brownish; 50–70 μ; spores 8, hyaline, simple, ovoid, 9–14 × 5–9 μ.

Growing on the bark of many kinds of trees and shrubs, especially on *Alnus, Salix,* and *Betula.* A circumpolar arctic and boreal species which is more common north of the range of *P. olivacea* (vide Ahti 1969), ranging southward to New England, the Great Lakes states, and Alberta.

Localities: 1, 2, 3, 8, 9, 11, 14.

19. Parmelia panniformis (Nyl.) Vain. Thallus foliose, the lobes tight or loose on the substratum, narrow, 1–1.5 mm broad at most, branching irregularly but forming small cushion-like masses; upper side olive brown, rough or smooth, lacking soredia, the center covered with tiny, flat lobules almost appearing as isidia; underside black, dull, rough with scattered black rhizinae; medulla white, K+, C-, KC-, P-; reported contents (Culberson 1970) atranorin, salazinic acid, possibly olivetoric acid and imbricaric acid, and an unidentified substance. Apothecia rare, not seen. European material reported to be to 2.5 mm broad; the margin paler than the thallus and crenate; disk blackish brown, dull, epruinose; spores ovoid, 9–12 × 5 μ.

A species of acid rocks. Probably circumpolar arctic and boreal, in North America ranging south to New York, Montana, and Washington.

Localities: 8, 11.

USNEACEAE

Fruticose lichens, cylindrical in build, simple or branching, of radial internal organization, corticate; lacking a prothallus, usually attached basally to the substratum. Apothecia, when known, with disk surrounded by thalloid margin.

1   Thallus hollow or with a cottony medulla containing several cartilaginous strands, or uniformly cottony, or with a solid center but no central strand   2

Thallus with a central strand surrounded by looser medullary tissue and an external cortical layer   1. *Usnea*

2 (1) Thallus with center solid or with cottony medulla interspersed with cartilaginous strands, usually much branched   3

Thallus with hollow centers, simple or little branched   9

3 (2) Thallus with solid center, bone white, growing in cold wet seepages
2. *Siphula*

Thallus with cottony medulla, with or without interspersed cartilaginous
strands   4

4 (3) Thallus with medulla of uniformly cottony hyphae, basically terete, at
least on ultimate branches   5

Thallus with medulla of cottony hyphae interspersed with cartilaginous
strands, basically flattened or angular throughout   8

5 (4) Cortex composed of longitudinally oriented hyphae   6

Cortex composed of radially oriented hyphae   3. *Cornicularia*

6 (5) Thallus lacking pseudocyphellae, dark brown to black, appressed and
attached to rock or gravel; cortex with superficial cells differentiated;
spores frequently present, 8 per ascus, hyaline, simple   4. *Pseudephebe*

Thallus often with pseudocyphellae, yellow green to brown or black, erect
or pendent, loose over the substratum if on soil or attached to trees; cortical cells not differentiated; spores rare, 8 or 2–4 per ascus, hyaline or
brown, simple   7

7 (6) Thallus on ground, rarely on lower tree branches, yellow green or brown
to black with pale base, pseudocyphellae always present, conspicuously
raised; medullary hyphae knobby ornamented; containing either usnic
acid or alectorialic acid; if spores are present (rare) they are 2–4 in ascus
and brown   5. *Alectoria*

Thallus epiphytic on trees, brown, pseudocyphellae present or not, adpressed to only slightly raised; medullary hyphae not ornamented, lacking usnic acid, species in this range lacking alectorialic acid; spores rare,
8, per ascus, hyaline   6. *Bryoria*

8 (4) Cross section flattened   7. *Ramalina*

Cross section round to angular   8. *Evernia*

9 (2) Thallus white with pointed tips   9. *Thamnolia*

Thallus greenish yellow to brownish, with blunt tips   10. *Dactylina*

1. USNEA P. Br. ex Adans.

Thallus fruticose, erect or hanging, usually attached to the substratum
by a holdfast; branching sometimes dichotomous, more commonly with
a central axis from which longer or shorter side branches project, the
main axis often dark at the base, smooth, papillate, tuberculate, or with
soralia or isidia, the lateral or peripheral branches often differing from

the main axis, sometimes the main branches are foveolate; corticate, the medulla with an outer layer which may be more or less solid or loosely arachnoid, and an inner, central cartilaginous strand which is very diagnostic. Apothecia lateral or appearing terminal, discoid, with thalloid margin concolorous with the thallus, usually fibrillate; the disk pale, sometimes pruinose; epithecium pale; hypothecium hyaline; paraphyses coherent, branched, septate; asci clavate; spore 8, hyaline, simple thin walled, ellipsoid or almost spherical. Pycnidia lateral, immersed or slightly projecting, pale or dark, fulcra exobasidial; pycnoconidiospores spindle shaped, acicular, or cylindrical, straight. Algae: *Trebouxia*.

1 Plants on trees; soredia in non-capitate small masses, the soralia not also
   producing fibrils   2
   Plants on rocks; soredia forming capitate masses with radiating fibrils
                                                                      3. *U. scholanderii*

2 Tips elongate; medulla dense; axis larger than medulla (CMA 110:180:380)
                                                                      1. *U. glabrescens*
   Tips short; medulla lax; medulla larger than axis (CMA 50:300:200)
                                                                      2. *U. lapponica*

1. **Usnea glabrescens** (Nyl.) Vain. Thallus suberect, caespitose, divergently branched, to 10 cm long and wide, the tips elongate, pale ashy green, subglabrous, the base short, thick, stiff, with a broad black basal zone, toward the apices sparsely subdichotomously branched or sympodial, more densely so toward the base, with many fibrils at right angles; the main branches curving-ascending, the tips straight, to 1.5 mm thick, continuous or little fractured, the main branches with minute, quite dense papillae, the elongate tips smooth, glabrous, sorediate; the soredia in small, sharply delimited soralia which are deeply eroded into the branches; white or yellow white, the tips little branched, hair-like, smooth, tapering, slender; medulla dense, K+ yellow becoming red in typical material, K- in ssp. *glabrella* Mot. to which our material belongs. Cortex 110 μ, medulla 180 μ, axis 380 μ. Lacking apothecia.

Growing on *Picea, Abies, Betula*. A probably circumpolar, boreal species. Locality: 14.

2. **Usnea lapponica** Vain. Thallus usually about 4 cm long, subpendulous, pale yellowish green, the base dark brownish straw colored, dull, distinctly tapering from the base which is thick, the mid-thallus 1 mm, the tips tapering, the thick branches terete, rarely fractured, with short subcylindrical, sparse, acute papillae; the lateral branches to 1 cm long, repeatedly branched, larger and smaller intermixed, irregularly curved,

smooth, slightly subangular and deformed, dull, paler than the main branches, the tips not elongated; soralia on the lateral branches, white, erose, sharply defined; medulla lax, K-. Cortex 50 μ, medulla 300 μ, axis 220 μ.

Growing on *Picea, Pinus, Betula*, rarely on rocks. A probably circumpolar, boreal species.

A species very similar to *U. substerilis* Mot. but differing in the short instead of elongate papillae, the soredia in erose well defined soralia instead of on raised, tuberculate soralia which deform the branchlets. It is treated as a synonym of *U. sorediifera* var. *substerilis* (Mot.) Keissl. by Keissler (1960).

Localities: 11, 14.

3. **Usnea scholanderii** Llano Thallus erect to subpendulous, to 6 cm long and 3 cm wide, soft, tufted, straw colored to yellowish green, the base rusty brown, constricted, densely irregularly branched to subsympodial, apically attenuate, subfibrillose, the cortex minutely papillose at base, subpapillate to smooth apically, strongly sorediate, the soredia farinose in small soralia which unite and form globose soralia 1–2 mm wide with many fibrils radiating from them; medulla K-, soralia K± rusty-brown. Cortex 30–45 μ, medulla 180 μ, axis 150 μ. Apothecia unknown.

Growing on rocks. Known only from a specimen collected at Lake Peters on the North Slope by P.F. Scholander, 1948, and described by Llano (1951). Dr A.W. Herre has examined the type specimen and noted that it was too matted in preservation to be observed very well, and that the radiating fibrils of the original description were fungal hyphae of a mold. I have not seen material of this species.

2. SIPHULA Fr.

Thallus fruticose, erect, sparingly branched, sparingly attached at the base to the substratum by rhizinae or the rhizinae lacking because of death of the lower part as the podetia grow up through mosses; podetia longitudinally sulcate, swollen at the nodes, simple or sparingly branched to caespitose; cortex of paraplectenchymatous hyphae; medullary layer of compact hyphae which are more or less longitudinally oriented; algae in a glomerulate layer in outer part of the medullary layer. Apothecia unknown in northern species. Algae: *Protococcus*.

1. **Siphula ceratites** (Wahlenb.) Fr. Podetia to 7 cm tall, 2 mm broad, sparingly dichotomously branched, the branches ascending and swollen at the nodes, white when fresh, darkening in the herbarium to grayish

or yellowish gray, longitudinally furrowed, lacking soredia or isidia; internal structure as above: C-, K+ orange, KC+ purple, P-: containing siphulin and siphulitol. Apothecia not known.

Growing in cold seepages, sometimes under water, below permanent or nearly permanent snowbanks. A circumpolar high arctic species, growing south to British Columbia in the west, to Labrador in the east.

Localities: 4, 5, 6.

3. CORNICULARIA Ach.

Fruticose, erect lichens with irregularly branching terete to somewhat angled podetia; cortex with irregularly or radially, not longitudinally oriented hyphae. Apothecia terminal or subterminal; margins with teeth or short branchlets or crenulate; spores 8, hyaline, simple, small, ellipsoid. Algae: *Trebouxia*.

1   Medulla C+ red, containing olivetoric acid; branches elongate tapering, with raised white pseudocyphellae   1. *C. divergens*
        Medulla C-, containing protolichesterinic acid; branches short and spinescent   2

2 (1) Thallus with few usually flattened branchlets in addition to the main branching; pseudocyphellae deeply concave   2. *C. aculeata*
       Thallus with many small terete side branchlets; pseudocyphellae flattened   3. *C. muricata*

1. **Cornicularia divergens** Ach. Thallus fruticose, erect, stiff, fragile, very shiny, dark brown, chestnut brown, terete or moderately flattened, 5–8 (–10) cm tall, dichotomously branched, with pseudocyphellae; medulla of white cottony hyphae, K-, C+ red, KC+ red, P-; containing olivetoric acid. Apothecia lateral and making the tips geniculate, adnate; margin concolorous with the thallus, crenulate to toothed; disk to 10 mm broad, chestnut brown, smooth, shining; epithecium brownish; hypothecium hyaline; spores 8, hyaline, simple, ellipsoid, 7–8 × 3.5–4 μ.

Common among mosses and on gravelly tundras. A circumpolar arctic-alpine species, also found in South America.

Localities: 1, 2, 3, 4, 7, 8, 9, 14.

2. **Cornicularia aculeata** (Schreb.) Ach. Thallus fruticose, erect, very stiff and brittle, broadly dichotomously branched, usually only a few cm tall, 1 or 2 cm in poor conditions, the branches angular, scantily pseudocyphellate, the pseudocyphellae usually concave, 1 mm thick on main axes, the smaller branches becoming pointed, often flattened, lacking soredia or isidia; medulla very thinly cobwebby to hollow, white, K-, C-,

KC-, P-; containing protolichesterinic acid. Apothecia rare, not seen by writer; terminal or in the upper axils, to 3 mm broad, the margin concolorous with the thallus, ciliate or toothed; disk flat to slightly convex, brown, shining; epithecium brownish; hypothecium yellowish; spores 6–8, hyaline, simple, ellipsoid, 6–8 × 4 μ.

Growing in open dry gravelly tundras as well as over high humus soils. An arctic and boreal species which is circumpolar, reported as far south in the Appalachian Mts as Mt LeConte, Tenn. (Mozingo, *Bryologist* 57: 31 [1954]) and into California (Herre, *Bryologist* 47: 86 [1944]). It is known from the Andes in South America and from Antarctica.

Localities: 7, 8, 9, 10, 11, 13, 14.

3. **Cornicularia muricata** (Ach.) Ach. Thallus fruticose, erect, richly branched, 1–2 cm tall, dichotomously branched, the branches terete with many short, thorn-like lateral branches, the pseudocyphellae flat rather than concave as in *C. aculeata*, containing protolichesterinic acid. Apothecia not seen but reportedly as in *C. aculeata*.

Growing over soils and humus as the previous. An arctic and boreal species with uncertain range in North America.

Locality: 13.

4. PSEUDEPHEBE Choisy

Thallus fruticose, sometimes becoming compacted centrally, more or less appressed to the substrate and attached by hapters over the area covered; branching isotomic-dichotomous, the branches terete but tending to become dorsiventrally compressed in one species, even or uneven, brown to black, dull to shiny; cortex of longitudinally oriented hyphae which become prosoplectenchymatous at the surface, medullary hyphae not ornamented. Apothecia frequent to abundant, lateral; thalloid margin concolorous with the thallus, sometimes ciliate, disk brown to black; asci clavate; spores 8, ellipsoid, hyaline, simple, 7–12 × 6–8 μ. Pycnidia common.

1 Branches terete, even, internodes elongate, becoming bushy toward the center of the thallus   1. *P. pubescens*
  Branches becoming dorsiventrally compressed, uneven, internodes short, closely attached over the entire surface by the hapters   2. *P. minuscula*

1. **Pseudephebe pubescens** (L.) Choisy (*Alectoria pubescens* (L.) Howe) Thallus fruticose, loosely lying on the substratum and forming small cushions, irregularly branching, the branches slender, cylindrical, 0.2–0.5 mm thick, blackish brown to olive brown, slightly paler below

when lying horizontally, matted on the ground, smooth, dull to slightly shining, lacking soredia or isidia; cortex of longitudinally oriented hyphae; medulla white, K-, C-, KC-, P-; no lichen substances known. Apothecia uncommon, adnate, to 3 mm broad; margin of same color as thallus, smooth to crenulate; sometimes with hair-like bordering fibrils; disk flat to convex, blackish brown, smooth, shining, epruinose; epithecium dark brownish; hypothecium hyaline; spores 8, hyaline, simple, ellipsoid, 9–12 × 5.5–8 μ.

Growing on rock surfaces, gravels, sometimes mixed among mosses in heath tundras. A circumpolar arctic-alpine species ranging into Quebec in the east and southwards in the west to New Mexico.

Localities: 8, 11, and Lake Peters.

2. **Pseudephebe minuscula** (Nyl.) Brodo & Hawksw. (*Alectoria minuscula* Nyl.) Thallus similar to *A. pubescens* but closer to the substratum, the tips more often attached by haptera, and the ultimate branches flattened. The lobes are to 0.5 mm broad, with knotty swellings, blackish brown and shining above, lacking soredia or isidia, paler below, underside with sparse black rhizinae; medulla white, K-, C-, P-; no lichen substances known. Apothecia uncommon, adnate, to 3 mm broad; margin concolorous with the thallus, smooth to crenulate or with projecting cilia; disk flat, blackish brown, dull, epruinose; epithecium brown; hypothecium hyaline or pale; hymenium 50–90 μ; spores 8, hyaline, simple, spherical or ellipsoid, 8–10 × 6.5–7 or 5–7 μ.

Growing on rock outcrops, on boulders, and on gravels. A circumpolar arctic-alpine species ranging south to Quebec in the east and to Colorado and California in the west.

Localities: 3, 8, 9, 11.

5. ALECTORIA Ach.

Thallus fruticose, erect, dichotomously or sympodially branched, the branches thin to several mm thick, round or rarely flattened, smooth, pitted or wrinkled; pseudocyphellae always present, soralia in some species, brown, black, yellow green or yellow; the cortical hyphae longitudinally oriented, conglutinate; medulla or loosely interwoven hyphae usually with knobby ornamentation. Apothecia with thalloid margin, common in some species, rare in others, lateral or appearing terminal, sessile; epithecium brown or yellowish; hypothecium hyaline or brownish; paraphyses little branched, coherent; spores 2–4 per ascus, simple, hyaline to brown, ellipsoid. Pycnidia rare.

Algae: *Trebouxia*.

1 Thallus yellow with the tips darkened, containing usnic acid and diffractaic or
   alectoronic acid   1. *A. ochroleuca*
   Thallus brown, brownish gray with darkened tips and paler base; lacking
   usnic acid, containing alectorialic acid (C+ pinkish red)   2. *A. nigricans*

1. **Alectoria ochroleuca** (Hoffm.) Mass. Thallus fruticose, erect or prostrate, stiff, to 15 cm tall but usually much less; branching sympodial,
rounded or somewhat flattened especially at the axils, the main
branches to 3 mm broad, smooth to slightly ridged or tuberculate, with
numerous longitudinally oriented, raised, whitish pseudocyphellae
0.5–1 mm long; lower parts yellowish or brownish yellow, turning
browner in old herbarium specimens; tips curved, pointed, black to blue
black, the dark part very variable in length; K-, C-, KC+ yellow (or red in
a few specimens), P-; reported to contain usnic, and diffractic acids,
possibly also barbatic acid (Krog 1968). Apothecia rare, lateral, the
podetial tip becoming reflexed with the thalloid margin; disk to 7 mm
(one 15 mm reported by Keissler), flat to slightly concave or slightly
convex, dark brown to blackish, appearing slightly roughened; ascus
cylindrical, the tip thickened; spores 2–4, simple, becoming darkened,
ellipsoid, 28–46 × 14–28 μ.

Growing on soil and humus in large swards in heath tundras espe-
cially. A circumpolar arctic alpine species which also ranges into the
Andes in South America; in North America the southern limits are
Labrador to Alberta.

Localities: 1, 3, 8, 9, 11, 14.

2. **Alectoria nigricans** (Ach.) Nyl. Thallus fruticose, forming large
entangled mats, podetia to 10 cm tall, the main branches to 0.8 mm thick,
dichotomous, more or less sympodial, the lower parts dying, dirty whitish or brownish, the upper growing parts dark chestnut brown to
blackish brown, dull or pruinose, rarely shining, cylindrical or deformed,
pseudocyphellae common, lacking soredia; cortex K+ yellow, C+ red, P+
yellow then orange; containing alectorialic acid. Apothecia extremely
rare, not seen by the writer, known from Scandinavia, Labrador, Newfoundland, and St Lawrence Island; 2–9 mm broad; the disk slightly
convex, brown to blackish brown; spores 2, hyaline or pale brownish,
20–35 × 12–20 μ.

Growing in large swards over gravel and on humus, and as more
depauperate specimens at the edge of frost boils; occasionally as an
epiphyte on the lower branches of spruce below the snow accumulation
level. This is a circumpolar, arctic-alpine species which ranges south to

Labrador and Miquelon in the east, and the Olympic Mountains in Washington in the west.

Localities: 1, 3, 4, 6, 7, 8, 9, 11, 12.

6. BRYORIA Brodo & Hawksw.

Thallus fruticose, erect, caespitose, decumbent, or pendent; branching usually terete, angular or foveolate in a few species, greenish gray to brown, dark brown or black; with lateral spinules in some, soralia absent to abundant, tuberculate or fissural; pseudocyphellae absent or present, fusiform, white, yellow or brownish; cortex of longitudinally oriented conglutinate hyphae, smooth or rough at the surface; medullary hyphae not ornamented with knobs. Apothecia lateral, rare in some species; thalloid margin concolorous with the thallus; disk reddish brown to dark brown; asci clavate, thick walled, arrested bitunicate; spores 8, ellipsoid, hyaline, simple. Pycnidia rare.

1	Sorediate	2
	Lacking soredia	5

2 (1) Soralia P-, greenish black to white; plants short, more or less erect with brown divergent branches; on trees   1. *B. simplicior*
Soralia P+ red, containing fumarprotocetraric acid   3

3 (2) On rocks and soil; the main branches much larger than the secondary, to 0.5 mm thick, often foveolate or flattened; soralia usually sparse
2. *B. chalybeiformis*
Epiphytic; main branches only slightly thicker than the secondary, to 0.4 mm thick, nearly terete   4

4 (3) Main branches very slender, most under 0.2 mm thick, ± uneven, uniformly dark olivaceous black, especially at tips; soralia fissural, abundant; cortex P-   3. *B. lanestris*
Main branches 0.2–0.4 mm thick, fuscous brown, usually pale at base; soralia tuberculate and fissural; cortex sometimes P+ red
4. *B. fuscescens*

5 (1) Medulla P+ red, containing fumarprotocetraric acid; plants erect, divergent branching; cortex shining; brown fissural pseudocyphellae usually conspicuous; growing on ground   6
Medulla P-   7

6 (5) Main branches shiny, black, with numerous gray to pale brown perpendicular branches and apices; spinules abundant; medulla P+ red at least in parts   5. *B. tenuis*
Main branches not black with pale side branches and apices; thallus more or less uniformly cervine brown to dark brown; medulla P+ red
6. *B. nitidula*

7 (5) Olivaceous brown to olive black; medulla C-; pseudocyphellae absent
2. *B. chalybeiformis*
     Shining red brown; medulla C+ red; white pseudocyphellae conspicuous;
main branches ± even   *Cornicularia divergens*

1. **Bryoria simplicior** (Vain.) Brodo & Hawksw. (*Alectoria simplicior* (Vain.) Lynge) Thallus small fruticose, 2–3(–5) cm long, forming small rather dense tufts to slightly hanging, sometimes attached to the substrate at the tips as well as the base, brownish black to black, dull or slightly shining, dichotomously or nearly dichotomously branched, the angles between the branches acute, making the plants appear brush-shaped, the main branches to 0.3 mm thick, the tips thinner, straight or slightly curved, round or slightly deformed, sorediate, the soralia fissural, sharply delimited, round or oval, young soredial masses convex, the soralia becoming crateriform as the soredia are shed, lacking isidia among the soredia, cortex and medulla C-, K-, KC-, P-, Soralia P-. Apothecia not known.

Growing on conifers (*Picea, Larix, Pinus*) and *Betula*. This is a circumpolar northern and middle boreal species, characteristic of the boreal forests of Canada and Alaska. It is in the spruce forests in the Brooks Range, and G.A. Llano collected it at Umiat (locality 3).

2. **Bryoria chalybeiformis** (L.) Brodo & Hawksw. (*Alectoria chalybeiformis* (L.) Röhl.) Thallus fruticose, prostrate on rocks or partly hanging, to 10–15 cm long, very dark, olive black or gray black, not uniformly brown, dull, rarely shining, irregularly submonopodial or subdichotomous, the angles of branching broad, narrower in hanging plants; main branches to 0.5 mm thick, irregularly curved, usually irregularly deformed, slightly flattened or cylindrical, longitudinally wrinkled or slightly sulcate; soredia irregularly distributed over the branches, becoming tuberculate then emptying and appearing more fissure-like; medulla white; K-, C-, KC-, P-, but the soredia P+ red. Apothecia unknown.

Growing on rocks, occasionally on humus. A circumpolar species, arctic-alpine; the North American range not well known.

Localities: 3, 7.

3. **Bryoria lanestris** (Ach.) Brodo & Hawksw. (*Alectoria lanestris* (Ach.) Gyel.) Thallus fruticose, short to several cm long, forming small tufts on twigs, slender, 0.1–0.2(–0.3) mm broad, dark olive brown to black, dichotomously or subdichotomously branched with many branch angles obtuse but then hanging downward, the branches of uneven diameter

(thicker and thinner), sorediate, the soralia fissure-shaped; the cortex P-; the medulla K-, C-, P-; the soredia P+ red; containing fumarprotocetraric acid. Apothecia not seen.

Growing on conifers, including *Picea, Larix,* and *Abies.* A boreal transcontinental species, in our range reaching limits with the spruce at Mancha Creek.

Locality: 14.

This species differs slightly from the very similar *B. fuscescens* in being darker, dark olive brown to black instead of paler brown, dark at base instead of paler, and with unevenly thickened instead of very evenly thickened branches. Both species have the medulla P- and the soredia P+ red. *B. fuscescens* may have the cortex P+ red in some specimens. The similar dark, shining, more tufted and dichotomous western species, *B. glabra,* also has the soredia P+ red along with evenly thickened branches.

4. **Bryoria fuscescens** (Gyel.) Brodo & Hawksw. (*Alectoria fuscescens* Gyel.) Thallus fruticose, pendent, short to several cm long, 0.2–0.4 mm thick, even in thickness, fuscous brown, paler at the base, dichotomously to subdichotomously branched, the branches angling obtusely but then pendent, sorediate, the soralia tuberculate and fissural; cortex sometimes P+ red; medulla K-, C-, P-; soredia P+ red, containing fumarprotocetraric acid. Apothecia not seen.

Growing on trees, sometimes on wood and rocks. A possibly circumpolar species known from Europe and North America. It is known from the Mackenzie delta region (Brodo, personal communication) but not yet from the North Slope.

5. **Bryoria tenuis** (Dahl) Brodo & Hawksw. (*Alectoria tenuis* Dahl) Thallus erect to decumbent, 4–6(–12) cm tall, forming tufts, branching isotomic dichotomous basally but also anisotomic toward the tips, the angles usually obtuse, branches terete, even in diameter, basal branches becoming black, apical branches pale brown to brown, always paler than the basal parts, perpendicular lateral spinules present, soredia absent, pseudocyphellae abundant to scanty, fissural, usually dark and inconspicuous; inner cortex and medulla K-, C-, KC-, P+ red at least in parts. Apothecia rare, lateral; thalloid margin of same color as thallus; disk concave to convex, yellowish brown to reddish brown, 1.5 mm; hymenium 70 μ; spores 8, subglobose to ellipsoid, hyaline, simple, 7–9.5 × 5–7 μ.

This species grows on mossy rocks and in rock crevices in the tundras and also on mossy tree bases in the coniferous forest. It is known from Europe and North America where it appears to be quite uncommon with scattered arctic and montane localities.

Locality: 1.

6. **Bryoria nitidula** (Th. Fr.) Brodo & Hawksw. (*Alectoria nitidula* (Th. Fr.) Vain. [incl. *A. irvingii* Llano]) Thallus fruticose, tufted, forming intricately entangled mats from 2 or 3 to 10 cm tall, dichotomously branched with the branches sympodial, the angles of branching wide, the tips tapering gradually, cylindrical, occasionally a little flattened, the branches 0.5–0.7 mm thick, the main branch 1–2 mm at most, occasionally deformed and swollen in parts, the branches sometimes forming loops, dark brown or blackish at the base, the branches olive brown to dark brown, shining, brown pseudocyphellae conspicuous, K-, C-, KC-, P+ red; containing fumarprotocetraric acid. No apothecia or soredia known.

Common on gravel or heath tundras in well drained situations, on both humus and bare soils, occasionally on boulders. A circumpolar arctic-alpine species.

Localities: 1, 2, 3, 8, 9, 11.

7. RAMALINA Ach.

Thallus fruticose, erect to hanging, attached to the substratum by a holdfast; flattened or cylindrical, the entire surface corticate with more or less longitudinally oriented hyphae or else hyphae at right angles to the length; an inner mechanical layer of cartilaginous strands close under the outer cortex; the medulla of arachnoid hyphae; soredia common, pseudocyphellae common; the surface often longitudinally rugose. Apothecia terminal or lateral, flat or concave; margin thalloid, concolorous with the thallus; disk pale, greenish, pruinose or not; epithecium pale; hypothecium hyaline; paraphyses unbranched, clavate, coherent; asci clavate; spores 8, hyaline, ellipsoid to spindle-shaped, straight or curved, 2- or 4-celled. Algae: *Trebouxia*.

1 Thallus fistulose (partially hollow) esorediate or with few soredia, bony white to pale yellow when fresh; containing usnic and divaricatic acids
1. *R. almquistii*
Thallus not fistulose, wih abundant lateral or terminal soredia, glaucescent in color; containing usnic, evernic, and obtusatic acids   2. *R. pollinaria*

1. **Ramalina almquistii** Vain. Thallus fruticose, erect, to 65 mm tall, in tufts, the lobes mainly terete, partly slightly flattened, fistulose, not or only slightly foraminous at the base, the main branches to 3 mm, usually more slender, branching irregularly, the apices slender; bony white to yellowish, surface smooth, shiny; lacking soredia or with some soredia, cartilaginous layer uniform in thickness; medulla fistulose, arachnoid between; K-, C-, KC- or +, P-; containing usnic and divaricatic acids, plus

possibly other substances. Apothecia to 4 mm broad, subterminal or partly lateral; margin concolorous with the thallus, the back smooth; disk pale brownish, bare; spores oblong with rounded apices, straight, 2-celled, 11–13 × 4–5 μ.

Growing on soil among other lichens and mosses. This is a member of the Beringian element in the flora of Alaska which ranges along the north side of North America as far east as King William Peninsula, NWT.

Localities: 1, 7, 8, 9, 10, 13.

2. **Ramalina pollinaria** (Westr.) Ach. Thallus erect, low, soft, papery, somewhat shining, gray green, brownish green, the base blackish; not fistulose, flattened, the short tips more or less divided and rounded, narrowed; soredia abundantly produced, not in well defined soralia, on the surface, the edges or the tips of the lobes, the latter dissolving into fine powder; the outer cortex thin, the inner becoming a net of cartilaginous strands; the medulla not fistulose, loosely arachnoid; K-, C-, KC-, P-; containing usnic, evernic, and obtusatic acids. Apothecia rare, not present in Alaskan material; mainly terminal, to 10 mm broad but usually less; margin of same color as thallus, the reverse smooth or furrowed; disk concave to flat or convex, brown, usually pruinose; epithecium brownish; hypothecium brownish; spores 8, hyaline, elongate, spindle-shaped, 2-celled, straight, 10–15 × 4–5 μ.

Growing on bark and on rocks. A circumpolar boreal and temperate species of poorly defined range in North America.

Locality: 11

8. EVERNIA Ach.

Fruticose, erect or hanging, the hanging species attached to the substratum by a holdfast, irregularly or dichotomously branching, the lobes round or flat, with thin cortex of radially oriented cells; lacking cartilaginous strands in the medulla which is composed of arachnoid hyphae. Apothecia not seen in the species treated here.

1 On soil, fragile, lacking soredia	1. *E. perfragilis*
On trees, soft and pliable, very sorediate	2. *E. mesomorpha*

1. **Evernia perfragilis** Llano (*Evernia arctica* (Elenk. & Sav.) Lynge) Thallus fruticose, erect, in tufts, 2–5 cm tall, brittle and fragile, breaking into pieces when dry, the main branches to 1.5 mm broad, angularly compressed, dichotomously branching, smooth or roughened to rugose, the tips attenuate, lacking soredia or isidia, pale yellowish; the cortex chondroid, to 55 μ thick, the cells more or less longitudinally arranged,

paraplectenchymatous; medulla white, arachnoid, I-, K-, C-, KC-, P-; containing usnic and divaricatic acids. Apothecia and pycnidia not known.

Growing mainly on calcareous soils and gravels. This species is distributed from Novaya Zemlya across the high arctic in Siberia, Alaska, and the Northwest Territories to Baffin Island.

Localities: 1, 9, 10, 11, 14.

2. **Evernia mesomorpha** Nyl. Thallus fruticose, much branched, erect, projecting from the substratum, partially hanging, to 10 cm long but shorter in the arctic, the branches irregular or dichotomous, angularly rounded to partly flattened but not dorsiventral, to 1 mm thick, yellow green to yellowish brown, the tips darker, with an abundance of yellowish to gray coarse soredia in small spots over the thallus, medulla white; K-, KC- or pale + yellow, P-; containing divaricatic acid and usnic acid. Apothecia not seen in arctic material, described (Keissler 1960) as adnate, lateral, with thin thalloid margin; disk to 6 mm broad, slightly concave, chestnut brown; epithecium pale brown; hypothecium hyaline; spores 8, hyaline, spherical, simple, thick walled, 3-4 μ in diameter.

Growing on the bark and twigs of trees, including *Picea, Larix, Betula,* exceptionally on rocks beyond the tree line. A circumpolar, boreal and temperate species reaching its range limits with the trees.

Locality: 14.

9. THAMNOLIA Ach.

Thallus fruticose, erect or decumbent, simple or little branched, the decumbent podetia forming erect branches which multiply the thallus, rhizinae seldom formed, the base usually dying and growth proceeding at the apices; cylindrical, the tips pointed; white or cream colored; cortex paraplectenchymatous, of more or less longitudinally oriented hyphae; medulla thin, of longitudinally oriented hyphae; the interior hollow. Apothecia have been described in conflicting reports which seem to be based on lichen parasites. Algae: *Trebouxia.*

1 Thallus UV-, K+ yellow, P+ orange to red, containing thamnolic acid
1. T. vermicularis

Thallus UV+, K+ weakly yellowish, P+ yellow; containing squamatic and baeomycic acids
2. T. subuliformis

1. **Thamnolia vermicularis** (Sw.) Schaer. Thallus erect or decumbent, occasionally forming tufts by branching from the base or along decumbent podetia; cylindrical, tips pointed; white or cream white becoming

browned or staining paper brown in the herbarium, smooth, seldom with short side-pointed branches, very variable in size from tiny to 12 cm tall and 8 mm broad; thallus UV-, K+ yellow, P+ orange to red; containing thamnolic acid.

Growing on many types of tundras from frost boils, gravels, to willow thickets and heaths, the more luxuriant specimens under more sheltered conditions. A circumpolar and also southern hemisphere arctic and alpine species. In North America it is more northerly than *T. subuliformis*, ranging from Baffin Island, across the Northwest Territories, and then south to Washington and Oregon in the mountains.

Localities: 1, 2, 3, 4, 5, 6, 7, 8, 11, 13.

2. **Thamnolia subuliformis** (Ehrh.) W. Culb. Thallus as in the preceding, of erect, simple to sparingly branched, pointed, cylindrical podetia which are white but do not change color nor stain the herbarium paper on long storage in the herbarium; UV+ yellow, K- or K+ pale yellow, P+ yellow; containing squamatic and baeomycic acids.

Growing under similar conditions to the preceding, more common than it in the more sheltered situations. A circumpolar arctic-alpine species which is much more common in North America than *T. vermicularis* and ranges farther south, to New York in the east and to Colorado in the west. It also occurs in the southern hemisphere but is less common there than *T. vermicularis* which replaces it in frequency southwards.

Localities: 1, 3, 7, 8, 9, 10, 11, 12, 13, 14.

10. DACTYLINA Nyl.

Thallus of erect, fruticose podetia, sparingly dichotomously branched, lacking rhizinae, corticate, the cortex of columnar, radiate, dense hyphae, prosoplectenchymatous, the medulla of pachydermatous hyphae close under the cortex, the interior becoming hollow or loosely arachnoid, algae dispersed or in small glomerules in the outer medulla. Apothecia on tips of lateral branches; margin thalloid; epithecium brownish; hypothecium hyaline to yellowish; paraphyses unbranched, capitate; asci clavate; spores 8, simple, hyaline, thick walled, spherical to slightly ellipsoid. Pycnidia immersed, globose; sterigmata exobasidial; pyconconidiospores straight or slightly curved. Algae: *Trebouxia*.

1   Plants little branched or unbranched, turgid, hollow, inflated, yellow or
     brownish yellow, C+, gyrophoric acid present   2
     Plants well branched, yellow or brownish yellow or partly violet pruinose,
     hollow or with cobwebby hyphae filling the center, C-, lacking
     gyrophoric acid.   3

2 (1) Medulla P+ orange red, physodalic acid present; cortex C+ pink, medulla C-, the gyrophoric acid limited to the inner cortex above the algal layer
1. *D. beringica*

Medulla P-, physodalic acid absent, cortex and medulla both C+ red, the gyrophoric acid distributed through the inner cortex and the entire medulla
2. *D. arctica*

3 (1) Plants sparingly dichotomously branched, yellow or yellowish green, the branches with few lateral branches, filled with cobwebby hyphae; pycnidia rare; medulla P-, acetone extract UV-, containing usnic and protolichesterinic acids
3. *D. madreporiformis*

Plants dichotomously or sympodially branched, yellowish or brownish, usually with a light violet pruina toward the tips; the branches commonly muricate knobbed with short lateral branchlets, hollow at least in part; pycnidia common; medulla P+ red or P-, acetone extract UV+ yellowish or UV-, containing usnic acid + or - physodalic and physodic acids
4. *D. ramulosa*

1. **Dactylina beringica** Thoms. & Bird Podetia like those of *D. arctica*, finger-like, unbranched to little branched, regenerating along their length when decumbent, yellowish green to brownish, the base dying and turning dark brown, inflated and hollow, more or less shining, smooth, the tips blunt or acute, the medulla P+ orange red, C-, the cortex C+ pink in inner part, containing usnic and gyrophoric acids in the cortex, physodalic and physodic acids in the medulla.

Growing on soils among mosses and on humus. An amphi-Beringian species ranging in Siberia and in North America east to the mouth of the Back River.

Localities: 6, 8, 9, 14, and Lake Peters.

2. **Dactylina arctica** (Hook.) Nyl. Podetia erect, finger-like, unbranched or little branched, when decumbent regenerating along the length of the podetium and then forming tufts, varying from very small to 7 cm tall and 14 mm broad, pale yellow green, brownish in exposed situations, the base dying and becoming dark brown, inflated, hollow, more or less shining, usually smooth, sometimes slightly foveolate, lacking soredia or isidia, the apices rounded or obtuse-pointed; K-, C+ red, (rarely C-), KC+ yellow, P- or P+ red orange; containing gyrophoric and usnic acids, accessory physodalic and barbatic acids. Apothecia rare, on tips of lateral branches, to 5 mm broad; margin concolorous with the thallus, slightly crenulate; disk chestnut brown, shining, epruinose; epithecium brownish; hypothecium hyaline; hymenium 100 μ; spores 8, spherical, hyaline, the wall thick, 4-6 μ.

Growing in dry or moist tundras, fruiting above late snow banks. A circumpolar arctic-alpine species in North America ranging south to Quebec, Labrador, and Washington.

Localities: 1, 2, 3, 4, 5, 6, 7, 8, 9, 10, 11, 12, 13, 14, plus Teshepuk Lake, Schrader Lake, and Chandler Lake.

3. **Dactylina madreporiformis** (Ach.) Tuck. Podetia fruticose, to 3.5 cm tall, 2 mm in diameter, soft rather than fragile, inflated, cylindrical, foveolate, branching sparsely dichotomously from near the base to form tufts, tips rounded; yellow to straw colored or greenish yellow, often sun browned, lacking soredia, isidia, or pruina; medulla white, more or less solid, becoming arachnoid to the center but not hollow; K-, C-, KC+ yellow, P-; containing usnic and protolichesterinic acid. Apothecia rare, at tips of branches; margin thick, concolorous with the thallus; disk to 4 mm broad, pale chestnut brown, shining; epithecium brownish; hypothecium hyaline; spores 8, simple, hyaline, more or less spherical, 10 μ in diameter.

Growing in calcareous tundras from very dry to quite moist. A circumpolar, arctic-alpine species which ranges southward in the Rocky Mountains in North America.

Localities: 3, 7, 8, 11.

4. **Dactylina ramulosa** (Hook.) Tuck. Podetia fragile, erect, to 2 cm tall, the base dying and growing from the apices, subterete or flattish, more or less sympodial with few branches but many short, divergent, muricate, black tipped side branches, usually forming tufts when undisturbed, yellowish straw colored, brown in strong light, usually with a whitish or violet pruina, lacking soredia or isidia, only slightly if at all foveolate, the interior hollow with very loose arachnoid hyphae scattered in the cavity; K-, C usually + red, P+ red orange or P-; containing gyrophoric acid, usually physodalic acid, usnic acid in about 50%, and accessory barbatic acid. Apothecia on tips of side branches, to 3 mm broad; margin coarsely crenulate or folded, thick; disk flat, smooth, shining, chestnut colored; epithecium brownish; hypothecium brownish or yellowish; spores hyaline, simple, spherical, 5-6 μ.

A calciphilous lichen which grows in a wide variety of tundras from very dry *Empetrum-Arctostaphylos* heaths to tussocks in moist *Carex* and *Eriophorum* tundras. Best fruiting adjacent to late snow patches. A circumpolar, arctic-alpine species which ranges south into Alberta in the Rocky Mountains.

Localities: 4, 6, 8, 9, 10, 11, 14.

BUELLIACEAE

Thallus crustose to squamulose, the margins lobate or not, poorly differentiated into layers or with a cortex and medulla, the thallus is attached by the lower surface to the substratum. The apothecia may have either a proper or thalloid margin. The spores are dark, 2-celled to muriform, non-halonate. Algae: *Protococcus* (*Trebouxia?*).

1 Apothecia with proper margin	1. *Buellia*
Apothecia with thalloid margin	2. *Rinodina*

1. BUELLIA De Not.

Thallus crustose, granulose, areolate, sometimes effigurate at the margins, a black or white hypothallus sometimes present, cortex poorly developed. Apothecia adnate or immersed, sometimes between the areoles: disk flat to convex, sometimes scabrid or pruinose, the proper margin of the same color as the disk, the exciple continuous with the hypothecium, and both dark, red brown, yellow brown or black, often carbonaceous; upper part of hymenium brown, yellow, or greenish, lower part hyaline; paraphyses thin, septate, the tips usually brown-capitate, the apical cell enlarged; asci clavate; spores 8 (or polysporous), brown 1–3-septate or muriform, non-halonate, the walls uniformly thickened or with the septum or apex or both thickened, usually ellipsoid.

1		Spores 3-septate to submuriform	2
		Spores 1-septate	4
2	(1)	On calcareous rocks	1. *B. alboatra*
		On moss, humus, or other lichens	3
3	(2)	On rotting moss and humus	2. *B. papillata*
		Parasitic on *Xanthoria* and *Caloplaca*	3. *B. nivalis*
4	(1)	Apothecia sessile, not immersed	5
		Apothecia immersed, aspicilioid	12
5	(4)	Parasitic on thallus of Baeomyces, thallus lemon yellow, K+ yellow	
			4. *B. scabrosa*
		Free living on other substrates, thallus black, brown, or white	6
6	(5)	On mosses, soil, or wood	7
		On rocks	10
7	(6)	On mosses or soil	8
		On wood	9
8	(7)	Thallus of white lobes	5. *B. elegans*
		Thallus of indeterminate white masses	2. *B. papillata*

9 (7) Spores large, over 16 μ long; hymenium inspersed with oil drops
6. *B. disciformis*

Spores medium, less than 17 μ long; hymenium not inspersed; thallus scant, areolate, or verrucose   7. *B. punctata*

10 (6) Hypothecium hyaline; thallus whitish ashy   11

Hypothecium dark; thallus black or ashy black; spores often 1-celled, medulla I-   10. *B. moriopsis*

11 (10) Medulla I+ blue; thallus thin to lacking   8. *B. vilis*

Medulla I-; thallus thick, areolate-verrucose   9. *B. notabilis*

12 (4) Medulla I-, K+ red   11. *B. immersa*

Medulla I+ blue, K-   12. *B. malmei*

1. **Buellia alboatra** (Hoffm.) Tuck. (*B. margaritacea* (Sommerf.) Lynge) Thallus white to ashy, effuse or forming a thick, chinky-areolate crust, hypothallus not developed. Apothecia adnate 0.2–0.6 mm broad, disk black, flat or becoming convex, sometimes white pruinose, margin thin; hypothecium red brown; exciple red brown; hymenium brown, granular but not inspersed with oil drops, 85–100 μ; asci clavate; spores 8, oblong-ellipsoid to ovoid-ellipsoid, muriform, 3–5-septate transversely, 1–2-septate longitudinally, occasionally 3-septate, brown, 12–22 × 6–9 μ.

Growing on calcareous rocks. A circumpolar arctic and boreal to temperate species.

Locality: 1.

2. **Buellia papillata** (Sommerf.) Tuck. Thallus crustose, membranaceous to a granulose or verrucose white crust, lacking hypothallus; K+ yellow, P-. Apothecia 0.3–1.0 mm broad, disk black, flat to convex, margin black, thin, persistent; hypothecium brown black, exciple brown black; hymenium brown above, hyaline below, not inspersed, 70–95 μ; spores 8, colorless or green at first then soon dark brown, 1-septate, occasionally 3-septate (=*B. geophila* (Sommerf.) Lynge) fusiform-ellipsoid, frequently curved, the walls uniform, 19–38 × 9–13 μ.

Growing over mosses and soil. A circumpolar arctic and boreal species which ranges south into Washington in the west.

Localities: 1, 7, 8, 10, 11, 14.

3. **Buellia nivalis** (Bagl. & Car.) Hertel (*Polyschistes nivalis* (Bagl. & Car.) Keissl. Thallus at first white, parasitic on *Xanthoria* and *Caloplaca*, later perhaps free (*Buellia margaritacea?*). Apothecia at first immersed in the host, becoming adnate, to 0.5 mm broad; the disk becoming convex and immarginate, blue pruinose and rough, finally blackening; the margin black and extending under the apothecium; the epithecium dark; the hypothecium blackish brown; the hymenium hyaline; paraphyses

brown at the tip; asci clavate; spores 6–8, elliptic or kidney-shaped, 3-septate to muriform with 1 longitudinal wall, 15–20 × 9–12 μ.

Parasitic on *Xanthoria* and *Caloplaca*. Known from the Alps and Scandinavia in Europe, Novaya Zemlya, Venezuela, and reported from Alaska (Hertel, *Mitt. Bot. München* 12: 113-52. 1975).

Locality: 1. Det. Hertel.

4. **Buellia scabrosa** (Ach.) Körb. Thallus crustose, delimited, parasitic on thalli of species of *Baeomyces*, granulose-verrucose, constricted below, to glebulose, yellow green. Apothecia centrally located, often concentrically arranged, adnate; the disk flat, black, scabrid; the exciple thin, black, disappearing; epithecium greenish; hypothecium blackish brown; hymenium 75 μ; asci clavate; spores 8, brown, 1-septate, ellipsoid, the center constricted, 12–16 × 6–8 μ.

Growing on the thallus of species of *Baeomyces*. A circumpolar arctic-alpine species, known from Northwest Territories, Yukon, Newfoundland, and Alaska.

Localities: 3, 8, 9.

5. **Buellia elegans** Poelt Thallus of white lobules or radiate lobes; lacking hypothallus; K+ yellow, P-. Apothecia at first immersed, soon becoming adnate, the broken-through thallus persisting around the margin; disk flat to slightly convex, black, dull; margin thin, black, persistent; epithecium brown; hypothecium and exciple brown; hymenium 70–85 μ, I+ blue; spores brown, 1-septate, slightly constricted, wall thin, uniform, oblong-ellipsoid, 13–18 × 6–9 μ.

Growing on soil. A species known in Europe and in North America where it is known from the arctic and the Great Plains.

Locality: 11.

6. **Buellia disciformis** (Fr.) Mudd Thallus white to orangish, smooth to chinky-areolate but not granulose, with black hypothallus; K- or yellowish, P-. Apothecia to 1 mm broad; adnate; the disk flat to convex, black; exciple continuous with the hypothecium, brown black; the margin thin, black, persistent; epithecium brown; hypothecium brown black, projecting downwards into the thallus, hymenium brown above, hyaline below, 70–100 μ, inspersed with oil drops; asci narrowly-clavate; spores 8 (rarely 16), brown, 1-septate (rarely appearing 3-septate), containing oil drops, fusiform-ellipsoid, the walls thin, uniform, 16–30 × 6–11 μ.

Growing on bark and wood. A circumpolar arctic to temperate species with a very wide range over much of North America.

Localities: 3, 14.

7. **Buellia punctata** (Hoffm.) Mass. Thallus variable, scurfy, lacking, chinky, to verruculose, ashy, greenish, or brownish, P-, K+ yellowish. Apothecia adnate, to 0.5 mm broad, flat becoming convex; disk black, dull; margin thin, black, disappearing; epithecium brown; hypothecium brown to brown black, continuing into the exciple which is brown black; hymenium 40–80 μ, hyaline below, brown above, not inspersed with oil droplets; asci clavate; spores 8 or 12–24; 1-septate, not constricted, the walls uniform, ellipsoid, 7–16 × 4–8 μ.

Growing on barks, old wood, less often on rocks or soil. A circumpolar boreal to temperate species with a range over most of North America.

Localities: var. *punctata* with 8 spores per ascus: 7, 8, 9, 13, 14; var. *polyspora* (Willey) Fink with many spores per ascus: 3, 13, 14.

8. **Buellia vilis** Th. Fr. Thallus obsolete to more or less areolate, the areoles mixed with rock particles, ashy, lacking hypothallus, medulla K+ yellowish, P-, I+ blue. Apothecia to 0.8 mm broad, adnate, flat to slightly convex; disk black, dull, margin raised, persistent, black; epithecium brown; hypothecium hyaline; exciple hyaline inside, the outer part brown, continuous with the hypothecium; hymenium 55–65 μ, upper part brown, lower part hyaline, not inspersed with oil droplets, I+ blue; hypothecium and exciple also I+ blue; spores 8, 1-septate, not constricted, brown, the walls thin, uniform, ellipsoid, 13–17 × 5–6 μ.

Growing on rocks. A probably circumpolar, boreal species.

Locality: 11.

9. **Buellia notabilis** Lynge Thallus crustose, irregular, not well limited, to 1 mm thick, verrucose areolate with deep chinks, the areolae to 1 mm broad, angular, bullate, with secondary chinks, white or ashy white, with a narrow black hypothallus, K-, P-. Apothecia commonly contiguous and becoming angular, to 1 mm broad, deep in thallus but not aspicilioid; disk flat or depressed, black, scabrid; margin black, thin becoming crenulate to disappearing; epithecium dark; hypothecium hyaline; hymenium 90 μ, upper part dark lower hyaline; asci clavate; spores 8, 1-septate, the center slightly constricted, the walls thickened and the lumena subangular, brown, 14–18 × 7–8 μ.

Growing on calcareous rocks. Described from northeast Greenland. New to Alaska.

Locality: 11.

10. **Buellia moriopsis** (Mass.) Th. Fr. Thallus brown or black, on a black hypothallus, of irregular, sometimes glebulose areoles, continuous or dispersed, more or less shining; medulla K-, P-. Apothecia adnate or

between the areoles, to 0.8 mm broad, flat to convex, black when dry, greenish when wet, dull; margin black, thin, disappearing; epithecium green; hypothecium red brown or violet shaded, with an extension down through the thallus to the hypothallus; exciple green at edges, brown inside; hymenium 70–85 μ, green above, hyaline below; asci clavate; spores 8, 1-septate, greenish then dark brown, not constricted, the walls thin, uniform, ellipsoid, 10–15 × 6–9 μ.

Growing on rocks. A probably arctic-alpine species which ranges into Washington and New Hampshire in the mountains.

Localities: 3, 8, 11.

11. **Buellia immersa** Lynge Thallus over a conspicuous black hypothallus which tends to be chinky and plicate to radiate, the thallus of minute areoles to 0.5 mm broad, round or angular, ashy white, smooth, subshining, contiguous to the center of the thallus, dispersed to the margins, I-, C-, K+ red. Apothecia minute, to 0.45 mm broad immersed in the areoles, approximately level with the surface, the margins a little raised; disk black, flat or slightly concave, rough; the margin thick, black; epithecium greenish brown; hypothecium brownish; exciple brown; hymenium 65–70 μ, brownish above, hyaline below; spores 8, 1-septate, the center slightly constricted, the walls thin, unevenly thickened, ellipsoid, 13–17 × 8 –10 μ.

Growing on rocks. Known from Novaya Zemlya and west Greenland. New to Alaska.

Locality: 9.

12. **Buellia malmei** Lynge Thallus thinly crustose, ashy, chinky areolate, the areolae to 0.45 mm broad, flat, angular; the medulla I+ blue, K-; black hypothallus poorly developed. Apothecia immersed in the areolae, to 0.4 mm broad, concave to nearly flat, the margin entire, thick, slightly higher than the thallus; hypothecium hyaline; hymenium 55 μ, upper part bluish; paraphyses coherent, the tips thickened, the cells constrictedly septate; spores 2-celled, dark, the center constricted, the walls evenly thickened, 11–15 × 6.5–8 μ.

Growing on acid rocks. It is a circumpolar species known from Bear Island, Novaya Zemlya, northeast Greenland, and west Greenland.

Locality: 8.

2. RINODINA (Ach.) Gray

Crustose lichens with thallus thick or thin to disappearing, chinky, areolate, granules, rarely isidiate, sometimes lobate at the margins, sometimes with a dark hypothallus, gray, brown, or ochraceous. Apothecia innate or sessile, margin concolorous with the thallus or the disk, containing algae, entire, crenulate, to disappearing; a cellular cortex usually present; proper exciple hyaline or rarely brown, prosoplectenchymatous; hypothecium hyaline or rarely dark brown; hymenium hyaline; paraphyses capitate, the tips brown, forming a dark brown, red brown or black epithecium; asci clavate, spores 8, brown, 2- or 4-celled, the septum well developed or not at maturity. Pycnidia rare, pycnoconidia bacilliform. Algae: *Trebouxia*.

1   On rocks   2
    On soil, moss, humus, old wood, woody plants   6

2 (1) Thallus lobate at the margins, yellow green   1. *R. oreina*
    Thallus non-lobate, effuse, white or brown   3

3 (2) On calcareous rocks   4
    On acid rocks   5

4 (3) Thallus disappearing, white ashy; apothecia elevated; spores evenly pigmented, 11–23 × 10–12 μ 2. *R. occidentalis*
    Thallus gray or brownish; apothecia immersed to adnate; spores with dark band by the septum, 14–23 × 8–13 μ 3. *R. bischoffii*

5 (3) Thallus well developed, dark brown, areolate; spores 17–20 × 10–11 μ 4. *R. milvina*
    Thallus forming small pulvini, pale gray to brown; spores 12–15 × 5–7 μ 5. *R. cacuminum*

6 (1) On soil, moss, and humus   7
    On woody plants and old wood   10

7 (6) Thallus squamulose-lobate at the margins; spores 17–22 × 8–9 μ 6. *R. nimbosa*
    Thallus crustose, non-lobate; spores 25–35 × 11–17 μ 8

8 (7) Disk and margin whitish or bluish pruinose; apothecia usually small, 0.3–0.9(–1.5) mm   7. *R. roscida*
    Disk and margin epruinose; apothecia larger, 1–1.5 mm   9

9 (8) Disk flat; hymenium 100 μ; cortex I+ blue   8. *R. turfacea*
    Disk convex; hymenium 120 μ; cortex I-   9. *R. mniaraea*

10 (6) Spores large, 17–24 × 8–10 μ; thallus thin, granular to disappearing, gray brown, K-; apothecia not constricted   10. *R. archaea*
    Spores smaller, less than 20 μ; thallus various; apothecia constricted at base   11

11 (10) Lower cortex massive, more than 40 μ thick, hyaline   11. *R. laevigata*
    Lower cortex rarely expanded, not over 20 μ thick, indistinct and brownish   12

12 (11) Apical wall of spore convex inward; thallus gray brown or lacking
13

Apical wall of spore not convex inward; apothecia constricted; thallus very thin, brown gray over dark hypothallus; cortex to 25 μ, the cells to 7 μ, rounded; spores 13–17 × 6.5–8 μ 14. *R. septentrionalis*

13 (12) Thallus indistinct; cortex of margin of apothecium to 35 μ; the cells 3–5 μ, with thick walls; spores 17–20 × 7–8.5 μ 12. *R. lecideoides*

Thallus lacking; cortex of margin of apothecium 10–15 μ, cells 2–5 μ, with thin walls; spores 17–19 × 7.5–8.5 μ 13. *R. hyperborea*

1. **Rinodina oreina** (Ach.) Mass. (*Dimelaena oreina* (Ach.) Norm.) Thallus crustose, the margins lobate-radiate, the center areolate, greenish yellow to straw colored, the edges black, upper side flattened; K-, I+ blue; occurring in several chemical strains: (1) fumarprotocetraric and usnic acids (P+ red, C-), (2) gyrophoric acid and usnic acid (P-, C+ red), and (3) only usnic acid (P-, C-); zeorin may also occur. Apothecia immersed to adnate or sessile on the thallus; margin concolorous with the thallus, thick, persistent; disk flat to slightly convex, black, to 5 mm broad, usually much less; epithecium dark; hypothecium hyaline; paraphyses coherent, brown-capitate; spores brown, 2-celled, the center not constricted, ellipsoid, 9–12 × 4.5–8 μ.

Growing on acid rocks. A circumpolar, arctic-alpine and temperate species with very wide distribution.

Localities: 8, 11.

2. **Rinodina occidentalis** Lynge Thallus poorly developed to disappearing, granulose, the granules to 0.5 mm, discrete to forming areolate masses, whitish ashy, I-, K-. Apothecia raised over the soft rock surface, round; margin concolorous with the thallus, entire, persistent; disk to 0.7 mm, convex, scabrid, epruinose; cortex of the margin 15–17 μ; epithecium brown; hypothecium hyaline; hymenium 100–120 μ, upper part brown; paraphyses capitate, brown tipped, coherent, septate, 4–5 μ thick; spores 8, brown, the color evenly distributed, the walls uniform in thickness, the center slightly or not constricted, 2-celled, 19–23 × 10–12 μ.

Growing on calcareous rocks. Known from Greenland and Hudson Bay area; new to Alaska.

Localities: 1, 3, 8.

3. **Rinodina bischoffii** (Hepp) Körb. Thallus crustose, thin, often indistinct, minutely granulose to running together to form a scurfy or subareolate crust, ashy to brownish. Apothecia small, to 0.8 mm, adnate

or often immersed in holes in the substratum, flat to convex, the margin thin, concolorous with the thallus, disappearing; disk brown black, dull, scabrid, epithecium brown; hypothecium hyaline to yellowish; paraphyses coherent, brown-capitate; spores 8, 2-celled, the locules distant, a dark band at the septum, ellipsoid, 14–23 × 8–13 μ.

Growing on or in calcareous rocks. A circumpolar arctic, boreal, and temperate species with a very wide distribution, known from Greenland, Northwest Territories. New to Alaska.

Locality: 3.

4. **Rinodina milvina** (Wahlenb.) Th. Fr. Thallus crustose, fairly thick, effuse, non-lobate, granulose to verrucose-glebulose, dark brown to blackish olive, over a black hypothallus. Apothecia adnate to sessile; margin concolorous with the thallus, entire, persistent; disk flat, brownish black; epithecium brown; hypothecium brown; hymenium 125 μ, upper part brown; paraphyses brown capitate, 5 μ, coherent; spores 8, 2-celled, not or slightly constricted at the septum, the walls uniformly thin; brown, ellipsoid, 14–20 × 6–11 μ.

Growing on non-calcareous rocks. A circumpolar arctic-alpine species, known south to New Mexico. New to Alaska.

Locality: 11.

5. **Rinodina cacuminum** (Th. Fr.) Malme Thallus crustose, thin, verruculose or granulose to scant, brown to brownish gray, sometimes with a dark hypothallus, K-. Apothecia grouped, to 0.75 mm broad; margin concolorous with the thallus, entire to crenate, persistent; disk concave to flat, black or brownish black; exciple thick with thin cortex; epithecium brownish; hypothecium brown; hymenium 65–75 μ; paraphyses capitate, brown tipped, easily separable and in this differing from *R. milvina*; spores 8, 2-celled, not constricted at the center, the wall uniform in thickness, brown, ellipsoid, 12–15 × 5–7 μ.

Growing on acidic rocks. Probably circumpolar and arctic. Known from Scandinavia, Franz Josef Land, and Greenland; new to Alaska and North America.

Locality: 9.

6. **Rinodina nimbosa** (Fr.) Th. Fr. Thallus crustose, areolate centrally, lobate at the margins, the lobes to 1 mm broad, incised or subcrenate, sometimes overlapping, dull, sometimes slightly pruinose, brownish yellow or reddish brown, cortex cellular, the cells thin walled, K-, P-. Apothecia to 1 mm broad, partly immersed, flat; the margins concolorous with the thallus or sometimes pruinose; disk black, bare or pruinose;

epithecium brownish; hypothecium pale or dark brown; hymenium 90–100 μ, I+ blue or greenish blue as is hypothecium, paraphyses unbranched, free in KOH, tips brown-capitate; spores 8, 2-celled, slightly constricted, the walls uniform in thickness, greenish brown, 15–22 × 8.5–10 μ.

Growing on soil or humus, sometimes over mosses. Circumpolar and arctic-alpine, in North America south to Colorado and New Mexico.

Localities: 3, 13.

7. **Rinodina roscida** (Somm.) Arn. Thallus crustose, very thin, minutely granular, grayish white, K-, P-. Apothecia to 1.5 mm broad, sessile; margin thick or thin, chalky white, sometimes radiate-striate; disk concave to usually flat, black, usually densely white pruinose; cortex of exciple 12–15 μ; epithecium brown; hypothecium hyaline; hymenium 100–115 μ, upper part brownish; paraphyses coherent, tips brownish clavate, sometimes branched near the tips; spores 8, 2-celled, not constricted, the apical wall convex inward, septum 5–7 μ, greenish brown, broadly ellipsoid, 27–35 × 11–12 μ.

Growing on humus, soil, rabbit or ptarmigan dung. A common circumpolar, arctic-alpine species, known from Canada; new to Alaska.

Localities: 1, 9, 10, 11, 14.

8. **Rinodina turfacea** (Wahlenb.) Körb. Thallus crustose, variable, from thick and uneven to verrucose, of scattered verrucae, or disappearing; grayish white to brownish gray, K-, P-. Apothecia to 1.5 or 2 mm broad, sessile; margin thin, entire to crenulate or flexuous; disk flat or concave, black or brownish black, bare; epithecium brownish; hypothecium hyaline; hymenium 110–115 μ, the upper part brownish, the rest hyaline; paraphyses coherent, the tips brownish capitate; spores 8, 2-celled, not constricted, the apical wall convex inward, the central septum 5.5–8 μ, brown or olive brown, broadly ellipsoid, 26–35 × 11–14 μ.

Growing on mosses, rotting wood, decaying vegetation, rotting lichens. A common circumpolar, arctic-alpine lichen which ranges south to Washington, Alberta, and Saskatchewan.

Localities: 1, 2, 3, 4, 5, 6, 7, 8, 11, 13.

9. **Rinodina mniaraea** (Ach.) Körb. Thallus crustose, thin to thick, granular, continuous, dark reddish brown, sometimes tinged violet, with thin only occasionally visible dark hypothallus, K-, P-. Apothecia innate to sessile, to 1 mm broad; margin thin, disappearing; disk flat to becoming convex, reddish brown to dark brown, bare or slightly pruinose; exciple indistinct, with thin cortex or cortex lacking; epithecium brownish; hypothecium hyaline; hymenium 100–115 μ, upper part yellowish, it

and hypothecium I+ blue; paraphyses coherent, tips brown capitate; spores 8, 2-celled, not constricted, apical walls convex inward, septum 5-7 μ, brown, ellipsoid, 25-32 × 10-17 μ.

Growing on soil, humus, rotting wood. A circumpolar arctic-alpine species of wide distribution in North America.

Localities: 2, 9, 11.

10. Rinodina archaea (Ach.) Vain. em. Malme Thallus crustose, effuse, granular, subareolate to disappearing, brown or brownish gray, hypothallus indistinct, K-, P-. Apothecia appressed, to 0.65 mm broad; margin prominent, concolorous with the thallus; exciple with cortex lacking at sides, thin below; disk flat, brownish black, bare; epithecium brownish; hypothecium hyaline; hymenium 75-80 μ, upper part brownish, lower hyaline; paraphyses coherent but separating in KOH, brownish capitate; spores 8, 2-celled, not constricted, the apical walls not thickened inward but the walls thickened toward the septum, septum 3-5 μ, brownish green, ellipsoid, 17-24 × 8-10 μ.

Growing on barks and old wood. A circumpolar arctic and boreal species of uncertain range in North America due to difficulties in identifications and confusions with related species.

Localities: 1, 3.

11. Rinodina laevigata (Ach.) Malme Thallus very thin, of grayish brown to brown hypothallus on which are dispersed paler thin, irregular slightly raised thallus portions. Apothecia appressed, widely attached, to 0.6 mm broad; disk flat, brownish black; margin thin, smooth, cortex of margin 25-35(-70) μ thick, hyphae radiating; hypothecium hyaline; hymenium 85 μ, upper part yellowish brown, rest hyaline, I+ blue; paraphyses capitate; asci clavate; spores 18-22 × 7-9 μ, the septum 3-5 μ, the apical wall generally convex inwards, the lumina angular, the spore often with one side flat or concave.

On bark of deciduous woody plants as *Populus, Salix, Alnus*. Range currently uncertain. Reported from Scandinavia and North America.

Locality: 8.

12. Rinodina lecideoides (Nyl.) Vain. Thallus not apparent except under apothecia. Apothecia to 0.5 mm broad, constricted at base; margin thin to thick, pale to dark brown, smooth; disk flat to concave, dark brown; exciple with 12-15 μ cortex; epithecium brownish; hypothecium thin, 10-20 μ, hyaline; hymenium 70-80 μ, upper part brownish, mainly hyaline; paraphyses coherent, apices brown capitate; spores 8, 2-celled, apical walls convex inward, septum 2-3.5 μ, brown; elongate-ellipsoid, one side often flattened, 17-20 × 7-8.5 μ.

On wood, rarely on barks. A probably circumpolar species subalpine and boreal. Little known or collected. New to Alaska.

Locality: 14.

13. **Rinodina hyperborea** Magn. Thallus lacking, hypothallus lacking. Apothecia small, to 0.5 mm broad; margin thin, continuous, persistent, brownish ashy; cortex of exciple 10–15 μ; epithecium brownish; hypothecium narrow, 10–35 or 50 μ; hyaline; hymenium 65–70 μ, upper part brown, rest hyaline; paraphyses separating in KOH, brown capitate; spores 8, 2-celled, apical walls either straight across or convex inward, septum 2–3.5 μ, dark brown, central lamella dark, ellipsoid, 17–19 × 7.5–8.5 μ.

Growing on twigs and bark of *Alnus, Picea, Betula*. Known from Finland and Siberia. Possibly circumpolar boreal. New to Alaska.

Localities: 1, 8, 9.

14. **Rinodina septentrionalis** Malme Thallus crustose, of convex verrucae, over a brown or indistinct hypothallus, grayish or brownish gray, K-, P-. Apothecia adnate, constricted at base; margin thick, slightly crenulate, concolorous with the thallus, finally thin and smooth to disappearing; exciple cortex thin, 8–12 μ; epithecium brownish; hypothecium hyaline; hymenium 65–85 μ, upper part brownish yellow; paraphyses separating in KOH, tips subcapitate, brownish; spores 8, 2-celled, not constricted, apical walls not convex inward in old spores but sometimes in young, septum 2–3.5 μ, brown, ellipsoid, 16–20 × 6.5–8 μ.

Growing on bark of *Alnus* and *Salix*. Known from Fenno-Scandia and Siberia. New to Alaska.

Locality: 9.

PHYSCIACEAE

Foliose lichens attached to the substratum by rhizinae; dorsiventral, with the upper cortex of interwoven hyphae in *Anaptychia* and paraplectenchymatous in *Physcia* and *Pyxine*; lower cortex present or absent; medulla of leptodermatous or mesodermatous hyphae. Apothecia adnate to sessile, with either proper or thalloid exciple, hypothecium dark or hyaline; asci clavate; paraphyses simple, usually clavate at tipe; spores 8, 2-celled (3-septate or submuriform in extraterritorial material), brown, with thick walls. Pycnidia immersed; fulcra endobasidial, pycnoconidiospores short and straight or elongate and curved. Algae: *Trebouxia*.

Although *Anaptychia* is also represented in the Arctic, only *Physcia* has been collected in our region.

1. PHYSCIA (Schreb.) DC.

Thallus foliose, appressed to ascending, deeply incised, attached by rhizinae; lobes flattened, narrow, upper cortex of paraplectenchyma, medulla white or red, lower cortex present or absent; sides sometimes with elongate cilia. Apothecia laminal, sessile or short pedicillate; margin lecanoroid, containing algae; disk orbicular, brown or black, sometimes pruinose; paraphyses coherent, sometimes slightly branched, the tips capitate; asci clavate; spores 8, 2-celled, brown, the walls thickened. Pycnoconidiospores short, 2-3 μ, straight in species represented in the arctic.

1 Thallus white or whitish gray, upper side K+ yellow (atranorin)   2
  Thallus dark ashy brown or grayish brown, olive brown or brownish black, upper side K-   6

2 (1) Lobes ascending with long marginal cilia and helmet-shaped tips which are sorediate on the interior of the underside   1. *P. adscendens*
  Lobes lacking long marginal cilia, tips not inflated, open and hollow   3

3 (2) Lacking soredia, upper cortex white dotted, cortex and medulla both K+ yellow   2. *P. aipolia*
  With soredia   4

4 (3) Soralia laminal, capitate or punctiform, thallus white spotted above, cortex and medulla both K+ yellow   3. *P. caesia*
  Soredia marginal or laminal and marginal, thallus not white spotted above   5

5 (4) Lobes narrow, to 0.3 mm broad, with terminal labriform soralia and warty soralia on the upper surface; medulla as well as cortex K+ yellow   4. *P. intermedia*
  Lobes broader than 0.3 mm, distinctly labriform soralia, marginal soralia only, medulla K-   5. *P. dubia*

6 (1) Thallus epruinose, small to middle sized; spores small, less than 25 μ 7
  Thallus pruinose, middle-sized to large; spores large, 25-35 μ 9

7 (6) Lacking soredia or isidia   6. *P. endococcinea*
  With either soredia or isidia   8

8 (7) Sorediate, the soralia usually marginal, more or less labriform, on woody plants, rarely rocks   7. *P. orbicularis*
  Isidiate, on rocks   8. *P. sciastra*

9 (6) Thallus sorediate   9. *P. grisea*
  Thallus lacking soredia   10. *P. muscigena*

1. **Physcia adscendens** (Fr.) Oliv. em. Bitt. Thallus small foliose in small rosettes or of isolated lobes to 1.5 mm broad, irregularly pinnately branched, flat to convex, the tips forming a cup-shaped or helmet-

shaped cupule which is sorediate inside, the margins ciliate, the underside attached by rhizinae, upper side pale bluish gray, greener when wet,
lacking isidia, underside white, rhizinae pale; cortex K+ yellow, medulla
K-; containing atranorin. Apothecia uncommon, on upper side, sessile or
pedicillate, to 2 mm broad; margin entire or slightly crenate, concolorous
with the thallus; disk red brown to black brown, flat or concave, bare or
pruinose; epithecium yellow brown; hypothecium pale yellowish; hymenium 75–85 μ, hyaline; paraphyses simple or branched, septate, tips
brownish-capitate; spores 8, 2-celled, brown, apical walls thickened,
little constricted at the center, ellipsoid, 15–18 × 7–9 μ.

Growing on coniferous and deciduous trees, old wood, and rocks. A
circumpolar, boreal, and temperate species which ranges southward in
the mountains and along the California coast ranges.

Localities: 1, 3, 8.

2. **Physcia aipolia** (Ehrh.) Hampe Thallus foliose, lying close to substrate, marginal lobes separate, central lobes becoming conglomerate, to
1.5 mm broad, irregularly pinnate, upper side gray white to bluish white
and densely spotted with whitish spots (pseudocyphellae), dull, smooth
or roughened, sometimes pruinose; underside dull white to pale brownish with pale to brown rhizinae; cortex and medulla K+ yellow, containing atranorin. Apothecia usually present, to 2 mm broad, adnate on
upper surface; margin persistent, thick, smooth to crenulate; disk red
brown to black, usually very pruinose; epithecium brown yellow;
hypothecium hyaline; hymenium 90–130 μ, hyaline; paraphyses rarely
branched, tips brown clavate; spores 8, 2-celled, little constricted, the
apical walls and the septum thickened, brown, ellipsoid, 20–25 × 9–11 μ.

Growing on woody plants usually in open sun, rarely on rocks. A
circumpolar, arctic, boreal, temperate species with very wide range
covering Canada and the continental United States.

Localities: 1, 2, 3, 8, 9, 14.

3. **Physcia caesia** (Hoffm.) Hampe Thallus closely appressed to the
substrate, of irregularly pinnate lobes, convex, to 1 mm broad, upper
side blue gray or ashy, white spotted, with scattered or dense capitate
laminal soralia with dark gray, blue gray, or whitish gray soredia; underside pale or brownish with short dark rhizinae; cortex and medulla K+
yellow, containing atranorin and zeorin. Apothecia rare, sessile, to 1.5
mm broad, laminal; margin entire to crenulate, concolorous with the
thallus; disk brown black, bare or pruinose; epithecium brownish;
hypothecium hyaline to yellowish; hymenium 60–100 μ, upper part

brownish; paraphyses coherent, rarely branched, tips brown capitate; spores 8, 2-celled, more or less constricted, the apical walls and septum thickened, brown ellipsoid, 15–22 × 7.5–9.5 μ.

Growing on rocks, tree roots, old wood. A circumpolar arctic and boreal species which ranges south to New York in the east, to Arizona, New Mexico, and California in the west, also known from the Great Plains.

Localities: 1, 2, 3, 8, 9, 11, 13, 14.

4. **Physcia intermedia** Vain. Thallus small, the lobes narrow, to 0.5 mm broad with marginal labriform soralia and warty capitate soralia which are laminal, the latter often emptying and appearing as crateriform hollows; upper side dark ashy gray or gray brown, sometimes paler spotted, rarely slightly pruinose; underside pale with darkening rhizinae; cortex and medulla K+ yellow; containing atranorin. Apothecia to 2 mm, adnate; margin entire to crenate or sorediate; disk brownish black, bare or slightly pruinose; epithecium brownish; hypothecium yellowish; hymenium 80–100 μ, upper part yellowish, paraphyses coherent, brown capitate at tips; spores 8, 2-celled, the apical walls and the septum thickened, brown, ellipsoid, 16–22 × 8–11.5 μ.

Growing on rocks. A circumpolar, arctic and boreal species which ranges south in North America to Ontario, Wisconsin, New Mexico, and California.

Locality: 7.

5. **Physcia dubia** (Hoffm.) Lettau Thallus small, the lobes narrow at the base and broadening toward the tips, more or less dichotomous, flat to convex, sometimes reflexed at the tips, with labriform soralia at tips of lobes on short side branches; upper side white gray to ashy gray, dull, epruinose or rarely with slight pruina; underside pale or brownish with pale or brownish rhizinae; cortex K+ yellow, medulla K–; cortex containing atranorin. Apothecia uncommon, to 2.5 mm, sessile; margin entire or slightly crenate; disk black brown, bare; epithecium reddish brown; hypothecium hyaline to yellowish; hymenium 70–100 μ, hyaline with upper part brown red; paraphyses simple or branched, tips brown capitate; spores 8, 2-celled, apical walls and septum greatly thickened, brown, ellipsoid or slightly fabiform, 15–23 × 6–11 μ.

Growing on rocks, old bones, and antlers, occasionally weathered boards. A circumpolar arctic and boreal species which ranges south in North America to Maine, Wisconsin, and Colorado.

Localities: 1, 2, 3, 10, 11.

6. **Physcia endococcinea** (Körb.) Th. Fr. Thallus small, the lobes narrow, to 0.5 mm broad, dichotomous or irregularly pinnate; upper side pale to dark greenish brown; underside black with dark rhizinae; cortex and medulla both K-. Apothecia to 2 mm broad, sessile; margin entire to crenulate; disk dark brown; epithecium brownish; hypothecium hyaline to yellowish; hymenium 100–120 μ, hyaline below, yellow brown above; paraphyses branched toward the tips, brown capitate; spores 8, 2-celled, apical walls and septum thickened, brown, ellipsoid, 17–26 × 8–13 μ.

Growing on rocks in the arctic. A circumpolar species, uncommon in North America, where it ranges south to Connecticut, North Dakota, Colorado, and Washington. The material in our range is f. *lithotodes* (Vain.) Thoms. with a white medulla.

Localities: 1, 8, 11.

7. **Physcia orbicularis** (Neck.) Poetsch. Thallus growing in small rosettes which often coalesce, lobes irregularly pinnatifid, imbricated, flat to slightly convex, tips narrow or broadened; upper side olive green to gray brown, epruinose, with soralia which may be maculiform and laminal or more or less capitate at the tips of small lobes, the soredia varying from yellowish green to gray, blackish green, or black; underside black, paler at the margins, with short black rhizinae; cortex and medulla K-. Apothecia to 2 mm, sessile, occasionally with a corona of short rhizinae around the base; margin entire to crenate; disk red brown to black, flat, dull or shiny, bare; epithecium brown; hypothecium hyaline; hymenium 75–100 μ, hyaline below, red brown above; paraphyses coherent, simple or branched, tips red brown capitate; spores 8, 2-celled, apical walls and septum thickened, brown, ellipsoid, 20–23 × 8–10 μ.

Growing on a wide variety of substrates, rocks, tree trunks, and mosses. A circumpolar species, boreal and temperate, ranging over most of North America.

Locality: 14. Probably new to Alaska.

8. **Physcia sciastra** (Ach.) DR. Thallus small but coalescing to form larger mats, lobes narrow, 0.2–0.5 mm broad, irregularly branched, quite linear, tips sometimes broadened; upper side gray brown, dark brown, or black, the margins and upper surface isidiate with blackish isidia; underside black with black rhizinae; cortex and medulla K-. Apothecia rare, sessile, to 2 mm broad; margin entire to very crenate; disk red brown to black, dull, bare; epithecium dark; hypothecium hyaline; hymenium 100 μ, upper part brown black, lower hyaline; paraphyses branched, tips brown black capitate; spores 8, 2-celled, apical walls and septum thickened, brown, ellipsoid 15–24 × 8–12 μ.

Growing on rocks and over mosses. A circumpolar arctic and boreal species ranging south in the mountains and cool habitats to Maine, Minnesota, and Arizona.
Localities: 1, 2, 3, 9, 11.

9. **Physcia grisea** (Lam.) Zahlbr. (*Physconia grisea* (Lam.) Poelt) Thallus middle-sized to large, the lobes to 3 mm broad, close together to overlapping, irregularly pinnatifid, flat to concave, sometimes lobate-ascending, with marginal, sometimes laminal, labriform soralia with white, gray white or black soredia; upper side gray brown, gray green or brownish usually white or blue white pruinose, the pruina sometimes almost lacking; underside pale to black with pale to black rhizinae; cortex and medulla K-. Apothecia uncommon, laminal, sessile or adnate, to 1.5 mm broad; margin entire, crenulate, or sorediate; disk concave to flat, dull dark brown to black, bare or pruinose; epithecium dark; hypothecium hyaline to brownish; hymenium 120–160 μ, upper part brown, lower hyaline; paraphyses simple or branched, coherent, apices not capitate; spores 8, 2-celled, apical walls and septum very thick, center constricted, straight or slightly curved, brown black, 24–33 × 13–17 μ.
Growing on trees and on rocks, over mosses. A circumpolar boreal and temperate species.
Localities: 8, 11.

10. **Physcia muscigena** (Ach.) Nyl. (*Physconia muscigena* (Ach.) Poelt) Thallus middle-sized to large, to 12 cm broad, the lobes often to over 3 mm broad, tips rounded, flat to concave, sometimes very erect-squarrose (f. *squarrosa* (Ach.) Lynge); upper side chestnut brown or gray brown, either with the tips pruinose or heavily covered with white to blue white pruina (f. *lenta* (Ach.) Vain); in a rare form with cylindrical isidia (f. *isidiata* (Lynge) Thoms.); underside black with whiter margins, rhizinae black or gray black; cortex and medulla K-. Apothecia large, to 5 mm, adnate; margin entire, crenate, or with a circlet of lobules; epithecium brown; hypothecium hyaline; hymenium 100–125 μ, upper part brown, lower hyaline; paraphyses occasionally branched toward tips, brown capitate; spores 8, 2-celled lumena rounded, constricted, brown, ellipsoid, 22–30 × 12–17 μ.
Growing on humus and over mosses, on soil. An arctic-alpine circumpolar species which ranges south in North America to Michigan, Minnesota to Arizona in the west.
Localities: 1, 2, 3, 9, 10, 11, 12, 13, 14; f. *lenta*: 2, 11, 14; f. *squarrosa*: 3, 4; f. *isidiata*: 11.

TELOSCHISTACEAE

Crustose, foliose, to fruticose lichens with upper cortex or lacking this and in some genera with lower cortex, thallus yellow or whitish yellow, containing parietin and turning purple with KOH at least in epithecium; rhizinae present in some genera. Apothecia sessile or adnate or terminal in fruticose species; with thalloid or proper exciple; paraphyses coherent or free, commonly branched toward the apices, septate; asci clavate; spores 4 or 8, hyaline, ellipsoid, simple, 2-celled or polaribilocular, (the two cells separated by a ± thick septum and the lumena in rounded form.) Algae: *Trebouxia*.

1   Thallus crustose, lacking lower cortex   2
      Thallus foliose, with upper and lower cortex   5. *Xanthoria*

2 (1) Apothecia lacking algae either in the margin or below the hypothecium
                                                     3
      Apothecia with algae in the margin and below the hypothecium   4

3 (2) Spores unicellular   1. *Protoblastenia*
      Spores polaribilocular   2. *Blastenia*

4 (2) Spores unicellular   3. *Fulgensia*
      Spores polaribilocular   4. *Caloplaca*

1. PROTOBLASTENIA Stein.

Thallus crustose, granulose, not differentiated into layers. Apothecia minute to small, immersed in substratum to sessile; proper exciple only present and concolorous with the disk; disk concave to flat or convex; epithecium yellow, turning purple with KOH; hypothecium hyaline; hymenium hyaline, upper part brownish yellow; paraphyses unbranched; asci clavate; spores 8, hyaline, simple, ellipsoid.

1   Growing on rocks; apothecia orange to reddish orange   2
      Growing on soil; apothecia dark red, thallus thick, whitish
                                           2. *P. terricola*

2 (1) Thallus present   1. *P. rupestris*
      Thallus absent   1. *P. rupestris* var. *calva*

1. **Protoblastenia rupestris** (Scop.) Stein. Thallus thin, minutely granulose, disappearing in var. *calva* (Dicks.) Stein., ashy to whitish ashy. Apothecia sessile or immersed in the substratum in var. *calva*; disk flat to convex, yellow to reddish brown; margin thin, soon disappearing, colored like the disk; epithecium yellowish, K+ violet; hypothecium hyaline; paraphyses brownish above, unbranched; spores 8, simple, ellipsoid, hyaline, 8–14 × 5–7.5 μ.

Growing on calcareous rocks, usually in moist, shaded conditions, seepage areas on cliffs, etc. A circumpolar arctic, boreal, and temperate species with a wide range in North America.

Localities: 1, 2, 3, 8, 9, 11, 14; var. *calva*: 11, 14.

2. **Protoblastenia terricola** (Anzi) Th. Fr. Thallus thick, warty-areolate, white, K-. Apothecia sessile, to 1.5 mm broad, proper exciple concolorous with the disk; disk orange yellow to olivaceous brownish, K+ purple; epithecium yellowish; hypothecium hyaline to brownish; paraphyses yellowish above, hyaline below; spores 8, simple, ellipsoid, hyaline, 7–10 × 4–5 μ.

Growing on calcareous soil and humus. A probably circumpolar arctic-alpine species known from Europe, Siberia, Greenland, Spitzbergen, and Novaya Zemlya. Specimens from Ellesmere Island, and Waterton Lakes, Alberta are in the WIS herbarium. New to Alaska.

Locality: 11.

2. BLASTENIA Mass.

Thallus crustose, smooth, powdery, granulose or squamulose, not differentiated into layers. Apothecia adnate to sessile; proper exciple of same color as the disk, often disappearing; disk concave to convex, rust colored to orange or yellow, turning purple with KOH; epithecium brownish; hypothecium hyaline; hymenium hyaline below, brownish above; paraphyses unbranched, septate, brownish capitate; asci clavate; spores usually 8, hyaline, the cells polaribilocular, ellipsoid.

1 Spores 4 per ascus, large, 24–34 × 12–16 μ; apothecia red orange.

see *Caloplaca tetraspora*

Spores 8 per ascus, small, 12–17 × 6–8 μ; apothecia brown to reddish brown, or black   1. *B. exsecuta*

1. **Blastenia exsecuta** (Nyl.) Serv. (*B. melanocarpa* Lynge) Thallus granular to areolate or disappearing, whitish or bluish gray to grayish orange. Apothecia adnate to sessile, to 0.7 mm broad; disk black or blackish brown, flat, margin thin, black; exciple black green with radiating hyphae, violet in HNO_3; epithecium brownish green or brownish yellow; hypothecium hyaline or yellowish; hymenium 50–85 μ, I+ blue; upper part dark, K+ violet, C+ violet red brown, HNO_3 yellowish or violet; paraphyses only slightly enlarged at tips to 2 μ; spores 8, hyaline, polaribilocular, ellipsoid, 12–17 × 5–8.5 μ, septum 5–7 μ.

Growing on rocks. An arctic alpine species known from Europe and Greenland. New to Alaska, collected in Colorado by R.S. Egan.

Locality: 1.

3. FULGENSIA Mass. & DeNot.

Thallus crustose, granulose to lobulate toward the margins; more or less differentiated into layers with an upper cortex being slightly developed, in *F. bracteata* Poelt (1965) describes raising of the thallus to form lobules which he terms schizidia; a hypothallus often present around the border of the thallus proper. Apothecia adnate to sessile; proper exciple of same color as the disk, often disappearing; a paler thalloid margin present; disk concave to convex, orange to brownish orange, turning purple with KOH; epithecium brownish; hypothecium hyaline to yellowish; hymenium hyaline below, brownish above; asci clavate; spores 8, hyaline, simple, ellipsoid.

Although only one species has been collected on the North Slope another which is fully expected there is given in the key.

1 Thallus warty-squamulose, not effigurate-lobate, centrally discontinuous
1. *F. bracteata*

  Thallus bordered by radiate lobes, centrally continuous   *(F. fulgens)*

1. **Fulgensia bracteata** (Hoffm.) Räs. Thallus verrucose to forming small squamulose lobules which are more or less loose on the substratum (schizidia- of Poelt), more or less convex; orange yellow to orange brown sometimes with upper surface pruinose, scattered over a white hypothallus which shows between the lobules, no rhizinae, the underside attached directly to the substratum; a cortex slightly developed, medulla very loose. Apothecia adnate, scattered, to 2 mm broad; margin thick at first but becoming thinner and irregular, paler than the disk; disk flat, rust red to orange brown, K+ purple; epithecium brownish; hypothecium hyaline to yellowish; hymenium 60–70 μ, hyaline below, brownish above; paraphyses little branched, brownish capitate; spores 8, simple, hyaline, ellipsoid, 9–15 × 4–7 μ.
   Localities: 9, 10, 11, 14, and Lake Peters.

4. CALOPLACA Th. Fr.

Thallus crustose or lobulate to squamulose, granulose, chinky, areolate, rarely lacking, poorly developed upper cortex, lower lacking, algal layer continuous or glomerulate; K+ purple or K-. Apothecia adnate to sessile, small to middle-sized; thalloid margin usually yellow, black in some; disk flat rarely concave or convex, orange, yellow, brown, or black, K+ purple; epithecium yellowish or brownish to greenish black; hypothecium usually hyaline, rarely brownish; hymenium hyaline, upper part colored as in epithecium; paraphyses septate, apices darkened and capitate, unbranched, coherent or free; asci clavate; spores 8, 4 in one species, 2-celled polaribilocular, ellipsoid.

1   Thallus yellow or orange, apothecia yellow or orange, K+ reddish,
     foliose or lobate   2
     Thallus whitish, ashy, buff, brownish, or darker, K- or K+ if yellowish,
     lacking cortex below, crustose   5

2 (1) Corticate above and below   see *Xanthoria*
     Corticate above, ecorticate below   3

3 (2) Lacking soredia or isidia, thallus of radiating lobes   1. *C. murorum*
     Either sorediate or isidiate   4

4 (3) Center of thallus becoming a mass of soredia   2. *C. cirrochroa*
     Center of thallus granulose-isidiate   3. *C. granulosa*

5 (1) Apothecium black, margins dark colored; exciple blackish green
                                                     see *Blastenia exsecuta*
     Apothecia yellow, orange, or brown, disk K+ reddish   6

6 (5) Growing over mosses   7
     Growing on bark, wood, rocks, or parasitic   11

7 (6) Apothecia dark rusty red with concolorous margins   8
     Apothecia yellow or orange but not rusty red   9

8 (7) Spores 4 per ascus, large, 24–34 × 12–16 μ; apothecium red orange;
     hypothecium over 150 μ 4. *C. tetraspora*
     Spores 8 per ascus, small 13–17 × 6.5–7.5 μ; apothecium brick red to
     rusty brown; hypothecium about 35 μ 5. *C. cinnamomea*

9 (7) Apothecium with yellow disk, gray pruinose margin; hypothecium less
     than 40 μ 6. *C. stillicidiorum*
     Apothecium orange with orange margin; hypothecium over 40 μ
                                                                         10

10 (9) Apothecia large, to 1.5 mm; disk brownish orange; parathecium 35 μ or
     less wide at top; hypothecium under 60 μ high; spores 18–21 × 7–10 μ
                                                       7. *C. jungermanniae*
     Apothecia small, less than 0.7 mm; disk yellow to olive; parathecium
     more than 35 μ wide at top; hypothecium over 60 μ high; spores
     11–16 × 6–9 μ 8. *C. tiroliensis*

11 (6) Parasitic on *Placynthium*, thallus lacking, disk dark orange, margin paler
                                                          9. *C. invadens*
     Non-parasitic   12

12 (11) Growing on bark or wood   13
     On rocks   18

13 (12) Apothecia with orange disk and black margin   10. *C. pinicola*
     Apothecia with orange or rusty red disk and margin   14

14 (13) Apothecia rusty red to red brown   15
     Apothecia orange, orange red, to waxy yellow   17

15 (14) Thallus esorediate, white or gray   16
         Thallus sorediate with coarse soredia, yellowish, K+ purplish; spores
         15–23 × 10–14 μ 13. *C. discolor*

16 (15) Spores with narrow septum, 2–3 μ, spores narrow, 12–14 × 5–6 μ
                                                 11. *C. fraudans*
         Spores with broader septum, over 6 μ, spores broader, 13–17 × 8–9 μ
                                                 12. *C. ferruginea*

17 (14) Apothecia orange to orange red; thallus whitish, black, or lacking;
         spores 12–13 × 7.5–8.5 μ, septum 5–6 μ 14. *C. holocarpa*
         Apothecia orange to waxy yellow; thallus bluish gray; spores 13–17 ×
         7–8 μ, septum 6–8 μ 15. *C. cerina*

18 (12) Apothecia dark red brown, hypothecium over 50 μ; spores narrow,
         under 6 μ wide, with narrow septum under 3μ 11. *C. fraudans*
         Apothecia reddish rusty to orange red; hypothecium less than 50 μ;
         spores broader than 6 μ, with wider septum 5–6 μ 16. *C. festiva*

1. **Caloplaca murorum** (Ach.) Th. Fr. Thallus of radiate lobes placed close together, convex, to 1.5 mm broad, orange yellow, orange or brownish red, usually pruinose; K+ purple; the center of the thallus becoming more or less areolate, sometimes the thallus of scattered lobes, a thin yellow hypothallus sometimes present. Apothecia adnate, to 1 mm broad, flat or convex; margin concolorous with the thallus; disk orange to yellow orange or brownish red; epithecium yellowish, K+ violet; hypothecium hyaline; hymenium 50–80 μ, hyaline, yellowish above; paraphyses simple or branched above, tips capitate to 6–9 μ; spores polaribilocular, hyaline, ellipsoid, 8–16 × 3–7 μ, the septum broad, 5–8 μ.

     Growing on calcareous rocks. A circumpolar arctic to temperate species with wide range.

     Locality: 1.

2. **Caloplaca cirrochroa** (Ach.) Th. Fr. Thallus radiate, of narrow lobes to 0.5 mm broad, simple to irregularly branched, convex, crenulate toward the tips; upper side bright orange yellow or lemon yellow to brownish yellow, pruinose, with round lemon yellow soralia toward the bases of the lobes; K+ purple; the center becoming verrucose-areolate. Apothecia rare, to 0.5 mm broad; margin thin, concolorous with the thallus, entire; disk flat, orange yellow; hypothecium hyaline; hymenium 60–70 μ, hyaline; paraphyses slender, 1 μ with apices 3–4 μ, unbranched or slightly branched; spores polaribilocular, hyaline, ellipsoid, 10–18 × 4–7 μ, septum 4–6 μ.

Growing on shaded calcareous rocks, cliff overhangs, bird nesting cliffs. A circumpolar arctic to temperate species.

Localities: 2, 11, 14.

3. **Caloplaca granulosa** (Müll. Arg.) Jatta Thallus radiate, of contiguous lobes to 4 mm long and 0.5 mm broad, irregularly branched and with few to numerous globular isidia, the center of the thallus areolate and isidiate-granular; orange or yellow; K+ purple. Apothecia rare, to 1 mm broad; margin entire or crenulate, disappearing, concolorous with the thallus; disk flat, orange to bright yellow, K+ violet; epithecium yellow; hypothecium hyaline; hymenium 70–80 μ, hyaline; paraphyses simple or slightly branched, tips to 6 μ; spores 8, hyaline, polarbilocular, ellipsoid, 10–16 × 5–8 μ, septum 4–5 μ.

Growing on shaded calcareous rocks, especially those splashed by bird dung. A circumpolar arctic to temperate species widely ranging in North America.

Locality: 11.

4. **Caloplaca tetraspora** (Nyl.) Oliv. Thallus crustose, thin, granulose to disappearing, white to ashy, K-. Apothecia to 0.8 mm broad, at first with flat disk and a thin margin concolorous with the thallus, this soon disappearing and the disk becoming very convex so that the apothecium appears like that of a *Blastenia*; red rusty or rusty brown to darkening; epithecium golden brownish, K+ reddish; hypothecium hyaline; hymenium 80–100 μ, hyaline, the upper part brownish; paraphyses free, septate, branched, tips capitate to 5 μ; spores 4, polaribilocular, ellipsoid, hyaline, 17–34 × 10–16 μ, the septum 5–8 μ.

Growing on mosses and humus, rarely soil. A circumpolar arctic-alpine species in North America ranging south to Washington in the west.

Localities: 1, 2, 8, 11.

5. **Caloplaca cinnamomea** (Th. Fr.) Oliv. Thallus indeterminate, poorly developed, of scattered ashy granules or disappearing, K-. Apothecia to 1 mm, at first flat with more or less thick thalloid margin of same color as the thallus, soon the disk becoming convex and the margin thin to disappearing, rusty red or brick red to dark brown or olive brown, K+ purple; epithecium brownish, K+ violet; hypothecium yellowish; hymenium 85 μ, hyaline, upper part yellowish; paraphyses coherent, the apices thickened to 3–3.5 μ; spores 8, polaribilocular, hyaline ellipsoid, 13–17 × 6.5–7.5 μ, the septum 3.5–7.5 μ.

Growing over mosses and humus, sometimes on soil or bark. A circumpolar arctic-alpine species.

Localities: 3, 4, 5, 7, 8, 9, 11, 14.

6. **Caloplaca stillicidiorum** (Vahl) Lynge Thallus poorly developed to disappearing, granulose to small squamulose, ashy, K-. Apothecia adnate, flat; margin conspicuously differing from the disk, pale gray or blue gray pruinose, entire to flexuous; disk to 1.5 mm broad, yellowish yellow orange, or olive, more or less pruinose; epithecium yellowish, K+ reddish; hypothecium hyaline; hymenium hyaline, 35–80 μ, upper part yellowish; paraphyses septate, branched at upper ends, tips capitate to 5 μ; spores 8, polaribilocular, hyaline, ellipsoid, 10–17 × 5–10 μ, septum 5 μ.

Growing on humus, mosses, old organic materials, soil. A circumpolar arctic-alpine species, growing south into New England in the eastern states, Minnesota, and in the west to New Mexico.

Localities: 1, 2, 7, 8, 9, 11, 14.

7. **Caloplaca jungermanniae** (Vahl) Th. Fr. Thallus very thin, subgranulose to subsquamulose, whitish, K-. Apothecia adnate, flat, margin of same color as the disk, orange red or yellowish brown; disk to 1.5 mm broad, dull, epruinose; margin thick, brownish yellow; epithecium yellowish, K+ reddish violet; hypothecium hyaline, less than 60 μ high; hymenium hyaline, 80–110 μ; paraphyses septate, slender, tips not capitate; spores 8, polaribilocular, hyaline, ellipsoid, 14–23 × 7–13 μ, septum 3.5–4 μ.

Growing over mosses. A circumpolar arctic-alpine species which in North America ranges south to Colorado and Washington.

Localities: 8, 11.

8. **Caloplaca tiroliensis** Zahlbr. Thallus very thin, subgranulose to subsquamulose, whitish, K-. Apothecia adnate, flat, the thin margin of the same color as the disk, waxy yellowish, yellowish orange to olive blackish; disk to 0.5 mm broad; margin thin, concolorous with the disk; epithecium yellowish or brownish, K+ reddish violet; hypothecium over 60 μ; hymenium hyaline, 30 μ; paraphyses slender, not capitate; spores 8, polaribilocular, hyaline, ellipsoid, 11–16 × 8–9 μ, the septum 4–5 μ.

Growing over mosses and humus. A circumpolar arctic-alpine species.

Localities: 2, 4, 7, 8, 9, 13, 14.

9. **Caloplaca invadens** Lynge Thallus parasitic on other lichens, a thin yellow brown arachnoid layer. Apothecia scattered, to 1 mm broad, the

margin paler than the dark orange disk; disk flat, dull; margin thin; epithecium yellowish, K+ violet; hypothecium pale; hymenium hyaline, 100–130 μ; paraphyses little branched, slender; spores 8, polaribilocular, hyaline, ellipsoid, 13–16 × 8–10 μ, the septum 3–5 μ.

Growing over other lichens, *Lecanora* (*Aspicilia*), and in this area *Placynthium*. A species reported from Novaya Zemlya, Lapland. New to Alaska and North America.

Locality: 11.

10. **Caloplaca pinicola** Magn. Thallus a thin dark film or indistinct. Apothecia 0.5–0.6 mm broad; margin thin, entire, dull black; disk waxy yellowish to darkening olivaceous yellowish; epithecium yellow, K+ deep reddish purple; hypothecium 30–35 μ, grayish; hymenium 70 μ, upper part dark yellow, lower part hyaline; paraphyses coherent, septate, the tips capitate to 4–6 μ; spores 8, polaribilocular, hyaline, ellipsoid, 13–15 × 7–8 μ, the septum 2–3 μ.

Originally reported as growing on *Pinus*, found on humus and *Salix* in the arctic. Known from Colorado and Arizona. New to Alaska. Specimens from Labrador and Alberta are in WIS.

Localities: 1, 2, 8, 9.

11. **Caloplaca fraudans** (Th. Fr.) Oliv. Thallus thin, forming yellowish pale patches on the rock or invisible. Apothecia sessile, to 1.3 mm broad; disk flat, orange brown to yellowish red; margin thick, inflexed, subshiny, paler than the disk but deep orange; epithecium brownish yellow; hypothecium pale, 60–100 μ; hymenium 85–100 μ, the upper part brownish yellow, the lower part hyaline; paraphyses slender, septate, branched in upper part, the tips only slightly thickened; spores 8, polaribilocular, hyaline, ellipsoid, 10–17 × 5–6 μ, the septum 2–3.5 μ.

Growing on rocks and on bones. A circumpolar arctic species. New to Alaska.

Localities: 1, 2, 3, 8, 11.

12. **Caloplaca ferruginea** (Huds.) Th. Fr. Thallus thin, indefinite, smooth to partly granular, pale gray, K-. Apothecia to 1 mm broad, sessile, the base constricted; margin lighter than the disk, thin, rusty reddish; disk flat, dull, rust red to darkened rusty red; epithecium reddish brown, granular, K+ violet; hypothecium about 50 μ, hyaline; hymenium 85–90 μ, hyaline below, upper part reddish; paraphyses coherent, the tips not thickened-capitate; spores 8, polaribilocular, hyaline, ellipsoid, 13–17 × 8–9 μ, the septum 5–8.5 μ.

Growing on bark of broad-leaved woody plants such as *Populus, Salix,* seldom conifers. A probably circumpolar arctic and boreal species.
Localities: 1, 2, 13.

13. **Caloplaca discolor** (Willey) Fink. Thallus yellow or yellowish white, diffuse, chinky, uneven, partly granular, verrucose with verrucae which burst into coarsely granular rusty yellow soredia which resemble short isidia, K+ violet. Apothecia to 1.5 mm, sessile; margin of same color as disk, thin; disk flat to convex, dark rusty red; epithecium brownish yellow, K+ violet; hypothecium 100–120 μ, grayish; hymenium 85 μ, hyaline below, brownish yellow above; paraphyses branched above, slender, free, tips slightly thickened to 2.5 μ; spores 8, polaribilocular, hyaline, broadly ellipsoid, 15–23 × 10–14 μ, the septum 3–5 μ.
Growing on bark and old wood. Reported as a North American species in the northeastern states, west to Minnesota. Probably new to Alaska.
Localities: 10, 14.

14. **Caloplaca holocarpa** (Hoffm.) Wade Thallus thin to disappearing, grayish white, K-. Apothecia sessile, to 0.5 mm broad; margin thin, of same color as disk or paler, entire; disk flat, dark orange, dull; epithecium brownish yellow, K+ violet; hypothecium less than 50 μ, grayish; hymenium 85 μ, lower part hyaline, upper brownish yellow; paraphyses coherent, septate, branched above, tips capitate thickened to 3–4.5 μ; spores 8, polarbilocular, hyaline, ellipsoid, 12–13 × 7–8 μ, the septum 3.5–6 μ.
Growing on barks. A probably circumpolar boreal to temperate species with wide distribution in North America.
Localities: 8, 9, 10, 14.

15. **Caloplaca cerina** (Ehrh.) Th. Fr. Thallus smooth, uneven, or thin-areolate, ashy gray to dark greenish gray, K-, with thin bluish gray to blue black hypothallus sometimes present. Apothecia to 2 mm broad, sessile; margin thick, pale gray to bluish gray, persistent, entire, crenulate, or flexuose; disk concave to flat, yellow, orange yellow, or brownish yellow; epithecium yellowish, K+ violet; hypothecium 30–35 μ, grayish; hymenium 60–80 μ, upper part yellow, lower part hyaline; paraphyses coherent, slightly branched above, capitate, the tips cells to 3–4 μ; spores 8, polaribilocular, hyaline, broadly ellipsoid, 15–17 × 7–8 μ, the septum 7–8 μ.
Growing on a wide variety of deciduous trees. A circumpolar boreal to temperate species with very wide range in North America.
Localities: 1, 2, 8, 9.

16. **Caloplaca festiva** (Ach.) Zw. Thallus effuse to areolate, the areolae flat to uneven, often separated by cracks with the dark hypothallus showing through, edge areolae often with black margin, whitish gray, K-. Apothecia sessile with narrow base, to 1 mm broad; margin thick, becoming shiny, entire, of same color or slightly darker than the disk; disk flat, sometimes convex, bright rusty red to darkening to blackish rusty; epithecium rusty yellowish, K+ violet; hypothecium 50 μ, grayish; hymenium 70–80 μ, upper part yellow or rusty yellow, lower part hyaline; paraphyses distinct, not clearly septate, slightly branched, tips not thickened; spores 8, polaribilocular, hyaline, ellipsoid, 14–17 × 7–8 μ, the septum 5–7 μ.

Growing on rocks, acid or calcareous. A circumpolar arctic to boreal and temperate species. The North American range not well defined. New to Alaska.

Localities: 1, 3, 11.

5. XANTHORIA (Fr.) Th. Fr.

Thallus foliose or closely lobate to fruticose-appearing, dorsiventral, with cortex on upper and lower sides, the cortex of hyphae more or less oriented along the surface, upper side orange, K+ purple; lower side pale, attached to the substratum with rhizinae. Apothecia adnate to sessile, margin lecanorine; disk K+ purple (containing parietin); hypothecium hyaline; hymenium brownish or yellow above, hyaline below; paraphyses septate, branched, capitate; asci clavate; spores 8, polaribilocular. Fulcra endobasidial, pycnoconidiospores elongate. Algae: *Trebouxia*.

1	Thallus neither sorediate nor isidiate	1. X. *elegans*
	Thallus either sorediate or isidiate	2
2 (1)	Thallus isidiate, or with both soredia and isidia	3
	Thallus sorediate	4
3 (2)	Thallus short-fruticose, the lobe tips becoming divided into coarsely isidiate lobuli	2. X. *candelaria*
	Thallus foliose-flattened, the lobe tips with isidia which dissolve into soredia	3. X. *sorediata*
4 (2)	The soralia distinctly labriform, with granular isidia	4. X. *fallax*
	The tips of the lobes becoming dissolved into coarse soredia, not labriform	2. X. *candelaria*

1. **Xanthoria elegans** (Link.) Th. Fr. Thallus foliose but appearing crustose and closely applied to the substratum, lobate with radiating branching lobes, corticate above and below, attached by short rhizinae; the lobes convex, smooth or roughened, bright orange, K+ purple, elongate,

contiguous or spreading, sometimes deeply piled up in presence of abundant manuring, sometimes pruinose; the medulla partly hollow, partly filled with arachnoid hyphae. Apothecia common, sessile; margin concolorous with the thallus, entire to crenulate; disk to 2 mm broad, a little darker than the thallus, rough; epithecium granulose, brownish yellow, K+ violet; hypothecium hyaline; hymenium hyaline below, yellowish above; paraphyses septate, branched and entangled, the tips slightly thicker, to 3 μ; spores 8, polaribilocular, hyaline, ellipsoid, 11–16 × 6–8 μ.

Growing on rocks, especially calcareous but also on bird perch rocks and heavily manured places as bird cliffs, also on mosses and soil on occasion. A cosmopolitan lichen from Arctic regions to the tropics in both hemispheres.

Localities: 1, 2, 3, 8, 9, 10, 11, 14.

2. **Xanthoria candelaria** (L.) Arn. Thallus small, irregular to rosette-shaped, greenish yellow to yellow or orange yellow, pale below with sparse pale rhizinae; lobes flat and narrow, often sharply incised, erect, the margins sorediate and the apices becoming isidioid; medulla more solid than in the previous species; upper cortex K+ purple. Apothecia uncommon, to 2 mm broad, adnate; margin concolorous with the thallus, entire to sorediate; disk yellow to orange; epithecium brownish yellow, granular, K+ violet; hypothecium hyaline; hymenium hyaline below, yellowish above; spores 8, polaribilocular, hyaline, broadly ellipsoid, 9–16 × 5–9 μ.

Growing on calcareous rocks, especially under overhangs. A circumpolar, arctic-alpine to boreal or temperate species. Reported by Krog, 1962, from the Ogotoruk Creek area near the Bering Straits.

Localities: 7, and near Rogers-Post monument S. of Barrow.

3. **Xanthoria sorediata** (Vain.) Poelt Thallus small, lobes narrow, greenish yellow to orange yellow, flat to raised, in the center with isidia which break out into soredia at the tips; K+ purple. Apothecia rare.

Growing on calcareous rocks and underhangs. A circumpolar arctic and alpine species. New to Alaska.

Localities: 1, 3, 11, 14.

4. **Xanthoria fallax** (Hepp) Arn. Thallus foliose, forming small rosettes or covering large expanses with lobes raised at the edges; lobes more or less erect, short, rounded, with labriform soralia along the margins and tips; upper cortex yellow to yellow orange or darker orange, K+ purple; lower side pale with sparse pale to darker rhizinae. Apothecia not

common, to 2 mm broad; margin of same color as the thallus, entire; disk dark orange or orange; epithecium brownish yellow, K+ violet; hypothecium hyaline to yellowish; hymenium hyaline below, yellowish above; spores 8, polaribilocular, hyaline, ellipsoid, 11–16 × 4.5–8 μ.

Growing on rocks, bark, old wood, and humus. A circumpolar boreal to temperate species ranging over most of North America.

Localities: 11, 13, 14.

DIPLOSCHISTACEAE

Thallus crustose, the layering poorly differentiated, attached to the substratum by the hyphae of the underside. Apothecia round, immersed in the thallus or adnate, the disk deeply concave or flat, with a well developed proper exciple usually surrounded by a thalloid margin; paraphyses simple or branched; spores transversely septate (not in our area) or muriform and dark. Fulcra exobasidial. Algae: *Trebouxia*.

1. DIPLOSCHISTES Norm.

Thallus crustose, usually rather thick, of warty, uneven, crowded and piled up areoles, often with a paler marginal zone, bluish gray; upper cortex poorly developed, of matted hyphae; attached to the substratum by hyphae. Apothecia usually abundant, immersed in the thallus, thalloid margin usually higher than the interior proper exciple which is usually radiately fimbriated; disk deeply concave to flat, black, commonly thinly whitish pruinose; hypothecium brownish to brown; hymenium hyaline or brownish in upper part; paraphyses simple or with few branches toward the apices, indistinctly septate, slender, to 1.5 μ, the tips thickened to 3.5 μ; asci cylindrico-clavate; spores 2–8, usually 4, per ascus, becoming brown, muriform, 1–3-septate longitudinally, 5–7-septate transversely.

1. **Diploschistes scruposus** (Schreb.) Norm. Thallus crustose, thin to several mm thick, warty, chinky but not areolate, rounded uneven or rough, gray, blue gray or brownish gray, upper cortex poorly developed; thallus K- or partly K+ yellowish or brownish, Baryta water + violet, C+ red, KC+ red, P-, I+ blue; containing diploschistesic and lecanoric acids. Apothecia immersed in the thallus, to 2 mm broad; thalloid margin thick, of same height or slightly higher than the proper exciple, of same color as the thallus, crenulate; proper exciple dark brownish black, radially fraying toward the inner side and graying; hypothecium brown; hyme-

nium brownish above, hyaline below; spores 4–8, brown, ellipsoid, muriform, 2–3-septate longitudinally, 5–7-septate transversely, 20–45 × 10–20 μ.

Growing on rocks, calcareous or not, on bird perch rocks, on soil, and over mosses. A circumpolar species with exceedingly broad range, occurring over most of North America.

Localities: 3, 11, 14.

VERRUCARIACEAE

Thallus crustose to squamulose, in the crustose species from inside the rock substratum to with corticate or not layers above the substratum, in the squamulose species corticate above or both above and below. The fruit a perithecium more or less imbedded in the thallus; sometimes with periphyses in the upper part inside the mouth; with paraphyses which persist or disappear, in some genera with algae distributed in the hymenium; asci unitunicate. Algae: *Trebouxia* or *Pleurococcus*.

1	Thallus crustose, closely applied to or immersed in the substratum	2
	Thallus squamulose	7
2 (1)	Paraphyses soon gelatinizing or disappearing	3
	Paraphyses persistent	6
3 (2)	Spores simple	1. *Verrucaria*
	Spores several-celled to muriform	4
4 (3)	Perithecia lacking algae in the hymenial layer	5
	Perithecia containing algae in the hymenial layer	4. *Staurothele*
5 (4)	Spores 2–4(–6)-celled with only transverse septa	2. *Thelidium*
	Spores muriform, with both transverse and longitudinal septa	
		3. *Polyblastia*
6 (2)	Spores 3-septate	5. *Geisleria*
	Spores muriform	see *Microglaena*
7 (1)	Spores simple, hyaline; hymenial layer lacking algae	6. *Dermatocarpon*
	Spores muriform, brown; hymenial layer containing algae	
		7. *Endocarpon*

1. VERRUCARIA Schrad.

Thallus crustose, within the rocks and undifferentiated or on the surface of the rocks and then corticate. Perithecia sunken in the thallus or in thalline warts, or half projecting, or adnate; if covered by an outer layer this is called the involucrellum; the perithecial wall (excipulum) in sections may appear to be in two sections which are darkened (dimidiate), or it may be entirely dark (complete) or entirely pale; the ostiole is usually small and inconspicuous; a group of sterile hyphae, the periphyses, radiates within the ostiole; the hypothecium is hyaline to brownish; the hymenium hyaline or brownish; the paraphyses soon gelatinizing and indistinct; spores 8, simple, hyaline, ellipsoid to broadly ellipsoid.

1   Thallus within the rock; the perithecia half immersed   2
    Thallus surficial; membranaceous to scurfy, continuous, or areolate   3

2 (1) Perithecia lacking thalline involucrellum; spores 30–36 × 11–18 μ
                                  1. *V. devergens*
    Perithecia with thalline involucrellum; spores 15–26 × 8–13 μ
                                  2. *V. rupestris*

3 (1) Thallus membranaceous to scurfy   4
    Thallus areolate, forming a thick chinky to almost squamulose crust;
       perithecia embedded in the areolae   5

4 (3) Thallus scurfy, involucrellum reaching only halfway up the perithecium;
       spores 18–23 × 8.5–13 μ   3. *V. muralis*
    Thallus continuous to chinky, blackish brown, gelatinous when moist;
       spores 16–22 × 8–10 μ   4. *V. aethiobola*

5 (3) Growing on dry rocks   6
    Growing in sites frequently wet by fresh water   7

6 (5) Thallus dark brown to black, chinky areolate; perithecia imbedded; spores
       20–27 × 9–12 μ   5. *V. nigrescens*
    Thallus ashy gray; perithecia projecting, not covered by thalloid involucrellum; spores 16–27 × 7–14 μ   6. *V. obnigrescens*

7 (5) Thallus continuous, only partly areolate, dark gray to blackish brown,
       thin; spores 24–36 × 11–17 μ   7. *V. margacea*
    Thallus areolate, yellowish gray, flat, moderately thick; spores 18–28 ×
       6–10 μ   8. *V. devergescens*

1. **Verrucaria devergens** Nyl. Thallus forming a whitish color on the rocks, immersed in the rock, sometimes with a black bordering line. Perithecia half immersed in the substratum, 0.25–0.3 mm broad, hemispherical, the excipulum entirely black or brownish black, even in thickness or the tip thickened, the ostiole with small, not sunken mouth; hymenial gelatin I+ violet; spores elongate, ellipsoid, 30–36 × 11–18 μ.

    Growing on calcareous rocks. A probably circumpolar arctic species known from Novaya Zemlya and the Northwest Territories.

    Localities: 10, 13.

    The very similar *Verrucaria deversa* Vain. which is also common in the arctic has much smaller spores, 15–20 × 7–10 μ, and the mouth of the ostiole is sunken.

2. **Verrucaria rupestris** Schrad. Thallus forming an ashy white discoloration on the rocks, immersed, lacking hypothallus. Perithecia immersed in the rocks with only the tip showing, 0.4 mm broad, an involucrellum outside the black excipulum; the excipulum dimidiate, the upper part brownish black, lower pale, inner part of wall pale; ostiole commonly slightly sunken; hymenial gelatin I+ blue becoming violet brownish; spores ellipsoid oblong, 15–26 × 8–13 μ.

Growing on calcareous rocks. A circumpolar species of wide distribution, arctic to temperate.

Localities: 1, 2, 11, 14.

3. **Verrucaria muralis** Ach. Thallus thin, scurfy or quite smooth, to chinky diffract and mixed with the substratum, ashy gray, greenish gray to reddish; hypothallus not distinct. Perithecia partly immersed in the substratum, in verrucules, to 0.4 mm broad, conical, the base broadened, the base partly covered by an involucrellum, the exposed part subpruinose, with an inner hyaline layer which is complete and an outer reddish brown excipulum which is dimidiate with the sides slightly thickened; ostiole slightly sunken; hymenial gelatin I+ wine red; spores 18–23 × 8.5–13 μ.

Growing on calcareous rocks. A circumpolar species ranging from arctic to temperate and with very wide range in North America.

Locality: 11.

4. **Verrucaria aethiobola** Wahlenb. ex Ach. Thallus smooth to chinky-diffract, gelatinous when moist, ashy brown, brownish black, rarely reddish. Perithecia hemispherical with the base broadened 0.25–0.5 mm broad, the base with slight covering of thallus, the involucrellum bare, black, broadening laterally; excipulum with inner dark wall entire, outer very dark, dimidiate to nearly complete but thin at base; ostiole thin to broadening; hymenial gelatin I+ violet; spores 16–22 × 8–10 μ.

Growing on rocks which are inundated at least at times, mainly in stream beds. A circumpolar arctic to temperate species.

Localities: 9, 11.

5. **Verrucaria nigrescens** Pers. Thallus moderately thick to quite thick, chinky-areolate, the areoles to 1.5 mm broad, angular, smooth to uneven, dull, dark brown to black, grading into a black hypothallus. Perithecia imbedded in the thallus, only the mouth showing, forming broadly conical elevations in the thallus, the excipulum black brown, hemispherical, the basal part broadening and merging with the hypothallus, the ostiole minute and covered by the thallus, irregular; hymenial gelatin I+ wine red; spores 20–27 × 9–12 μ.

Growing on calcareous rocks. A circumpolar species with broad range from arctic to temperate, widely distributed in most of North America.
Locality: 9.

6. **Verrucaria obnigrescens** Nyl. Thallus forming a thin, dull crust of flat areoles to 0.2 mm broad, ashy gray, hypothallus lacking or thin and black. Perithecia 0.2–0.4 mm broad, on verrucules of the thallus, the base narrowly covered by the thallus, hemispherical; the excipulum subglobose, paler, exterior dark brown, dimidiate, thin above broadening to 50–52 μ thick at base; spores 16–27 × 7–14 μ.

Growing on rocks. A possibly circumpolar boreal species but the range is imperfectly known. Described from Kamchatka, new to Alaska.
Locality: 2.

7. **Verrucaria margacea** Wahlenb. Thallus a very smooth, continuous to slightly areolate thin dark gray to black brown crust. Perithecia scattered, sunken in hemispherical warts the excipulum dark brown to black in places; the hymenium I+ violet; spores broad, ovate-ellipsoid, 24–36 × 11–17 μ.

Growing on acid rocks in or near brooks. A circumpolar arctic-alpine species.
Locality: 8.

8. **Verrucaria devergescens** Nyl. Thallus continuous, only partly areolate, superficial on the substrate, yellowish gray, areolae to 0.3 mm thick, flat. Perithecia immersed in the areolae, with black tip slightly projecting, the excipulum brown black, thickened around the ostiole; hymenium I-; spores elongate ellipsoid, 18–28 × 6–10 μ.

Growing on acid rocks in wet situations. Too few collections are known to be able to state a range. Reported from Siberia and from the Alps in Europe.
Locality: 8.

2. THELIDIUM Mass.

Thallus crustose, not differentiated into layers, sometimes imbedded in calcareous rocks, at times on silicate rocks with the thallus superficial and becoming chinky or warty areolate. Perithecia small, more or less immersed; the wall (excipulum) entire or dimidiate, ostiole very small; paraphyses gelatinizing very soon; asci clavate; spores 8, hyaline, oblong ellipsoid, 1-3-septate.

1   Perithecia without separate thalloid covering, only 0.2 mm broad; spores
     mainly smaller than 20 μ (15–21 × 6–8 μ) thallus brownish gray
                                                        1. *T. acrotellum*
     Perithecia with separate thalloid covering, larger than 0.2 mm; spores
     larger, over 20 μ   2

2 (1) Perithecia 0.4–1.0 mm; thallus gray white; spores 19–32 × 9–14 μ
                                                        2. *T. pyrenophorum*
     Perithecia 0.3–0.5 mm; thallus gray brown to cherry brown or rust
     brown; spores 24–40 × 13–18 μ   3. *T. aeneovinosum*

1. **Thelidium acrotellum** Arn. Thallus very thin, membranaceous,
brownish gray when dry, greenish when wet, with poorly developed
hypothallus. Perithecia very small, 0.1–0.16 mm broad, flattened-
spherical, almost adnate, the base slightly immersed, the upper part
bare; excipulum pale below, brown dimidiate above; asci saccate; paraphyses soon gelatinizing, the hymenial gelatin I+ red; spores 6–8, hyaline, 1-septate, partly simple, ellipsoid, 15–21 × 6–8 μ.
     Mainly on calcareous rocks. Probably a circumpolar arctic-alpine spe-
cies but the range very incompletely known. New to Alaska, previously
reported from Newfoundland (Arnold 1899).
     Localities: 8, 14.

2. **Thelidium pyrenophorum** (Ach.) Mudd Thallus thin, rough, becoming chinky, or warty areolate, ocher or whitish or ashy, sometimes with
black lines bordering and running through the thallus. Perithecia with
the base immersed in the substratum, forming hemispherical verrucae;
thalloid margin half to entirely covering the perithecium; excipulum
0.4–1.0 mm, globose with the tip depressed, the upper part brown,
dimidiate or sometimes entirely dark; paraphyses gelatinizing, the
hymenial gelatin I+ red, the base of the asci I+ weakly blue; spores 8,
hyaline, ellipsoid, 1-septate, 19–32 × 9–14 μ.
     Growing on calcareous rocks. A circumpolar boreal to temperate
species distributed over much of North America.
     Locality: 11.

3. **Thelidium aeneovinosum** (Anzi) Arn. Thallus diffuse, thin, almost
membranaceous, continuous or finely chinky, grayish brown, cherry
brown, to nearly rust colored; upper cortex of a hyaline and a brown
layer present. Perithecia adnate, hemispherical, 0.3–0.5 mm broad, involucrellum extending with a thick, black layer extending to the base or
halfway down the perithecium, the margin spreading and thickening at
the base; the excipulum spherical, thin, entire; paraphyses becoming

gelatinized, the hymenial gelatin I+ red; spores 8, hyaline, ellipsoid, 1-septate, occasionally slightly curved, 24–40 × 13–18 μ.

Growing on calcareous rocks. An arctic and alpine species known from Europe, Greenland, and in Novaya Zemlya. New to North America.

Locality: 11.

3. POLYBLASTIA Mass.

Thallus crustose, sometimes within the rock substratum, when visible the thallus powdery to chinky, sometimes with a well developed hypothallus.

Perithecia adnate, bare or covered by thallus, in the rock-imbedded species the perithecia are sometimes buried in cavities in the rock; involucrellum sometimes present; excipulum pale or blackening; periphyses present; paraphyses soon becoming gelatinized; asci saccate; spores 1–8 per ascus, becoming muriform, hyaline or darkening, round to elongate ellipsoid.

1	On moss and earth	2
	On rocks	4
2 (1)	Spores hyaline	3
	Spores brown; thallus dark, gelatinous; spores 18–28 × 7–9 μ	
		3. *P. gothica*
3 (2)	Thallus gray or white; spores 15–30 × 9–14 μ	1. *P. sendtneri*
	Thallus black, gelatinous; spores 28–38 × 12–18 μ	2. *P. gelatinosa*
4 (1)	Spores dark, 63–84 × 35–45 μ; thallus thick, white	4. *P. theleodes*
	Spores hyaline	5
5 (4)	Spores 45–65 × 20–28 μ; thallus thin, gray white; perithecia immersed in the thallus verrucae	5. *P. sommerfeltii*
	Spores smaller, less than 45 μ; thallus not visible to scant; perithecia not immersed in thallus	6
6 (5)	Spores 25–27 × 13–17 μ	6. *P. hyperborea*
	Spores 30–42 × 18–23 μ	7. *P. integrascens*

1. **Polyblastia sendtneri** Kremp. Thallus a cartilaginous, 0.5 mm thick, gray, white, or pale reddish gray crust with a black hypothallus. Perithecia to 0.3 mm broad, at first immersed, later a little projecting from the thallus, the top flattened, black or blue black, in that case covered with a hyaline layer; excipulum hyaline to brown black, spherical, the upper half or more with a black involucrellum outside; periphyses elongate and quite thick; hymenial gelatin I+ red or occasionally blue; spores 8, hyaline, ellipsoid, containing 8–16 irregularly arranged cells, 15–30 × 9–14 μ.

Growing over mosses and soil. A circumpolar arctic-alpine species. New to Alaska; previously reported from Ellesmere Island by Darbishire (1909).

Localities: 10, 14.

2. **Polyblastia gelatinosa** (Ach.) Th. Fr. Thallus thin to moderately thick, gelatinous when wet, uneven in globules, the upper cortical layer brown. Perithecia half immersed in the thallus with slightly flattened summits, 0.2 mm broad; excipulum spherical, entirely hyaline or with the outer part darker, surrounded by a black layer; hymenial gelatin blue with I and rapidly turning red; spores 8, hyaline, ellipsoid, muriform with 1-2 longitudinal septa, 6-7 septa transversely, 28-38 × 12-18 μ.

Growing over moss and humus in moist sites. A circumpolar arctic and boreal species, rarely collected. Reported by Fink (1935) from New Hampshire and by Darbishire from Ellesmere Island (1909). New to Alaska.

Locality: 5.

3. **Polyblastia gothica** Th. Fr. Thallus thin, gelatinous, green or dark blackish. Perithecia to 0.2 mm broad, partly immersed in the thallus, involucrellum and excipulum black brown; asci saccate; spores brown, elongate elliptical, transversely several-septate to muriform with 5-7 transverse, 1 longitudinal wall, 18-28 × 7-9 μ.

Growing on humus and mosses. A circumpolar arctic species, reported from Greenland and Ellesmere Island.

Locality: 8.

4. **Polyblastia theleodes** (Somm.) Th. Fr. Thallus crustose, thick and warty-areolate to scurfy, or nearly lacking, whitish, sometimes reddish tinged; upper cortex of 30-40 μ, hyaline and poorly structured. Perithecia 0.8-1.0 mm broad, in warts in the thallus with the flattened and deepened ostiole more or less free, becoming adnate on the thallus and only the base inserted; excipulum spherical, pale, darkening near the ostiole and sometimes the outer layer also darkening; surrounded by a dark involucrellum which is lacking below the perithecium; periphyses very long and slender; spores 8, soon blackish brown, ellipsoid, muriform with many cells, 63-84 × 35-45 μ.

Growing on rocks, usually with calcareous content. A circumpolar probably boreal species.

Localities: 1, 8, 11, 14.

5. **Polyblastia sommerfeltii** Lynge Thallus thin, gray white, irregularly chinky. Perithecia immersed in thallus warts, at first only the ostiole, later 1/3 of the perithecium visible, 0.75 μ broad; excipulum brown or black, the upper part fused with the involucrellum, the lower part free, the involucrellum thick, black; periphyses elongate, slender; hymenial gelatin I+ wine red; spores 8, remaining hyaline, muriform with many cells, ellipsoid, 45–65 × 20–28 μ.

Growing on calcareous rocks. A circumpolar arctic species. New to Alaska.

Localities: 1, 2.

6. **Polyblastia hyperborea** Th. Fr. Thallus not distinguishable. Perithecia at first immersed in the rock, later becoming hemispherical, 0.2–0.3 mm broad, black, bare, with swollen summit; excipulum below pale brownish; involucrellum thick, black, dimidiate, spreading away from the perithecium at the base; hymenial gelatin I+ red; spores 8, hyaline, muriform with many cells, 25–27 × 13–17 μ.

Growing on calcareous rocks. A circumpolar arctic species.

Localities: 1, 2, 14.

7. **Polyblastia integrascens** (Nyl.) Vain. Thallus moderately thick, finely chinky-areolate, ashy gray or whitish. Perithecia with the base immersed, hemispherical, 0.3–0.4 mm broad, black, bare or with the involucrellum over the entire surface; excipulum spherical, black brown, thinner at the base; hymenial gelatin I+ red; spores 8, hyaline, ellipsoid, muriform with many cells, 30–42 × 18–23 μ.

Growing on calcareous rocks. A possibly circumpolar arctic species.

Locality: 3.

4. STAUROTHELE Norm.

Thallus crustose, smooth, warty-areolate, chinky, radiate-areolate, or embedded in the rock substratum; corticate; usually lacking hypothallus but this present in some species. Perithecia embedded in the areolae or adnate to sessile; terminal ostiole present; periphyses present; paraphyses becoming gelatinous and the hymenial gelatin reacting with IKI; asci saccate; spores 1–8, hyaline or brown, muriform with many cells; algae present in the hymenial layer. Algae: *Stichococcus mirabilis* Lagerheim in both hymenium and thallus (Ahmadjian and Heikkilä 1970).

1	Thallus thick, areolate; hymenial algae elongate	2
	Thallus thin, varnish-like to chinky; hymenial algae globose	3. *S. fissa*

2 (1) Thallus continuous, thick, deeply chinky or warty-areolate; the perithecia
deep in the areolae   1. *S. clopima*
Thallus with the areolae arranged in radiating separated lines; the perithecia with only the base immersed   2. *S. perradiata*

1. **Staurothele clopima** Th. Fr. Thallus quite thick, chinky areolate or
areolate brownish black, dull. Perithecia deeply sunken in the areolae,
0.2–0.3 mm broad, the tip slightly projecting, the ostiole narrow;
perithecial wall globose, the upper projecting part black brown, including fused involucrellum and excipulum, the rest hyaline; hymenial
gelatin I+ blue; hymenial algae usually elongate; spores 2, brown, ellipsoid, many celled muriform 30–52 × 12–25 μ.
  Growing on many kinds of rocks, including calcareous. A circumpolar
arctic-alpine species.
  Localities: 2, 3.

2. **Staurothele perradiata** Lynge Thallus crustose forming small circles,
striate-plicate, toward the margins becoming radiate lobate, the lobes
very discrete at margin, confluent toward the center, dull, dark brownish black. Perithecia numerous, 0.2–0.3 mm, verruculose-globose, protuberant, the base immersed in the thallus, top of same color as the
thallus; excipulum thick, dark brown, lower part pale; hymenium I+
blue; spores 2, at first hyaline, becoming dark, oblong, broadly rounded,
muriform, 3–5-septate longitudinally, 9–11-septate transversely, 30–62
× 14–22 μ.
  Growing on rocks. Known from Greenland, the Northwest Territories, Saskatchewan, North Dakota, and Arizona. New to Alaska.
  Locality: 11.

3. **Staurothele fissa** (Tayl.) Zwackh. Thallus thin, continuous or finally
chinky, often varnish-like, shining, dark brown to brown black, paraplectenchymatous, upper cells strongly browned, algae distributed
through the entire thallus. Perithecia hemispherical or broad based,
immersed in 0.3–0.5 mm broad warts on the thallus, with the mouth
projecting and black when dry, brown when wet; amphithecium spherical, to 0.36 mm in diameter, pale, the thalloid covering brown to black
brown; the hymenial algae subspherical, frequently with 2 cells united;
hymenium I+ blue turning violet; spores 2, muriform many-celled,
30–50 × 14–22 μ.
  Growing on granitic rocks. In high mountains and the arctic in
Europe and Greenland.
  Locality: 8.

5. GEISLERIA Nitschke

Thallus crustose, undifferentiated. Perithecia erect, immersed in thallus or bare; excipulum hyaline; paraphyses persistent, branched; asci cylindrico-clavate; spores 8, hyaline, pointed at each end, 3-septate. Algae: *Trebouxia*.

1. **Geisleria sychnogonioides** Nitschke Thallus forming a thin granular-scurfy ashy gray layer with an indistinct hypothallus. Perithecia ovate with narrowing toward the ostiole, adnate, 0.25–0.3 mm broad, brownish, or reddish when wet; hymenium I-; paraphyses slender, branched; asci cylindrico-clavate; spores 8, hyaline, pointed at each end, 3-septate, 16–25 × 6–9 μ.

Growing on bare soil. Known as boreal and alpine in Europe. New to North America.

Locality: 4.

6. DERMATOCARPON Eschw.

Thallus squamulose to umbilicate; with upper cortex and usually a well developed lower cortex, heteromerous. Perithecia immersed in the thallus with only the ostiole showing or a small portion of the top; with perithecial wall hyaline or the upper part black and dimidiate, or entirely dark; periphyses abundant, slender; lacking algae in the hymenium; the paraphyses soon gelatinizing or rarely branching and interwoven; asci cylindrico-clavate to saccate; spores 8 (rarely 16), simple, hyaline, ellipsoid. Algae: *Pleurococcus*.

1 Perithecium entirely black; squamules ashy gray, whitish pruinose.

1. *D. cinereum*

Perithecium hyaline below, only the tip dark; squamules brown or reddish brown, epruinose 2

2 (1) Squamules 3–5 mm broad, free at the margins. 2. *D. lachneum*

Squamules less than 3 mm broad, totally adnate to the substratum

3. *D. hepaticum*

1. **Dermatocarpon cinereum** (Pers.) Th. Fr. Thallus adnate squamulose, chinky between the squamules in the center, gray brown, ashy gray, with whitish gray pruina when young, this disappearing, the squamules 0.2–1.5 mm long, 0.2–0.4 mm broad, the margins entire, over a black hypothallus. Perithecia immersed in the thallus the summit forming 0.2 mm broad warts; perithecium brown to black; periphyses numerous; paraphyses gelatinizing, the hymenial gelatin I+ pale violet, at the base of

the perithecium blue; asci cylindrico-clavate; spores 8, biseriate, simple, hyaline, ellipsoid, 16–23 × 6–7 μ.

Growing on bare soil or humus, usually calcareous. A circumpolar arctic, boreal, and alpine species, but reported with very wide range in North America.

Locality: 14.

2. **Dermatocarpon lachneum** (Ach.) A.L. Smith (*D. rufescens* (Ach.) Th. Fr.) Thallus squamulose with the squamules becoming free at the edges, overlapping, shining red brown, 7–9 mm long, 3–5 mm broad; corticate above and below, over a black hypothallus. Perithecia immersed in the thallus, only the ostiole showing, perithecial wall hyaline or pale, only dark at the edge of the mouth; periphyses abundant, 30–40 μ long; paraphyses gelatinizing, hymenial gelatin I+ brownish red; asci cylindrical; spores 8, hyaline, simple, ellipsoid, 11–18 × 6–9 μ.

Growing on calcareous soils. A circumpolar arctic and alpine species. New to Alaska.

Localities: 1, 9, 11, 14.

3. **Dermatocarpon hepaticum** (Ach.) Th. Fr. Thallus adnate-squamulose, the squamules forming an almost continuous crust with chinks between, the squamules usually less than 3 mm broad, flat or slightly convex, rounded or angular, light or dark reddish brown, dark margined, over a dark hypothallus, corticate above and below. Perithecia immersed in the thallus, only the ostiole showing; perithecial wall hyaline with only the edge at the ostiole dark; periphyses abundant, 25 μ long; paraphyses gelatinizing, the hymenial gelatin I-; asci cylindrical; spores 8, hyaline, simple, ellipsoid, 11–16 × 5–7 μ.

Growing on calcareous earth. A circumpolar, boreal, and temperate species ranging over most of North America.

Locality: 11.

7. ENDOCARPON Hedw.

Thallus squamulose, the squamules adnate to ascending; upper side corticate paraplectenchymatous, lower cortex also present; attached to the underside by rhizinae. Perithecia immersed in the thallus; periphyses present; hymenial algae present; asci saccate; paraphyses soon gelatinizing, hymenial gelatin I+; spores 2, becoming brown, muriform with many cells, ellipsoid. Algae: *Stichococcus diplosphaeria* (Bialosuknia) Chodat, in both hymenium and thallus (Ahmadjian and Heikkilä 1970).

1. **Endocarpon pusillum** Hedw. Thallus squamulose-areolate, adnate, the areolae contiguous or dispersed, 0.5–3.5 mm broad, flat, crenulate or subentire, upper side brown or pale brownish to reddish brown, dull, underside black, attached by hyphae to the substratum. Perithecia immersed in the thallus, only the ostioles showing, 0.25–0.4 mm broad, the perithecial wall spherical, brown; periphyses 15 μ long, the hymenial algae usually spherical; asci saccate; paraphyses gelatinizing, the hymenial gelatin I+ violet; spores 2, soon reddish brown to brown, ellipsoid, muriform with many cells, 29–55 × 18–23 μ.

Growing on calcareous soils. A circumpolar boreal to temperate species which ranges over most of North America.

Locality: 2.

ARTHOPYRENIACEAE

Crustose lichens, the thallus poorly developed, within the substratum to a granular undifferentiated layer on the surface. Fruit a perithecium, periphyses usually present; paraphyses persistent; asci of the bitunicate type, the outer wall thin, an inner wall extending to form a cylindrical sac carrying up the ascospores and releasing them one at a time through an indistinct elastic pore at the apex of the inner wall.

1. MICROGLAENA Körb.

Thallus crustose, smooth, granulose, chinky, areolate, or squamulose, not differentiated into layers. Perithecia small, more or less immersed in the areolae, or adnate, the perithecial wall (excipulum) hyaline to brown, complete or dimidiate; no involucrellum; periphyses present; pseudoparaphyses persistent, branched; asci cylindrical; bitunicate; spores 2–8 in ascus, hyaline, yellowish, or brown, ellipisoid, muriform with many cells. Algae: *Trebouxia*.

1. **Microglaena muscorum** (Fr.) Th. Fr. Thallus thin, ashy gray granules dispersed over a thin white hypothallus. Perithecia with the base immersed, 0.15–0.3 mm broad, hemispherical, reddish- or brownish-black; the perithecium hemispherical, reddish brown, brownish black; pale below; pseudoparaphyses persistent, sparingly branched and interwoven; hymenium I-; asci cylindrical; spores 2, 4, 6, or 8 per ascus, at first 6–7-septate transversely, becoming muriform, hyaline, elongate ellipsoid, 60–86 × 10–26 μ.

Growing over mosses and humus. A probably circumpolar arctic-alpine and boreal species.

Locality: 1.

CALICIACEAE

Thallus crustose, poorly developed, inconspicuous to disappearing, not differentiated into layers. Apothecia borne on the tips of stipes, cup-shaped to top-shaped, usually with a proper exciple; asci cylindrical but soon dissolving, leaving the spores in a network of the branching and interwoven paraphyses; spores 8, simple or 1-7-septate, rarely submuriform.

1. MYCOCALICIUM Vain.

Thallus a thin mycelium discoloring the substratum, usually lacking algae. Apothecia on well developed stipes; asci cylindrical; network from paraphyses produced or not; spores 8, uniseriate, simple, brown or pale brownish. Algae: *Cystococcus*.

1. Mycocalicium parietinum (Ach.) Vain. Thallus very thin, making white patches of discoloration on the substratum; stipes short, 0.3-0.8 mm tall, brownish black; capitulum top-shaped, 0.13-0.38 mm broad, black, exciple black; hypothecium brown, lacking paraphysal hairs which disappear, hymenium pale, I+ yellow; asci cylindrical; spores 8, uniseriate, simple, ellipsoid, 6-11 × 3-6 μ.

Growing on old wood. A circumpolar boreal to temperate lichen ranging over much of North America.

Locality: 3.

CYPHELIACEAE

Crustose lichens with the thallus effigurate or not at the margins, lacking cortex, little differentiated. Apothecia sessile or adnate; with either proper or thalloid exciples or both.

1. CYPHELIUM Ach.

Thallus crustose, smooth to granulose, verrucose, or areolate, a few species marginally lobed; apothecia immersed in the thallus or becoming adnate, forming a mazaedium; thalloid margin present, sometimes a proper one also; hypothecium brown; hymenium with few unbranched to little branched, slender paraphyses; asci cylindrical; spores uniseriate, simple, 2-celled, or submuriform, brown, more or less rough walled. Algae: *Trebouxia*.

1 Apothecium forming a cup-shaped black excipulum; thallus coarsely verrucose, ashy gray; spores 10–12 × 4–7 μ   1. *C. inquinans*
  Apothecium without external persistent excipulum; thallus chinky to areolate, greenish yellow or yellow; spores 14–24 × 7–11 μ.   2. *C. tigillare*

1. **Cyphelium inquinans** (Sm.) Trev. Thallus granulose to chinky or areolate, ashy to greenish gray, continuous or scattered; K-, C-. Apothecia sessile, small, to 1 mm broad, surrounded by the cup-like exciple, black or sometimes slightly pruinose, sometimes with thin pruinose or dark thalloid margin surrounding it; the mazaedium spilling over the margins; paraphyses slender, 1.5–2 μ, unbranched or slightly branched; spores 8, 2-celled, brown, broadly ellipsoid or oblong, constricted in the middle, 12–20 × 7–12 μ according to Fink, 10–12 × 4–7 μ according to Weber.

A species of the western coastal states, inland to Idaho and Colorado, and north through British Columbia to Alaska and Northwest Territories. In addition to the Alaskan specimens reported here I have specimens from Reindeer Station, Northwest Territories, Thomson 15520, 15616. New to Alaska.

Locality: 14.

2. **Cyphelium tigillare** (Ach.) Ach. Thallus crustose, areolate with rounded areolae, chinky, greenish yellow or yellow, sometimes darkening; containing rhizocarpic acid. Apothecia sunken in the thallus or in projecting areolae, usually common, bordered only by the thalloid margin; a thin indefinite proper black margin sometimes visible inside; disk black, epruinose, flat or the mazaedium piling up on top in a black heap; paraphyses agglutinate, 1 μ, spores 8, 2-celled, brown, broadly ellipsoid or oblong, constricted in the middle, 14–24 × 7–11 μ.

Growing on old wood of many kinds. A circumpolar boreal to temperate species with a range in North America south to Connecticut, Illinois, Kansas, and California. New to Alaska.

Locality: 14.

SPHAEROPHORACEAE

Thallus fruticose, terete or flattened, erect, heteromerous with cortex, algal layer, and central medulla which is solid or subfistulose with thick walled but not conglutinate hyphae. Apothecia terminal and immersed in the tips of the branches; with thick thalloid margin; opening widely; paraphyses branching-anastomosing; asci with thin membranes which soon dissolve, leaving a mass of spores and forming a mazaedium with the paraphyses; spores brown, simple or 1-septate (in *Tholurna*). Algae: *Cystococcus*.

1. SPHAEROPHORUS Pers.

Thallus fruticose, terete or flattened, erect or procumbent, heteromerous, solid to subfistulose, sometimes with a hypothallus which forms small erect papillae or hypothallus absent; cortex chondroid, with small lumena. Apothecia in the tips of the branches, the covering a thalloid globose head which bursts without a definite ostiole; lacking proper exciple; the mazaedium and spores exposed as a black mass at the lacerate tip; spores 8, uniseriate, globose or nearly so, simple, brown.

1 Medulla I+ blue, thallus partly foveolate, not very fragile   1. *S. globosus*
   Medulla I-, thallus fragile, non foveolate   2. *S. fragilis*

1. **Sphaerophorus globosus** (Huds.) Vain. Thallus erect, branching sympodially or dichotomously, the branches round, partly impressed foveolate, smooth, the bases brown stained, the upper parts bluish gray to sun browned; medullary layer I+ blue, containing sphaerophorin, fragilin, squamatic acid and either hypothamnolic or thamnolic acid. Apothecia rather rare, in the globose tips of branches, dehiscing lacerately; hypothecium reddish above, pale below; paraphyses poorly developed, thin; asci cylindrical; spores 8, uniseriate, simple, globose or nearly so, the walls rugulose papillate, 7–15 μ.

Growing on humus, soil and over rocks. A common circumpolar arctic-alpine species. Although the Pacific Coast rainforest material which has many small lateral branches has been described as belonging to this species as var. *gracilis* (Müll. Arg.) Zahlbr., it differs in its ecology as well as morphology and is distributed in Asia as well as the west coast. It has been named as *Sphaerophorus tuckermanii* Räs. and should be separated from the typical arctic species.

Localities: 1, 2, 3, 6, 7, 8, 9, 11, 13.

2. **Sphaerophorus fragilis** (L.) Pers. Thallus erect, in close tufts or tussocks on the rock substratum, branching sympodial or dichotomous, the branches rounded, bluish gray or ashy gray, browned in strong light, smooth; cortex chondroid; medullary layer I-; containing sphaerophorin and fragilin (and hypothamnolic acid in European material [Rehm

1971]). Apothecia rare, in the globose tips of branches; hypothecium brown; upper part of the hymenium blue black; paraphyses slender, the tips branched; asci cylindrical; spores 8, uniseriate, turning from blue into black, the walls rugulose-papillate and very dark, spherical, 7–16 μ.

Growing on rocks and rock outcrops. A circumpolar arctic-alpine species.

Localities: 3, 4, 6, 7, 8, 11.

BASIDIOLICHENES

1. CORISCIUM Vain.

Thallus foliose, squamulose, of small cochleate lobes, the upper cortex paraplectenchymatous, the algal layer thick, glomerulate, the medulla loose, the lower cortex discontinuous, lacking rhizinae, attached by a weft of hyphae. Algae: *Coccomyxa*.

The fruiting bodies are uncertain, the basidiomycete *Omphalina hudsonica* (Jenn.) Bigelow is listed as being the fruiting body in Hale and Culberson (1970), and is claimed to be *Omphalina luteolilacina* (Favre) Henderson by Heikkilä and Kallio (1966). The first suggestion that the accompanying *Omphalina* could be the fruiting body of the lichen was made by Gams (1962).

1. Coriscium viride (Ach.) Vain. Thallus small, squamulose, of rounded cochleate lobes, circular or becoming slightly lobate with rounded lobes, to 5 mm broad, usually less, the margins inrolled and white, the surface blue green, slightly pruinose, rhizinae lacking but the underside attached to the substratum by a loose network of hyphae; upper cortex paraplectenchymatous, several cells thick, uneven in thickness, algae distributed through the medulla in glomerules, the lower cortex plectenchymatous to paraplectenchymatous, thin, the hyphae anastomosing in part into the hyphae penetrating the ground, toward the edges the lower cortex is more continuous.

On mossy or woody humus in partly shaded sites. A circumpolar species known from northern and central Europe, Siberia, southeast Greenland, and in North America from Baffin Island, Labrador, Ungava Peninsula, the Northwest Territories, and Alaska, south to the White and Adirondack Mountains.

Localities: 2, 3, 4, 5, 6, 7, 11.

Experimental or other proof of the association of this lichen and the accompanying fungus is much needed.

Glossary

Acicular Needle-shaped, pointed at both ends.

Amphithecium The part of a lecanorine apothecium outside the proper exciple and usually containing algae.

Anisotomic branching Unequal branching in which one branch becomes a more or less distinct main branch from which more slender side branches arise.

Apothecium (pl. *apothecia*) A disk- or saucer-shaped ascocarp.

Areolate Broken up into angular patches.

Ascocarp The fruiting structure of Ascomycete fungi, in turn containing asci which produce ascospores.

Ascospore A spore produced in an ascus.

Aspicilioid With the apothecia sunken in the thallus as in the section *Aspicilia* of the genus *Lecanora*.

Biatorine With the upper part of the hymenium and the exciple pale.

Botryoid With rounded clusters like bunches of grapes.

Bullate With high, rounded swellings.

Capitate With a rounded or head-like shape.

Capitulum A head-like structure.

Carbonaceous With black, dull, more or less brittle structure.

Cartilaginous With translucent, more or less stiff tissues.

Cephalodium (pl. *cephalodia*) Gall-like structures containing different algae than the rest of the thallus. They may be external or internal.

Cortex The outer protective layers of the thallus, sometimes cellular in character.

Corymbose With branches coming up to the same general level.

Crenate With small rounded teeth or scallops.

Crenulate With very fine rounded teeth or scallops.

Cyphellate (noun, *cyphella*) With small rimmed cup-like depressions in the thallus, usually corticate within.

Decorticate Lacking a cortex.

Dichotomous Forked in pairs of branches.

Dimidiate Referring to sections of the perithecium in which the wall appears as two dark lateral areas.

Ecorticate Lacking a cortex.

Effigurate With elongate lobing toward the outer part of a thallus.

Entire Not divided.

Epispore A gelatinous outer layer on the spores; sometimes also called a halo.

Epithecium The uppermost part of the hymenial layer in the ascocarp consisting of the apices of the paraphyses imbedded in a gelatinous matrix which is often colored or inspersed with tiny granules.

Euthyplectenchymatous Referring to hyphae with the cells more or less parallel to the surface.

Exciple The outer layer of the thecium forming the proper margin of a lecideine type of apothecium or internal to the amphithecium in a lecanorine type of apothecium. Also applied to the inner wall of a perithecium type of fruit.

Excluded Eliminated, as when the disk of an apothecium swells over and causes the margin or exciple to disappear.

Farinose Finely granular, like flour.

Fastigiate In bundles.

Glaucescent Of a bluish green color.

Glebulose With rounded cushion-like areolae.

Halonate With a 'halo' or gelatinous epispore.

Hapters Fungal threads or hyphae which attach the underside of the thallus to the substratum.

Hemiangiocarpic Used with reference to the apothecium in which there is an exciple plus a pseudoexcipulum formed from thallus tissue during ontogeny. Characteristic of Peltigeraceae and Stictaceae.

Heteromerous Referring to a type of thallus in which the algae are in a distinct layer in the upper part.

Homiomerous Referring to a type of thallus in which the algae are distributed throughout, often in a rather gelatinous matrix.

Hymenium The layer of the fruiting structure in which the asci containing the spores are produced; paraphyses are also present in the tissue.

Hypothallus A layer of fungal tissue next to the substratum and below the thallus proper.

Hypothecium A layer of tissue immediately below the hymenium and separate from the exciple from which it is distinguished with difficulty.

Involucrellum An exposed cap or covering over the perithecium and its exciple, usually black and carbonaceous.

Isidium (pl. *isidia*) A coral-like outgrowth of the thallus with cortical outer tissues; it may function in vegetative reproduction.

Isotomic branching A type of branching in which the branches are of equal diameter.

Isthmus The narrow canal between the two cells of a polarilocular spore.

Lecanorine Pertaining to the margin of an apothecium in which the amphithecium contains algae.

Lecideine Pertaining to an apothecium which has no amphithecium containing algae but in which the exciple forms the margin of the apothecium; a 'proper margin.'

Lumen A cell cavity of spore or of hyphae.

Mazaedium A powdery mass of spores and of disintegrated asci and paraphyses in the apothecia of Caliciaceae.

Muriform Referring to a division of spores by both longitudinal and transverse walls.

Palisadeplectenchyma A layer of cells, conglutinate and vertically oriented in columns.

Paraphysis (pl. *paraphyses*) A sterile hypha among the asci in the hymenium.

Paraplectenchymatous With the hyphae densely coherent but with large lumina and a cellular appearance.

Periphyses Sterile hyphae in the Verrucariaceae located to the inside of the ostiole and projecting into the cavity of the perithecium.

Perithecium A flask-shaped ascocarp characteristic of the fungal order Sphaeriales, sessile or else usually sunken into the thallus tissue.

Phyllocladium (pl. *phyllocladia*) A small granular or scale-like growth on the pseudopodetia of some lichens, such as *Stereocaulon*.

Placodioid Referring to a type of thallus which is crustose in the center and lobed and plicate at the margins.

Podetium (pl. *podetia*) A stalk formed by the vertical growth of the apothecial tissues, usually the hypothecium and stipe. It secondarily becomes covered with algal tissues and cortex as in *Cladonia*.

Polaribilocular Referring to spores characterized by having two lumina separated by a thick septum through which a narrow canal passes.

Polytomous branching Branching with several branches at the same level.

Proper margin, Proper exciple A rim around the fruit or apothecium making up the entire margin in lecideine types of fruits or an inner rim within the thalloid margin in lecanorine types of fruits. It may also be applied to the inner wall of perithecia.

Prosoplectenchymatous With thick walled hyphae which have very minute lumina.

Pseudocyphellae Minute dots or pores through the upper or lower cortex, with the medullary hyphae extending to the surface through the cortex. They differ from the cyphella in lacking a definite rim.

Pseudopodetium (pl. *pseudopodetia*) A stalk formed by a vertical growth of the thallus tissue rather than of apothecial tissues.

Pycnidium (pl. *pycnidia*) A small flask-shaped structure in which pycnoconidia are produced.

Pycnoconidium (pl. *pycnoconidia*) Minute spores produced in the pycnidia.

Rhizine (pl. *rhizinae*) A hyphal extension from the lower cortex and attaching the thallus to the substratum.

Scleroplectenchyma Hyphal tissue with thick, indurate walls.

Scrobiculate With a pitted appearance.

Squarrose Rough with projecting scales.

Subhymenium A layer of hyphal tissue immediately below the hymenium and above the hypothecium.

Sulcate Having long narrow channels or grooves, fluted.

Sympodium A branching pattern in which there is a main branch from which lateral branches arise.

Terete Round in cross section.

Tetrachotomous Divided into four equal branches.

Thalloid margin, Thalloid exciple A rim surrounding the disk of the fruit and containing algal as well as fungal cells, usually having the same color as the thallus.

Tholus An upper portion of the ascus forming a thick cap beyond the spores and reacting IKI+ blue.

Trabeculae Flattened, plate-like or shreddy extensions of the lower cortex in some of the Umbilicariaceae.

Umbilicate Attached by a central point.

Verruca (pl. *verrucae*) A wart-like protuberance.

Verruculose With warty protuberances.

References

Ahmadjian, V., and H. Heikkilä. 1970. The culture and synthesis of *Endocarpon pusillum* and *Staurothele clopima*. *Lichenologist* 4: 259–67

Ahti, T. 1961. *Taxonomic studies on reindeer lichens* (Cladonia, *subgenus* Cladina). Ann. Bot. Soc. Zoo. Bot. Fenn. 'Vanamo' 32

– 1969. Notes on brown species of *Parmelia* in North America. *Bryologist* 72: 233–9

Ahti, T., G.W. Scotter, and H. Vänskä. 1973. Lichens of the Reindeer Preserve, Northwest Territories, Canada. *Bryologist* 76: 48–76

Anderson, R.A. 1965. Additions to the lichen flora of North America. I. *Bryologist* 68: 54–63

– 1967. Additions to the lichen flora of North America. II. *Bryologist* 70:339–43

Arnold, F. 1899. Lichenologische Fragmente 36 (Labrador and Newfoundland). *Oesterr. Bot. Z.* 1899: 1–25

Blomberg, O.G. and J.B.K. Forssell. 1880. Lichens. In *Enumerantur plantae Scandinaviae*, 1–116. Lund

Bocher, T.W. 1954. *Oceanic and continental vegetational complexes in southwest Greenland*. Medd. om Grønl. 148 (1)

Culberson, C.F. 1970. Supplement to 'Chemical and Botanical Guide to Lichen Products'. *Bryologist* 73: 177–377

Culberson, W.L. and C.F. Culberson. 1965. *Asahinea*. A new genus in the Parmeliaceae. *Brittonia* 17: 182–90

– 1968. The lichen genera *Cetrelia* and *Platismatia* (Parmeliaceae). *Contr. U.S. Natl. Herb.* 37(7): 449–558

Cummings, Clara. 1904. The lichens of Alaska. In Harriman, *Alaska Expedition 5: Cryptogamic Botany*, 1–424

Darbishire, O.V. 1909. Lichens collected during the second Norwegian Polar Expedition in 1898–1902. In *Report of the Second Norwegian Arctic Expedition in the 'Fram'*, No. 21: 1–68

Degelius, G. 1954. *The lichen genus* Collema *in Europe: morphology, taxonomy, ecology.* Symb. Bot. Upsal. 13(2)

– 1974. *The lichen genus* Collema *with special reference to the extra-European species.* Symb. Bot. Upsal. 20(2)

Egan, R.S. 1970a. Alpine lichens from Mt Audubon, Boulder County, Colorado. *Bryologist* 73: 385-9
– 1970b. Additions to the lichen flora of New Mexico. *Bryologist* 73: 143-5
– 1971. Additions to the lichen flora of New Mexico. II. *Bryologist* 74: 387-90
Eigler, G. and J. Poelt. 1965. Flechtenstoffe und Systematik der lobaten Arten der Flechtengattung *Lecanora* in der Holarktis. *Oesterr. Bot. Z.* 112: 285-94
Erichsen, C.F.E. 1938. Neue arktische und subarktische bes. von Dr. E. Hultén und Prof. B. Lynge gesammelte Pertusarien nebst einer Bestimmungstabelle arktischer und subarktischer, über Erde und Moosen wachsender Pertusariaceae. *Ann. Mycol.* 36: 349-66
– 1941. Neue Pertusarien aus den Vereinigten Staaten von Nordamerika. *Ann. Mycol.* 39: 379-95
Fink, Bruce. 1935. *The lichen flora of the United States.* Ann Arbor: University of Michigan Press
Follmann, G., and S. Hunek. 1971. Mitteilungen über Flechteninhaltsstoffe LXXXVIII. Zur vergleichenden Phytochemie der Krustenflechtenfamilie Acarosporaceae. *Philippia* 1(2): 65-79
Gams, H. 1962. Die Halbflechten *Botrydina* und *Coriscium* als Basidiolichenen. *Oesterr. Bot. Z.* 109: 376-80
Hale, M.E., Jr., and W.L. Culberson. 1970. A fourth checklist of the lichens of the continental United States and Canada. *Bryologist* 73: 499-543
Heikkilä, H. and P. Kallio. 1966. On the problems of subarctic basidiolichens. I. *Ann. Univ. Turka A*, II, 36:9-35
Henssen, A. 1963a. A study of the genus *Vestergrenopsis. Can. J. Bot.* 41: 1359-66
– 1963b. The North American species of *Massalongia* and generic relationships. *Can. J. Bot.* 41: 1331-46
– 1963c. The North American species of *Placynthium. Can. J. Bot.* 41: 1687-1724
– 1963d. *Eine revision der Flechtenfamilien Lichenaceae und Ephebaceae.* Symb. Bot. Upsal. 18
– 1965. A review of the genera of the Collemataceae with simple spores (excluding *Physma*). *Lichenologist* 3: 29-41
– 1969. Eine Studie über die Gattung *Arctomia. Svensk Bot. Tidsk.* 63: 126-38
Hertel, H. 1967. *Revision einiger calciphiler Formenkreise der Flechtengattung* Lecidea. Beih. Nova Hedwigia 24
– 1968. Beiträge zur Kenntnis der Flechtenfamilie Lecideaceae I. *Herzogia* 1: 25-39
– 1969a. Beiträge zur Kenntnis der Flechtenfamilie Lecideaceae II. *Herzogia* 1: 321-9
– 1969b. Über Flechtenstoffe und Systematik einiger Arten der Gattungen *Lecidea, Placopsis* und *Trapelia* mit C+ reagierenden Thallus. *Willdenowia* 5: 369-83
– 1971a. Beiträge zur Kenntnis der Flechtenfamilie Lecideaceae IV. *Herzogia* 2: 231-61

- 1971b. Über holarktische Krustenflechten aus den venezuelanischen Anden. *Willdenowia* 6: 225-72
- 1973. Beiträge zur Kenntnis der Flechtenfamilie Lecideaceae V. *Herzogia* 2: 497-515

Howard, G.E. 1970. The lichen genus *Ochrolechia* in North America north of Mexico. *Bryologist* 73: 93-130

Imshaug, H.A. 1950. New and noteworthy lichens from Mt. Rainier National Park. *Mycologia* 42: 743-52
- 1957. Alpine lichens of western United States and adjacent Canada. *Bryologist* 60: 177-272

Johnson, A.W., L.A. Viereck, R.E. Johnson, and Herbert Melchior. 1966. Vegetation and flora of the Cape Thompson-Ogotoruk Creek area. In *Cape Thomson Report*, 277-354 Washington: US Atomic Energy Commission

Keissler, K. 1960. Usneaceae. In *Rabenhorsts Krypotgamen-Flora von Deutschland, Österreich und der Schweiz*, 9(5), part 4

Kristinsson, Hordur. 1969. Chemical and morphological variation in the *Cetraria islandica* complex in Iceland. *Bryologist* 72: 344-57

Krog, H. 1962. A contribution to the lichen flora of Alaska, *Ark. Bot.*, Ser. 2, 4: 489-513
- 1968. *The macrolichens of Alaska*. Norsk Polarinst. Skrift. 144
- 1973. *Cetraria inermis* (Nyl.) Krog, a new lichen species in the Amphi-Beringian flora element. *Bryologist* 76: 299-300

Lamb, I.M. 1947. A monograph of the genus *Placopsis. Lilloa* 13: 151-288
- 1961. Two new species of *Stereocaulon* occurring in Scandinavia. *Bot. Not.* 1961: 265-75

Lindahl, P. 1962. Taxonomical aspects of some *Peltigera* species. *Svensk Bot. Tidsk.* 56: 471-6

Llano, G.A. 1950. *A monograph of the lichen family Umbilicariaceae in the Western Hemisphere*. Washington: Office of Naval Research (Navexos P-831)
- 1951. A contribution to the lichen flora of Alaska. *J. Wash. Acad. Sci.* 41: 196-200
- 1956. New Umbilicariaceae from the Western Hemisphere, with a key to genera. *J. Wash. Acad. Sci.* 46: 183-5

Magnusson, A.H. 1929. *A monograph of the genus* Acarospora. Kgl. Svensk. Vetensk. Handl. 7
- 1933. *A monograph of the genus* Ionaspis. Acta Hort. Gothenberg 8: 1-46
- 1934. On the species of *Biatorella* and *Sarcogyne* in America. *Ann. Crypt. Exotic.* 7: 115-34
- 1939. *Studies on species of* Lecanora *mainly the* Aspicilia gibbosa *group*. Kgl. Svensk. Vetensk. Handl. 17
- 1947. Studies on non-saxicolous species of *Rinodina* mainly from Europe and Siberia. *Medd. Göteborgs Bot. Tradg.* 17: 191-338

- 1952. Lichens from Torne Lappmark. *Ark. Bot.*, Ser. 2, 2: 1–249

McCullough, H.A. 1965. Lichens of the Mendenhall Valley, southeastern Alaska. *Bryologist* 68: 221–6

Runemark, H. 1956. *Studies in* Rhizocarpon. 1. *Taxonomy of the yellow species in Europe*. Opera Botanica 2 (1); 2. *Distribution and ecology of the yellow species in Europe*. Ibid. 2 (2)

Scotter, G.W. and J.W. Thomson. 1966. Lichens of the Thelon River and Kaminuriak Lake regions, Northwest Territories. *Bryologist* 69: 497–502

Sierk, H.A. 1964. The genus *Leptogium* in North America north of Mexico. *Bryologist* 67: 245–317

Thomson, J.W. 1955. *Peltigera pulverulenta* (Tayl.) Nyl. takes precedence over *Peltigara scabrosa* Th. Fr. and becomes of considerable phytogeographic interest. *Bryologist* 58: 45–9

- 1968. *The lichen genus* Cladonia *in North America*. Toronto: University of Toronto Press

- 1970. Lichens from the vicinity of Coppermine, Northwest Territories. *Can. Field Nat.* 84: 155–64

Weber, W.A. and L.A. Viereck. 1967. Lichens of Mt. McKinley National Park, Alaska. *Bryologist* 70: 227–35

Wetmore, C.M. 1960. The lichen genus *Nephroma* in North and Middle America. *Mich. State Univ. Biol. Ser.* 1 (11): 371–452

- 1970. The lichen family Heppiaceae in North America. *Ann. Mo. Bot. Gard.* 57: 158–209

Taxonomic index

Descriptions are on pages in italics, synonyms are in italics.